Biology and Ecology of Groupers

Editors

Fabiana Cézar Félix-Hackradt

Marine Ecology and Conservation Laboratory
Centre for Environmental Sciences
Federal University of Southern Bahia
Porto Seguro, Bahia, Brazil

Carlos Werner Hackradt

Centre for Environmental Sciences
Federal University of Southern Bahia
Porto Seguro, Brazil

José Antonio García-Charton

Department of Ecology and Hydrology
University of Murcia
Murcia, Spain

CRC Press is an imprint of the
Taylor & Francis Group, an **informa** business
A SCIENCE PUBLISHERS BOOK

Cover credit: Cover illustration reproduced by kind courtesy of Javier Ferrer.

First edition published 2022
by CRC Press
6000 Broken Sound Parkway NW, Suite 300, Boca Raton, FL 33487-2742

and by CRC Press
2 Park Square, Milton Park, Abingdon, Oxon, OX14 4RN

CRC Press is an imprint of Taylor & Francis Group, LLC

Library of Congress Cataloging-in-Publication Data

Names: Hackradt, Fabiana Cézar Félix, 1981- editor.
Title: Biology and ecology of groupers / editors, Fabiana Cézar Félix Hackradt, Marine Ecology and Conservation Laboratory, Centre for Environmental Sciences, Federal University of Southern Bahia, Porto Seguro, Bahia, Brazil, Carlos Werner Hackradt, Centre for Environmental Sciences, Federal University of Southern Bahia, Porto Seguro, Brazil, Jose Antonio García Charton, Department of Ecology and Hydrology, University of Murcia, Murcia, Spain.
Description: First. | Boca Raton : CRC Press, 2022. | Includes bibliographical references and index.
Identifiers: LCCN 2021030467 | ISBN 9781482253092 (hardcover)
Subjects: LCSH: Groupers. | Groupers--Ecology.
Classification: LCC QL638.S48 B56 2022 | DDC 597/.736--dc23
LC record available at https://lccn.loc.gov/2021030467

ISBN: 978-1-4822-5309-2 (hbk)
ISBN: 978-1-032-19913-9 (pbk)
ISBN: 978-1-315-11887-1 (ebk)

DOI: 10.1201/b20814

Typeset in Palatino
by Radiant Productions

Preface

Our fascination with groupers stems from having carried out much of our research work in the waters of the Cabo de Palos - Islas Hormigas marine reserve (Murcia, SE Spain, W Mediterranean). Twenty-six years after the implementation of the protection measures, we can confidently say that this zone is one of the most successful marine protected areas in the Mediterranean in terms of the biomass of groupers. Myriads of large specimens of dusky, goldblotch and mottled groupers are found throughout the year, sheltered in its rocky shallows and islands, or performing all kinds of behaviours, unconcerned by the presence of divers. The core area of this marine reserve is full of large male "silverback" dusky groupers struggling to maintain a territory and attract females, while it is not at all uncommon to find juvenile groupers sheltering in the shallower rocky crevices. The biomass increase reached in this marine area is unparalleled in the entire Mediterranean Sea, rivalling the richest coral reef areas on the planet.

Inspired by the seminal work of pioneers such as Jo Harmelin, Alfonso Ramos and James Bohnsack, and encouraged by his mentor, prof. Ángel Pérez-Ruzafa, one of us (Jose) has been working with rocky reef fish assemblages in this exceptional place for more than three decades. During the first half of this long period, several national and European projects (e.g., ECOMARE, BIOMEX, EMPAFISH) allowed sharing experiences in this marine area; and in others across the Mediterranean Sea with countless European colleagues, whose mention here would make this text never-ending, and to whom we can only be grateful for their crucial influence on our vision of Mediterranean marine ecology. We also benefited from the expertise of the members of the GEM (Group d'Étude du Mérou), a French group of enthusiastic professional and amateur researchers and divers, genuinely concerned about the conservation status of the grouper populations on the French Mediterranean coastline.

It was in 2008 that Jose accepted to supervise the PhD theses of a couple of young Brazilian scientists (Fabi and Carlos, funded by individual fellowships provided by CAPES and CNPq, respectively), whom he had met through their master's thesis supervisor at the Federal University of Parana, Professor Fred Brandini, who had opened the doors of Brazilian marine science to him through his work with artificial reefs. This decision proved to be extremely wise, given the boundless enthusiasm and know-how they brought to the receiving research group, and the fruitfulness of this collaboration to now. Carlos, supported by the research project EcoMero - EcoGaroupa (funded by the Fundación BBVA), studied the mobility patterns of groupers in the reserve, as well as the variations in abundance and biomass in various Mediterranean marine reserves, in relation not only to the different protection measures, but also to other potential prey species and their habitat. For her part, Fabi, partly funded by the Spanish Agency for International Development Cooperation (AECID), studied the dynamics of the first stages of the life cycle of rocky reef fish species in several marine reserves in the southeast of the Iberian Peninsula, both post-larvae (sampled by light traps) and juveniles (censused by scuba diving) and their relationship with adult populations and their habitat. Our joint research also took us to the Brazilian coast, where we collaborated with several groups to study the effect of fisheries protection on grouper and comber populations, as well as the dynamics of coral reef colonisation by post-larvae and juveniles in those latitudes. The successful completion of both doctoral thesis periods in late 2012 was immediately followed by a short postdoctoral period, and the attainment of professorships at the then brand-new Federal University of Southern Bahia, where Fabi and Carlos are currently working.

Our scientific collaboration only grew from that moment on. The achievement, over the last few years, of several national (REDEMED - funded by MINECO - Spanish Ministry of Economy; ABHACO^{2}DE, funded by the Seneca Foundation, responsible for science funding in the region of Murcia) and European projects (RECOMPRA Staff Exchange Scheme, ITN-MMMPA) allowed the continuation of our collaboration with so many colleagues from Europe and beyond the ocean (mainly Argentina and Brazil). All this research would not have been possible without the work of the postdocs, doctoral students and masters' students who have continued working in our respective groups since then, and the important contribution of numerous visiting researchers.

As a result, our interest in the role of groupers in the functioning of ecosystems, and the tools for the management and conservation of their populations has only grown, and naturally led us to the idea of a book in which we present the current state of research on these extraordinary fish.

To this end, we have relied on an outstanding panel of prestigious leading researchers. Obviously, trying to translate a (otherwise untranslatable) Spanish saying, "neither are all who are, nor are all who are", meaning that, although those who enthusiastically agreed to be part of the cast of this book are undoubtedly the best representation of the scientific community dedicated to the study of the biology and ecology of these fascinating animals. Some of the experts contacted to be part of this initiative declined to participate due to other commitments. This has probably meant that some issues have been left out of the scope of the book (notably, everything to do with the connectivity, mobility, and habitat of grouper populations), which will have to wait for a successive edition of this work.

Our sincere thanks to all researchers who finally accepted the invitation to participate in this endeavour, and the Editorial Department of CRC Press for the kind and indulgent patience with which they have handled the extremely excessive length of the editing process of the book. We would also like to thank the reviewers of the different chapters for their contribution to improving the quality of the texts.

While writing and editing the manuscript, we have had to mourn the unfortunate loss of one of the co-authors, our great friend and colleague Professor Patrice Francour, to whom we dedicate this book.

November 2021

Fabiana C. Félix-Hackradt, Carlos W. Hackradt
Porto Seguro (Brazil)

José A. García-Charton
Murcia (Spain)

Contents

Preface iii

Section 1: Biology and Ecology of Groupers

1.1 Classification of Groupers 3
Ka Yan Ma and *Matthew T. Craig*

1.2 Early Life Development 23
Fabiana C. Félix-Hackradt

1.3 Sexual Patterns and Reproductive Behaviours in Groupers 54
Beatrice Padovani Ferreira, Simone Marques, Mario Vinicius Condini and *Min Liu*

1.4 Feeding Biology of Groupers 74
M. Harmelin-Vivien and *J.G. Harmelin*

1.5 Interspecific Relationships 102
Patrice Francour and *Giulia Prato*

Section 2: Conservation and Management

2.1 Fisheries Regulation: Groupers' Management and Conservation 119
D. Rocklin, I. Rojo, M. Muntoni, D. Mateos-Molina, I. Bejarano, A. Caló, M. Russell, J. Garcia, F.C. Félix-Hackradt, C.W. Hackradt and *J.A. García-Charton*

2.2 Grouper Aquaculture: World Status and Perspectives 166
Branko Glamuzina and *Michael A. Rimmer*

2.3 The Importance of Groupers and Threats to Their Future 191
Yvonne Sadovy de Mitcheson and *Min Liu*

Index 231

Section 1
Biology and Ecology of Groupers

CHAPTER 1.1

Classification of Groupers

Ka Yan Ma[1,#] and *Matthew T. Craig*[2,*]

Introduction

Biological classification aims to define groups of organisms. As the broader field of systematics has evolved, classification has developed into a system that is based on the evolutionary relationships among organisms. It is generally accepted that a natural classification is one that groups together all the descendants of a common ancestor into a taxon, a condition that is termed "monophyletic" (Hennig 1965). Organisms in a taxon share a common genetic inheritance, which may be expressed in a variety of observable and measurable characters that taxonomists use to develop classification schemes. Traditionally, characters used in fish classification are morphological, such as lateral-line scale count, fin shape and position, and various morphometric ratios. However, similar morphological characters may evolve independently in distantly related lineages due to convergent evolution, and classification based on these analogous characters would create unnatural, non-monophyletic taxonomic groups. Additionally, selection plays a large role in shaping the morphology of an organism, which further complicates evolutionary histories derived from these data. Molecular data have emerged as valuable tools that, along with morphology, can increase our understanding of the evolutionary relationships among lineages, thus enhancing our ability to create robust classification systems. In some cases, these data are thought to be relatively free of strong selective pressures and may not experience such frequent instances of convergent evolution.

[1] Simon F. S. Li Marine Science Laboratory, School of Life Sciences, The Chinese University of Hong Kong, Hong Kong SAR.

[2] National Oceanic and Atmospheric Administration, National Marine Fisheries Service, Southwest Fisheries Science Center, 8901 La Jolla Shores Drive, La Jolla, CA 92037, USA.
Email: makayana@gmail.com

[#] Current affiliation: School of Ecology, Sun Yat-sen University, Shenzhen, China.

[*] Corresponding author: matthew.craig@noaa.gov

Groupers (Epinephelinae *sensu* Craig and Hastings (2007)) are a species-rich group of marine reef fish whose classification has undergone many changes and remains controversial across various levels of the taxonomic hierarchy from family to species. Historically, their classification was based on morphology. In the past decade or so, a number of molecular phylogenetic studies, using a variety of markers and covering many taxonomic levels, have contributed significantly to clarifying the evolutionary relationships among groupers and creating a more coherent classification (e.g., Maggio et al. 2005, Ding et al. 2006, Craig and Hastings 2007, Craig et al. 2011, Zhuang et al. 2013, Schoelinck et al. 2014, Ma et al. 2016, Qu et al. 2017, Ma and Craig 2018). This chapter discusses the chronological development of grouper classification from the morphology-era to the molecular-era, from family to genera, and to species. We discuss the most comprehensive grouper phylogeny developed to date (Ma and Craig 2018) and its implications for taxonomy. By consolidating both morphological and molecular information, we present a revised grouper classification scheme that reflects evolutionary relationships of this diverse group of marine fish.

The Family for Groupers: Serranidae or Epinephelidae

Groupers have long been recognized as part of the large and diverse family Serranidae, which has traditionally been used as a convenient pigeon-hole for lower percoid species with ambiguous affinity and has received extensive systematic treatments (Fig. 1). In the first attempt to coherently organise the Serranidae, Jordan and Eigenmann (1890) defined six subfamilies: Anthiinae (the anthias), Epinephelinae (the groupers), Grammistinae (the soapfishes and podges), Latinae (the late perches), Percichthyinae (the temperate perches), and Serraninae (the basses) (Fig. 1a). Jordan (1923) elevated the groupers to family level (Epinephelidae), thus restricting the Serranidae to the subfamilies Anthiinae and Serraninae, and treated the monotypic genus *Niphon* as its own family, Niphonidae (Fig. 1b). The first cladistically-based classification of the family was by Gosline (1966), in which Serranidae was restricted to Jordan and Eigenmann's (1890) Anthiinae, Epinephelinae, and Serraninae (Fig. 1c). Kendall (1976, 1979) added to Serranidae the subfamily Grammistinae (treated as family Grammistidae by Gosline (1966)) based on the similar number and orientation of predorsal bones to those of the Epinephelinae (*sensu* Jordan and Eigenmann (1890)) (Fig. 1d). Later, Johnson (1983) agreed with Gosline's (1966) three-subfamily scheme, but also followed Kendall (1976, 1979) in placing members of Grammistinae into the family Serranidae (Fig. 1e). Johnson (1983) defined a monophyletic subfamily Epinephelinae based on the absence of an autogenous distal radial on the first dorsal-fin pterygiophore, and in doing so he included Kendall's Grammistinae as well as the enigmatic species *Niphon spinosus* as part of Epinephelinae. Johnson (1983, 1988) further divided his Epinephelinae into five tribes: Diploprionini (soapfishes), Gramminstini (soapfishes), Liopropromini (basslets), Niphonini

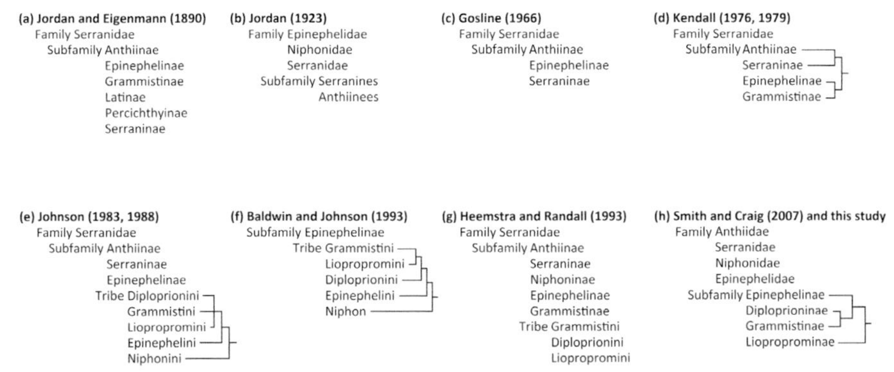

Fig. 1 Classification schemes of the family Serranidae proposed by (a) Jordan and Eigenmann (1890), (b) Jordan (1923), (c) Gosline (1966), (d) Kendall (1976, 1979), (e) Johnson (1983, 1988), (f) Baldwin and Johnson (1993), (g) Heemstra and Randall (1993) and (h) Smith and Craig (2007) and this study. Cladograms illustrating relationships among tribes based on morphology are shown in (d), (e), (f), and (h).

(*Niphon spinosus*), and Epinephelini (groupers) (Fig. 1e). Baldwin and Johnson (1993) cladistically demonstrated the monophyly of Johnson's Epinephelinae and analyzed the relationships among its tribes: *Niphon* was hypothesized to be most distantly related to all other epinehelines, followed by Epinephelini, and then Diploprionini, whereas Grammistini and Liopropromini, were thought to be closely related sister groups (Fig. 1f). By contrast, Heemstra and Randall (1993) proposed an alternative classification which restricted Epinephelinae to Johnson's (1983, 1988) Epinephelini, elevated the genus *Niphon* to subfamily Niphoninae, and united Diploprionini, Gramministini, and Liopropromini to subfamily Grammistinae (Fig. 1g). The classification schemes proposed by Johnson (1983, 1988), Baldwin and Johnson (1993), and Heemstra and Randall (1993), though substantially different, are all used today, causing considerable confusion (Fig. 1).

In the past two decades, several studies used DNA sequence data to scrutinize the evolutionary relationships within the Perciformes and among the wider percomorph group. Many of these studies challenged the monophyly of Serranidae (Smith and Wheeler 2004, Dettai and Lecointre 2005, Smith and Craig 2007, Lautredou et al. 2013, Schoelinck et al. 2014, Smith et al. 2018). While demonstrating the polyphyly of the "Serranidae", a common feature in all of these studies is a monophyletic Epinehpelinae (*sensu* Baldwin and Johnson (1993), but excluding *Niphon*). In light of their findings, Smith and Craig (2007) recommended restricting the Serranidae to members of Johnson's (1983, 1988) Serraninae and resurrecting Niphonidae and Epinephelidae, a scheme also supported by data in Craig and Hastings (2007) (Fig. 1h). Monophyly of the Epinephelidae is supported not only by molecular data, but also morphologically by the presence of three opercular spines in adults (Gosline 1966, Johnson 1983), and the extremely elongated, fleshy sheath-encased spine serially associated with the first dorsal-fin pterygiophore in larvae (Baldwin and Johnson 1993). While some studies that were designed to test higher-level relationships (above family level) recover a group comprising the three traditional "serranid" subfamilies (e.g., Near et al. 2012, 2013, Betancur-R et al. 2013, Betancur-R et al. 2017), these were based on limited taxon sampling within each family and not aimed at testing monophyly of families. Nonetheless, these have led some to be reluctant to accept the Epinephlidae as a valid family. Politics and the relative acceptance of genetic data have thus resulted in variable usage of the Epinephelidae as a valid family. Considering that all molecular phylogenies with adequate taxon sampling demonstrated the non-monophyly of the traditional "Serranidae" (Smith and Wheeler 2004, Dettai and Lecointre 2005, Smith and Craig 2007, Lautredou et al. 2013, Schoelinck et al. 2014, Smith et al. 2018), we follow Smith and Craig (2007) and Craig and Hastings (2007) and accept the family Epinephelidae, treating the tribes of Johnson (1983, 1988) as subfamilies (excluding Niphonini; treated as Niphonidae). The revised family Epinephelidae comprises all genera of the Epinephelinae *sensu* Baldwin and Johnson (1993) except *Niphon*, and can be divided into four subfamilies: Diploprioninae (diploprionine soapfishes), Gramminstinae (gramistine soapfishes), Liopropominae (basslets), and Epinephelinae (groupers). This scheme is compatible with all modern phylogenetic treatments of the group. There have been no studies published, to our knowledge, in which recognizing the Epinephelidae results in a non-monophyletic taxon.

The Groupers (Family Epinephelidae, Subfamily Epinephelinae)

Groupers are one of the most species-rich percoid clades, comprising 167 species traditionally divided into 15 genera: *Aethaloperca, Alphestes, Anyperodon, Cephalopholis, Cromileptes, Dermatolepis, Epinephelus, Gonioplectrus, Gracila, Mycteroperca, Paranthias, Plectropomus, Saloptia, Triso*, and *Variola* (Heemstra and Randall 1993). With the exception of works by Johnson (1983, 1988) and Baldwin and Johnson (1993), few systematic studies have been undertaken to resolve the phylogenetic relationships among these genera or to verify their monophyly. In contrast, since Craig et al. (2001) presented the first molecular analysis of groupers and challenged the monophyly of *Cephalopholis, Mycteroperca*, and *Epinephelus*, several molecular studies have attempted to scrutinize the relationships of the epinepheline genera. In this section, we will discuss the most comprehensive molecular phylogeny of the Epinephlinae to date (Ma and Craig 2018), followed by (dis-)agreements with morphology and resulting taxonomic implications.

The Phylogenetic Relationships of Grouper Genera

Craig et al. (2001) was the first to demonstrate the polyphyly of the three largest grouper genera (*Cephalopholis, Epinephelus*, and *Mycteroperca*) using DNA sequences of mitochondrial 16S rRNA. Later, Maggio et al. (2005) analysed the mitochondrial cytochrome *b* and 16S rRNA DNA sequences of

East Atlantic grouper species and again demonstrated the non-monophyly of *Epinephelus* and *Mycteroperca*. Ding et al. (2006) analysed 16S rDNA of 30 species of groupers in the China Seas with the same results. However, those studies assumed regional monophyly, a hypothesis already challenged by data in Craig et al. (2001).

A more comprehensive phylogenetic study of groupers was presented by Craig and Hastings (2007), who analysed two nuclear (TMO4C4 and histone H3) and two mitochondrial (16S and 12S) genes for 119 of the 163 (73%) epinepheline species described at that time. Their phylogeny recovered a monophyletic Epinephelidae and polyphyletic genera *Cephalopholis*, *Epinephelus*, and *Mycteroperca*. Despite the high taxon coverage and ample genetic data used in that study, even more comprehensive taxonomic sampling and additional molecular markers were deemed required to better resolve several phylogenetic ambiguities, including: (1) How is the monotypic genus *Gonioplectrus* related to other groupers? This genus, found only in the West Atlantic, was not analysed in previous studies, but based on the low number of dorsal fin spines, it was hypothesized to belong to the basal clades of the family (Heemstra and Randall 1993, Craig and Hastings 2007). (2) The phylogenetic position of the genus *Variola* was poorly resolved by Craig and Hastings (2007), but it was considered a probable member of the basal epinepheline clades. (3) The phylogenetic position of the genus *Hyporthodus* (erected by Craig and Hastings (2007) for the "deep bodied" groupers) was also ambiguous, and it was unclear whether the genus as currently defined contained other, unsampled species. (4) The fine scale phylogeny within the species-rich genus *Epinephelus* was unclear. Specifically, the "reticulated-grouper complex", which contains nine species characterized by having a "rounded caudal fin and close-set dark brown spots with pale interspaces forming a network on the body" (Heemstra and Randall 1993) appeared polyphyletic, though statistical support was lacking in Craig and Hastings (2007).

Following Craig and Hastings (2007), three molecular phylogenetic studies of groupers also questioned traditional morphological based classification schemes and examined the relationships of the epinepheline genera, including Zhuang et al. (2013) using complete mitochondrial genome, Schoelinck et al. (2014) based on five genes (COI, 16S, TMO-4C4, Rhodopsin, and Pkd1), and Qu et al. (2017) based on mitochondrial COI and ND2 genes. The recovered relationships were similar to Craig and Hastings (2007), however they sampled woefully few taxa and the aforementioned phylogenetic ambiguities were not fully resolved.

Ma and Craig (2018), who built on the study of Ma et al. (2016), presented the most comprehensive molecular phylogeny of groupers, which includes 147 of 167 (88%) known species and representatives of all known epinepheline genera (Figs. 1, 2; Ma and Craig 2018). Phylogeny was reconstructed with four genes, COI, 12S, 16S, and TMO4C4 sequences using Bayesian inference performed in BEAST v1.8.0 (Drummond et al. 2012). A detailed biogeographic analysis of these data may be found in Ma et al. (2016).

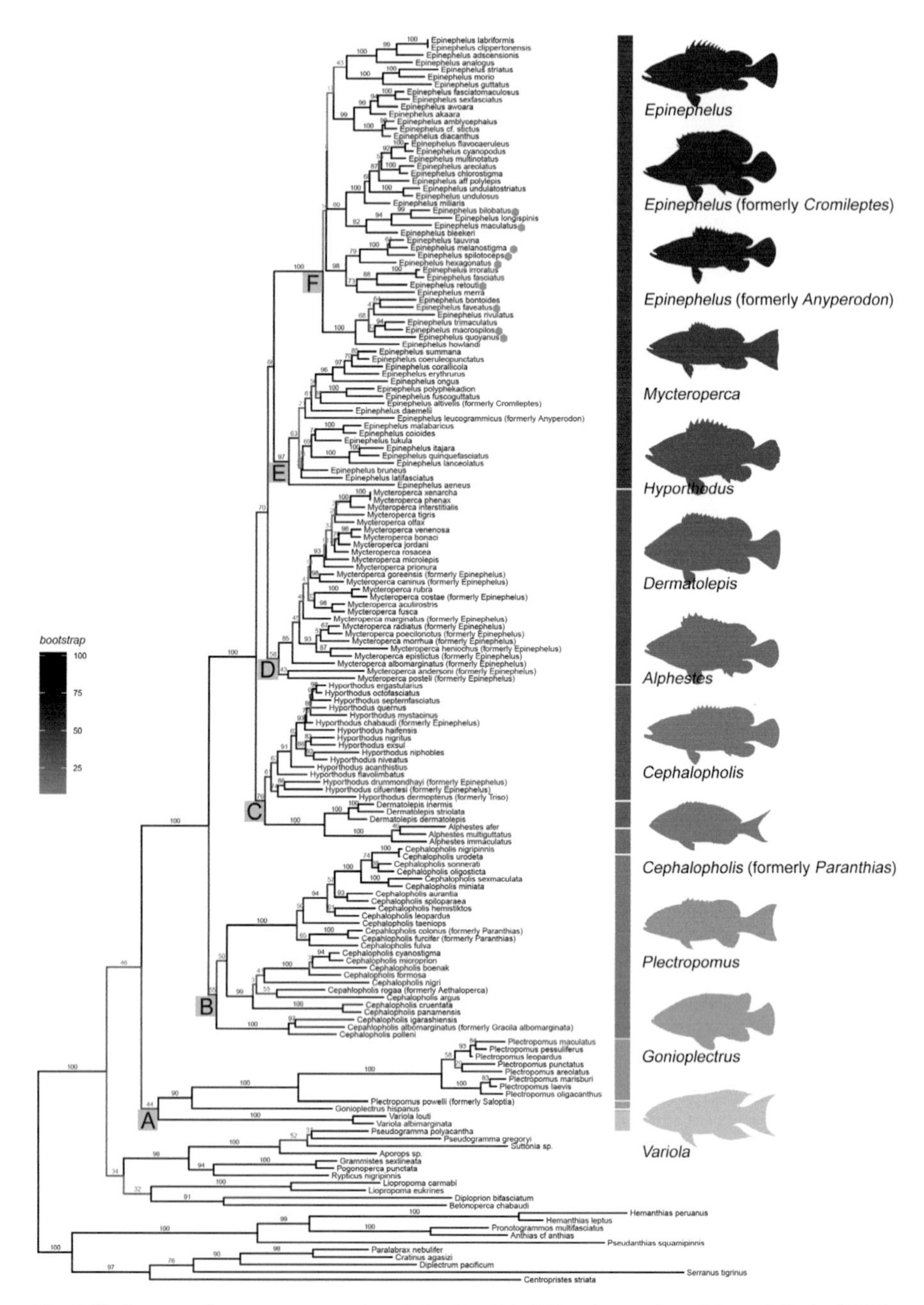

Fig. 2 Phylogeny of groupers reconstructed using RAxML redrawn from Ma and Craig (2018). Numbers on branches represent bootstrap supports. Epinepheline clades are labelled A to F. Members of the "reticulated-grouper complex" (Heemstra and Randall 1993) are indicated by hexagons.

Ma and Craig (2018) recovered several polyphyletic epinepheline taxa including *Cephalopholis*, *Epinephelus*, and *Mycteroperca*, confirming results from previous studies. It recovered six strongly supported clades that diverged sequentially (Fig. 2). The basal clade (A) contained *Variola*, *Gonioplectrus*, *Saloptia*, and *Plectropomus*. It was followed by Clade B, which contained *Cephalopholis*, *Gracila*, *Aethaloperca*, and *Paranthias*. Clade C comprised *Dermatolepis*, *Alphestes*, *Triso*, *Epinephelus chabaudi*, *E. cifuentesi* and *E. drummondhayi*, and *Hyporthodus*. Clade D consisted of *Mycteroperca* and 12 *Epinephelus* species. Clade E contained 17 *Epinephelus* species, *Cromileptes*, and *Anyperodon*, whereas the most speciose clade, Clade F, contained exclusively *Epinephelus* species.

Concordance between Morphological and Molecular Phylogenetic Inferences

Groupers have a rather large range of dorsal-fin hard spines with eight, nine, or 11 spines being the most common condition. Leis (1986) and Johnson (1988) concluded that the eight or nine spine condition is the pleisiomorphic state among groupers. The grouper phylogeny proposed by Ma et al. (2016) corroborates this hypothesis, as all genera with eight or nine dorsal spines (*Aethaloperca*, *Cephalopholis*, *Paranthias*, *Plectropomus*, *Saloptia*, and *Variola*) are basal to genera with 10 or more dorsal-fin spines (*Hyporthodus acanthistius* also has nine dorsal-fin spines, however this is considered autapomorphic in this species as it is clearly a member of *Hyporthodus*). This study also provides some interesting implications for the evolution of the dorsal-spine character in the Epinephelinae. From a strictly parsimonious viewpoint, genera with eight, nine, and 11 dorsal-fin spines would be presumed to be monophyletic. This suggests that this is not the case, as the nine-spine condition has apparently evolved twice.

Among the genera with low dorsal-fin spine counts is the monotypic genus *Gonioplectrus*, which was not included in previously published analyses. The single species, *G. hispanus*, is a deep-water (60–360 m) species inhabiting the West Atlantic. Although considered a very likely member of the Epinephelinae (Kendall and Fahay 1979, Johnson 1983), the placement of *Gonioplectrus* within the Epinephelinae has been uncertain due to a suit of apomorphic characters. *Gonioplectrus* has two well-developed supraneurals (versus only one in *Saloptia* and *Plectropomus*), 15 branched caudal-fin rays (versus 13 in *Saloptia* and *Plectropomus*), a prominent knob on the lower corner of the maxilla (absent in *Cephalopholis*, *Saloptia*, *Plectropomus*) and cranial differences (Heemstra and Randall 1993). Its low dorsal spine count (eight) and the presence of epineural ribs on vertebrae 1–9 led to speculation that it has close affinity to *Plectropomus* or *Saloptia* (with eight dorsal spines) or *Cephalopholis* (with nine dorsal spines). Ma and Craig (2018) support the hypothesis that *G. hispanus* is indeed a member of the Epinephelinae and is closely related to, but distinct from, *Saloptia*, *Plectropomus*, and *Cephalopholis*, and supports its monotypic status.

A close relationship between *Saloptia* and *Plectropomus* recovered in molecular phylogenetic analyses (Ma et al. 2016, Ma and Craig 2018) has long been recognized by Smith (1964), Leis (1986), and Heemstra and Randall (1993), because they are the only groupers with a single supraneural bone, low dorsal-fin spine count, three large antrorse spines on the lower edge of the preopercle, and only 13 branched caudal-fin rays. *Saloptia* has been treated as a distinct genus because of the "stronger" median fin spines than those in *Plectropomus* (Heemstra and Randall 1993), and the lack of enlarged canine teeth on the lower jaw. In view of the many synapomorphies shared by these two genera, the argument for retaining *Saloptia* as a distinct genus is relatively weak.

The genus *Variola* contains two Indo-Pacific coral reef-associated species *V. albimarginata* and *V. louti*. Three molecular phylogenetic studies (Craig and Hastings 2007, Schoelinck et al. 2014, Qu et al. 2017) placed *Variola* sister to all grouper genera except *Plectropomus* and *Saloptia* (which formed a clade basal to all the epinephelids), but with low to intermediate support. Zhuang et al. (2013), who analysed the mitochondrial genome of 22 grouper species, placed *Variola* in the same position with strong statistical support. This placement is puzzling as with nine dorsal spines, *Variola* would seemingly be closely allied with *Cephalopholis* (also possessing nine dorsal spines). Ma and Craig (2018) revealed that *Variola* is closely related to the lineage containing *Plectropomus*, *Saloptia*, and *Gonioplectrus*, but distinct from *Cephalopholis*, implying that the nine-spine condition has evolved twice in groupers. Leis (1986) found that in larval *Cephalopholis* and *Paranthias*, the first 8 dorsal fin spines are formed directly, while the ninth spine resulted from a transformation of the anteriormost dorsal soft ray. In grouper species with more than nine spinous rays, the anteriormost two soft rays of the larvae develop into spines (Kendall 1979). The development of dorsal-fin spination in *Variola* is unknown. The nine-spine condition, therefore, may not represent a homologous character state.

Cephalopholis, *Paranthias*, *Aethaloperca*, and *Gracila* form a monophyletic clade in all molecular phylogenetic analyses to date (Craig et al. 2001, Craig and Hastings 2007, Zhuang et al. 2013, Schoelinck et al. 2014, Ma et al. 2016, 2018, Qu et al. 2017), and morphological data suggests that they are close allies (e.g., Randall 1964, Smith-Vaniz et al. 1988). *Gracila albomarginata* was removed from *Cephalopholis* and placed in the monotypic *Gracila* in view of its shorter head and semipelagic behaviour (Randall 1964). *Aethaloperca rogaa* was removed from *Perca* and placed in a monotypic genus as it has a much deeper body and much steeper profile of the head (Randall 1964). The semipelagic behaviour of *Paranthias* is rare among epinephelines (shared with *Gracila*), and its distinct morphological features associated with its niche occupancy, such as deeply forked tail, high gill raker counts, and small teeth (Randall 1967), are convergent with those seen in basslets. However, these four genera shared a number of characters, including: (1) the presence of nine dorsal fin spines (shared only with *Hyporthodus acanthistius*) that are formed indirectly (Leis 1986, Kendall 1979); (2) the presence of trisegmental pterygiophores in the dorsal fin (bisegmental in remaining epinephelines) (Heemstra and Randall 1993); and (3) *Paranthias* and *Cephalopholis* share ctenoid midlateral scales, epineural ribs on vertebrate 1–9 (*Epinephelus* and *Mycteroperca* have epineurals on vertebrae 1–10) (Baldwin and Johnson 1993, Heemstra and Randall 1993). Therefore, there is strong morphological and molecular evidence which supports the close relationship among these genera.

Hyporthodus was resurrected for the "deep bodied" groupers (Craig and Hastings 2007). It is characterized by having a deeper body, drab brown or olive coloration with or without dark bars on the body, pelvic fins that insert immediately below or in front of the pectoral insertion, and an elongate, triangular foramen formed by the articulation between the cleithrum and the coracoids (type I articulation of Craig 2005). Currently, 11 species (plus three species *incertae sedis*) are assigned to *Hyporthodus* (Craig and Hastings 2007). The analysis of Ma and Craig (2018) included all *Hyporthodus* species, except *H. darwinensis* and *H. perplexus*, which have only a single holotype and showed that *Epinephelus chabaudi* should be regarded as a member of this genus. *Epinehelus chabaudi* shares the characteristic deep body and greyish coloration with bars of the remaining species.

Hyporthodus has been found to be closely related to *Triso*, *Dermatolepis*, and *Alphestes* in molecular phylogenetic analyses (Craig et al. 2001, Zhuang et al. 2013, Ma et al. 2016, 2018), suggesting the grouping of two *Epinephelus* species, *E. drummondhayi* and *E. cifuentesi*, into this clade (Clade C in Fig. 2), but their relationships with other members of this clade are obscure. More genetic data are required to properly resolve this phylogenetic ambiguity, and we would regard the status of these two species as *incertae sedis*. Morphologically, clade C members are generally deep bodied (depth contained less than 3 times in standard length), have nine anal fin rays (except *Hyporthodus exsul* which has eight, *Triso dermopterus* which has 10, and *Dermatolepis inermis* which has 8–10) while other epinephelids with 10 or more dorsal fin spines generally have eight or less anal fin rays (in most *Epinephelus* except *Cromileptes* and *Anyperodon*) or 10–13 (in *Mycteroperca*). *Alphestes* and *Dermatolepis* are morphologically similar. The mottled colour pattern with white spots in some temporary phases, dark spots in others, the rather smooth scales, and certain similarities in body proportion led Smith (1971) to suggest that they are close relatives.

Morphologically, the genus *Mycteroperca* appears to be closely related to *Epinephelus*, as they both have ten or eleven dorsal-fin spines and lack the trisegmental pterygiophores in the dorsal and anal fins and the knob at the lower corner of maxilla which are diagnostic characteristics of *Cephalopholis* (Heemstra and Randall 1993). *Mycteroperca* can be distinguished from *Epinephelus* by its parallel skull crests, well developed supraethmoid wall, long anal fin, greater number of anal fin rays (10–13 in *Mycteroperca* and mostly eight but ranges 7–10 in *Epinephelus*), head length (2.5–2.7 in standard length in *Mycteroperca* versus 2.1–2.5 in standard length in *Epinephelus*), and in caudal fin shape (truncate, emarginate, or distinctly concave in *Mycteroperca* versus generally round and rarely truncate in *Epinephelus*). Based on molecular phylogenetic analyses, 12 *Epinephelus* species nested within the clade of *Mycteroperca*, including *E. radiatus*, *E. epistictus*, *E. heniochus*, *E. poecilonotus*, *E. morrhua*, *E. andersoni*, *E. posteli*, *E. albomarginatus*, *E. marginatus*, *E. costae*, *E. goreensis*, and *E. caninus* (Craig and Hastings 2007, Zhuang et al. 2013, Ma et al. 2016). Yet, most of these *Epinephelus* species have 8–9 anal fin rays and a rounded caudal fin. Although morphological characters derived from their common ancestor are yet to be found, molecular data provide strong support for the monophyly of this clade of groupers, and hence it seems most prudent to include these 11 species in *Mycteroperca*.

Although the genus *Epinephelus sensu lato* appears polyphyletic in all molecular phylogenetic analyses conducted, most of its species (59) are recovered in the crown group—two sister clades (Clades E and F) in Ma and Craig (2018) (see Fig. 2, and also Ma et al. 2016, Schoelinck et al. 2014, Qu et al. 2017). Morphological differentiation among these two sister clades is yet to be identified, but they exhibit subtle differences in biology. Clade E contains *Epinephelus* and the monotypic *Anyperodon* and *Cromileptes*. The majority (at least 12 of 19) of this clade are large (maximum total length 75 to 250 cm), and most are restricted to coral reefs. On the other hand, Clade F contains only *Epinephelus* species. Clade F *Epinephelus* species are predominantly small (31 of 42), having maximum total length < 75 cm (Heemstra and Randall 1993), and most can inhabit non-coral reef environments. Ecological factors have likely played a major role in driving the divergence and radiation of these two species-rich clades of groupers, because 88% of cladogenesis events in this clade were non-allopatric (Ma et al. 2016). It is also important to note that the "reticulated-grouper complex" (Heemstra and Randall 1993), all of which are analysed phylogenetically in Ma and Craig (2018), is polyphyletic. The nine reticulated grouper species are scattered across three distinct clades along with "non-reticulated" groupers. Hence, body colour patterns may not be a good character for revealing phylogenetic relationships of epinephelid groupers.

The inclusion of the morphologically distinct *Anyperodon* and *Cromileptes* within clade E, dominated by *Epinephelus* species, is consistent with all molecular studies. *Anyperodon* is deemed closely related to *Epinephelus* as they share 11 dorsal-fin spines and the absence of trisegmental pterygiophores. It is treated as a distinct genus because of the elongated and distinctly compressed head and body and lacking palatine teeth. Nonetheless, the body width of *Anyperodon* is 2.3–2.8 times in its depth, while the range of body width for *Epinephelus* species is 1.8–2.8 in their depth (Heemstra and Randall 1993, Craig et al. 2011). Additionally, *Anyperodon*'s body depth is contained 3.1–3.7 times in standard length, while *Epinephelus* species' body depth ranges from 2.3–3.7 times in standard length (Heemstra and Randall 1993, Craig et al. 2011). Finally, the head length of *Anyperodon* is contained 2.3–2.5 in standard length, while in *Epinephelus* species, it is contained

2.1–2.8 times in standard length (Heemstra and Randall 1993, Craig et al. 2011). Therefore, *Anyperodon* is not more elongated or compressed than any other *Epinephelus* species. *Cromileptes* has a head shape (depressed anteriorly and elevated posteriorly) that is unique among groupers. However, it shares epineural ribs on vertebrae 1–10, two supraneural bones, and no trisegmental pterygiophores with *Epinephelus*. The unique characters of these *Anyperodon* and *Cromileptes* are curious autapomorphic specializations that are worthy of future investigation, however they are insufficient to exclude them from Epinephelus. Based on the genetic data, we deem it is most appropriate to synonymise these two genera with *Epinephelus*.

A Revised Taxonomy of Grouper Genera

Multiple molecular phylogenetic analyses have challenged past generic classification of groupers by clearly refuting the monophyly of the genera *Epinephelus*, *Mycteroperca*, and *Cephalopholis.* A revised classification of the Epinephelinae was therefore proposed by Craig and Hastings (2007) comprising nine genera: *Alphestes*, *Cephalopholis* (subsuming *Aethaloperca*, *Gracila*, and *Paranthias*), *Dermatolepis*, *Epinephelus* (*sensu stricto*), *Gonioplectrus*, *Hyporthodus* (subsuming *Triso dermopterus*, with the addition of three species previously placed in *Epinephelus*), *Mycteroperca* (with the addition of 12 species previously placed in *Epinephelus*), *Plectropomus* (subsuming *Saloptia*), and *Variola*. Molecular phylogenetic analyses corroborated the scheme of Craig and Hastings (2007) and provides additional revisions of this classification based on newly added taxa. A revised classification is showed in Appendix I.

Cryptic Speciation in Groupers

Many marine species, including groupers, have broad distribution, with some species ranging more than 10,000 km (Jablonski and Lutz 1983). With the use of genetic data, however, many marine species once thought to inhabit wide ranges are found to contain cryptic species (two or more distinct species that are morphologically indistinguishable and classified as the same species). A good example of cryptic species discovery in groupers is one that concerns the goliath grouper, which is one of the largest reef fishes and is considered critically endangered by the IUCN. The goliath grouper *Epinephelus itajara* (*sensu* Heemstra and Randall (1993)), was thought to have a relatively large range, occurring in sub-tropical and tropical waters of the East Pacific and the two Atlantic coasts (Heemstra and Randall 1993). No morphological differences have been identified to date between these widely distributed populations (Heemstra and Randall 1993). By analysing mitochondrial (16S rRNA and cytochrome *b*) and nuclear (S7) gene sequences, Craig et al. (2009) found that the goliath grouper populations in the Pacific and West Atlantic are genetically distinct and represent two distinct species. The species *Epinephelus quinquefasciatus* was therefore resurrected for Pacific goliath grouper.

More recently, genetics (complete mitochondrial genome) and morphological data provide ample evidence to validate the species status of *Epinephelus moara* (Liu et al. 2013). This species has long been synonymised with *E. bruneus* (Heemstra and Randall 1993, Craig et al. 2011). The former species inhabits the South and East China Seas, while the latter is a South China Sea endemic (Cheng and Zheng 1987). Interestingly, *E. moara* costs almost double in the wholesale market when compared to *E. bruneus*. Detailed morphological comparison found several diagnostic characters for *E. moara*, including white edges on the margin of fins, orange or greenish-yellow membranes/margin of dorsal and caudal fins, the shapes of preopercle, lachrymal, and urohyal bones, and the sub-branch number of pyloric caeca (Liu et al. 2013). The addition of this "cryptic species" is reflected in our revised grouper classification in Appendix I.

Ma et al. (2016) also discovered a potential new cryptic species of groupers-the two subspecies *P. pessuliferus pessuliferus* and *P. pessuliferus marisrubri* are found to be nested in two deeply diverged lineages of *Plectropomus* in the phylogenetic tree (Fig. 2). The distinction of the two subspecies was described by Randall and Hoese (1986): the Red Sea endemic *P. pessuliferus marisrubri* has more blue spots on the cheek (30–50 versus 5–12) than the Indo-Pacific *P. pessuliferus pessuliferus*, and pectoral fins of the former taxon are dark brown and become pale distally, while those of the latter taxon are uniformly pale and translucent. Currently, the two subspecies are considered near threatened in the IUCN Red List (Ferreira et al. 2008) and are overexploited in the Red Sea and Maldives. It is therefore crucial and urgent to carry out a more comprehensive morphological and morphometrical investigation of these *Plectropomus* taxa and to formally revise the taxonomy of these two commercially important groupers.

Epinephelus craigi (Frable et al. 2018), *E. fuscomarginatus* (Johnson and Worthington Wilmer 2019), *E. insularis* (Nakamura and Motomura 2021) and *E. kupangensis* (Tucker et al. 2016) are recent addition to the genus. *Epinephelus tankahkeei* (Wu et al. 2020), which has been historically treated as *E. chlorostigma*, was established based on both morphological and genetic differentiation. However, Nakamura and Motomura (2021) argued that *Serranus areolatus japonicus* originally described by Temminck and Schlegel (1843) but synonymized with *E. chlorostigma* by Randall and Heemstra (1991) should be reinstalled as a valid species, *E. japonicus*. And as *E. japonicus* and *E. tankahkeei* are indistinguishable morphologically and genetically, the former should be treated as a senior synonym of the latter.

As more genetic data become available, it is likely that more cryptic grouper species will be discovered, and more phylogenetic ambiguities of groupers will be resolved.

Acknowledgements

We would like to thank J.H. Choat, L. van Herwerden and P.A. Hastings for guidance and input in our research on groupers phylogenetics.

References

Baldwin, C.C. and G.D. Johnson. 1993. Phylogeny of the Epinephelinae (Teleostei: Serranidae). Bull. Mar. Sci. 52: 240–283.

Betancur-R, R., R.E. Broughton, E.O. Wiley, K. Carpenter, J.A. Lopez, C. Li, N.I. Holcroft, D. Arcila, M. Sanciangco, J.C. Cureton II, F. Zhang, T. Buser, M.A. Campbell, J.A. Ballesteros, A. Roa-Varon, S. Willis, W.C. Borden, T. Rowley, P.C. Reneau, D.J. Hough, G. Lu, T. Grande, G. Arratia and G. Orti. 2013. The tree of life and a new classification of bony fishes. PLOS Currents Tree of Life. 2013 Apr 18. Edition 1.

Betancur-R, R., E.O. Wiley, G. Arratia, A. Acero, N. Bailly, M. Miya, G. Lecointre and G. Orti. 2017. Phylogenetic classification of bony fishes. BMC Evol. Biol. 17: 1–40.

Cheng, X.T. and B.S. Zheng. 1987. Systematic Synopsis of Chinese fishes, 1. Science Press, Beijing, China (in Chinese).

Craig, M.T., D.J. Pondella, J.P.C. Franck and J.C. Hafner. 2001. On the status of the serranid fish genus Epinephelus: evidence for paraphyly based on 16SrDNA sequence. Mol. Phylogenet. Evol. 19: 121–130.

Craig, M.T. 2005. A Molecular Phylogeny and Revised Classification for the Groupers of the Subfamily Epinephelinae (Serranidae). Doctoral dissertation, University of California, San Diego, United States.

Craig, M.T., P. Bartsch, P. Wirtz and P.C. Heemstra. 2006. Redescription and validation of *Alphestes afer* as an amphi-Atlantic grouper species (Perciformes: Serranidae). Cybium 30: 327–331.

Craig, M.T. and P.A. Hastings. 2007. A molecular phylogeny of the groupers of the subfamily Epinephelinae (Serranidae) with a revised classification of the Epinephelini. Ichthyol. Res. 54: 1–17.

Craig, M.T., R.T. Graham, R.A. Torres, J.R. Hyde, M.O. Freitas, B.P. Ferreira, M. Hostim-Silva, L.C. Gerhardinger, A.A. Bertoncini and D.R. Robertson. 2009. How many species of goliath grouper are there? Cryptic genetic divergence in a threatened marine fish and the resurrection of a geopolitical species. Endanger. Species. Res. 7: 167–174.

Craig, M.T., Y.J. Sadovy de Mitcheson and P.C. Heemstra. 2011. Groupers of the World: A Field and Market Guide. NISC (Pty) Ltd., Grahamstown, South Africa.

Dettai, A. and G. Lecointre. 2005. Further support for the clades obtained by multiple molecular phylogenies in the acanthomorph bush. C. R. Biol. 328: 674–689.

Ding, S., X. Zhuang, F. Guo, J. Wang, Y. Su, Q. Zhang and Q. Li. 2006. Molecular phylogenetic relationships of China Seas groupers based on cytochrome b gene fragment sequences. Science in China Series C 49: 235–242.

Drummond, A.J., M.A. Suchard, D. Xie and A. Rambaut. 2012. Bayesian phylogenetics with BEAUti and the BEAST 1.7 Mol. Biol. Evol. 29: 1969–1973.

Ferreira, B.P., A.L.B. Gaspar, S. Marques, Y. Sadovy, L. Rocha, J.H. Choat, A.A. Bertoncini and M. Craig. 2008. *Plectropomus pessuliferus*. *In*: IUCN 2008. 2008 IUCN Red List of Threatened Species.

Frable, B.W., S.J. Tucker and H.J. Walker. 2019. A new species of grouper, *Epinephelus craigi* (Perciformes: Epinephelidae), from the South China Sea. Ichthyol. Res. 66: 215–224.

Gosline, W.A. 1966. The limits of the fish family Serranidae, with notes on other lower percoids. Proc. Calif. Acad. Sci. 33: 91–111.

Heemstra, P.C. and J. Randall. 1993. Groupers of the World. FAO Species Catalogue 16, Rome.

Hennig, W. 1965. Phylogenetic systematics. Annual Review of Entomology 10: 97–116.

Jablonski, D. and R.A. Lutz. 1983. Larval ecology of marine benthic invertebrates: paleobiological implications. Biol. Rev. 58: 21–89.

Johnson, G.D. 1983. *Niphon spinosus*: a primitive epinepheline serranid, with comments on the monophyly and intrarelationships of the Serranidae. Copeia. 1983: 777–787.

Johnson, G.D. 1988. *Niphon spinosus*, a primitive epinepheline serranid: corroborative evidence from the larvae. Jpn. J. Ichthyol. 35: 7–18.

Jordan, D.S. and C.H. Eigenmann. 1890. A review of the genera and species of Serranidae found in the waters of America and Europe. Bull. US. Fish. Com. VIII 1888: 329–441.

Jordan, D.S. 1923. A classification of fishes including families and genera as far as known. Stanford Univ. Publ. Univ. Ser. Biol. Sci. 2: 77–243.

Kendall, A.W. Jr. 1976. Predorsal and associated bones in serranid and grammistid fishes. Bull. Mar. Sci. 26: 585–592.

Kendall, A.W. Jr. 1979. Morphological comparisons of North American seabass larvae (Pisces: Serranidae). NOAA Tech. Rep. Circ. 428: 1–50.

Kendall, A.W. Jr. and M.P. Fahay. 1979. Larvae of the serranid fish *Gonioplectrus hispanus* with comments on its relationships. Bull. Mar. Sci. 29: 117–121.

Lautredou, A.C., H. Motomura, C. Gallut, C. Ozouf-Costaz, C. Cruaud, G. Lecointre and A. Dettai. 2013. New nuclear markers and exploration of the relationships among Serraniformes (Acanthomorpha, Teleostei): The importance of working at multiple scales. Mol. Phylogenet. Evol. 67: 140–155.

Leis, J.M. 1986. Larval development in four species of Indo-Pacific coral trout *Plectropomus* (Pisces, Serranidae: Epinephelinae) with an analysis of relationships of the genus. Bull. Mar. Sci. 38: 525–552.

Liu, M., J.L. Li, S.X. Ding and Z.Q. Liu. 2013. *Epinephelus moara*: a valid species of the family Epinephelidae (Pisces: Perciformes). J. Fish Biol. 82: 1684–1699.

Ma, K.Y. and M.T. Craig. 2018. An inconvenient monophyly: an update on the taxonomy of the groupers (Epinephelidae). Copeia.

Ma, K.Y., M.T. Craig, J.H. Choat and L. van Herwerden. 2016. The historical biogeography of groupers: Clade diversification patterns and processes. Mol. Phylogenet. Evol. 100: 21–30.

Maggio, T., F. Andaloro, F. Hemidac and M. Arculeoa. 2005. A molecular analysis of some Eastern Atlantic grouper from the *Epinephelus* and *Mycteroperca* genus. J. Exp. Mar. Biol. Ecol. 321: 83–92.

Nakamura, J. and H. Motomura. 2021. *Epinephelus insularis*, a new species of grouper from the western Pacific Ocean, and validity of *E. japonicus* (Temminck and Schlegel 1843), a senior synonym of *Serranus reevesii* Richardson 1846 and *E. tankahkeei* Wu et al. 2020 (Perciformes: Epinephelidae). Ichthyol. Res. 68: 263–276.

Near, T.J., R.I. Eytan, A. Dornburg, K.L. Kuhn, J.A. Moore, M.P. Davis, P.C. Wainwright, M. Friedman and W.L. Smith. 2012. Resolution of ray-finned fish phylogeny and timing of diversification. Proc. Natl. Acad. Sci. U.S.A. 109: 13698–13703.

Near, T.J., A. Dornburg, R.I. Eytan, B.P. Keck, W.L. Smith, K.L. Kuhn, J.A. Moore, S.A. Price, F.T. Burbrink, M. Friedman and P.C. Wainwright. 2013. Phylogeny and tempo of diversification in the superradiation of spiny-rayed fishes. Proc. Natl. Acad. Sci. U.S.A. 110: 12738–12743.

Qu, M., W. Tang, Q. Liu, D. Wang and S. Ding. 2017. Genetic diversity within grouper species and a method for interspecific hybrid identification using DNA barcoding and RYR3 marker. Mol. Phylogenet. Evol. 121: 46–51.

Randall, J.E. 1964. Notes on the groupers of Tahiti, with description of a new serranid fish genus. Pac. Sci. 18: 281–296.

Randall, J.E. 1967. Food habits of reef fishes of the West Indies. Stud. Trop. Oceanogr. 5: 665–847.

Randall, J.E. and D.F. Hoese. 1986. Revision of the groupers of the Indo-Pacific genus *Plectropomus* (Perciformes: Serranidae). Indo-Pacific Fishes No. 13, P. Bernice Bishop Museum, Honolulu. 31 pp.

Randall, J.E and P.C. Heemstra. 1991. Revision of the Indo-Pacific groupers (Perciformes: Serranidae: Epinephelinae), with descriptions of five new species. Indo-Pacific Fishes No. 20, P. Bernice Bishop Museum, Honolulu. 296 pp.

Schoelinck, C., D.D. Hinsinger, A. Dettai, C. Cruaud and J.L. Justine. 2014. A phylogenetic re-analysis of groupers with applications for ciguatera fish poisoning. PloS One 9: e98198.

Smith, C.L. 1971. A revision of the American groupers: *Epinephelus* and allied genera. Bull. Am. Mus. Nat. Hist. 146: 1–241.

Smith, J.L.B. 1964. A new Serranid fish from deep water off Cook Island, Pacific. Ann. Mag. Nat. Hist. 6: 719–720.

Smith, W.L. and W.C. Wheeler. 2004. Polyphyly of the mail-cheeked fishes (Teleostei: Scorpaeniformes): evidence from mitochondrial and nuclear sequence data. Mol. Phylogenet. Evol. 32: 627–646.

Smith, W.L. and M.T. Craig. 2007. Casting the percomorph net widely: the importance of broad taxonomic sampling in the search for the placement of serranid and percid fishes. Copeia 2007: 35–55.

Smith, W.L., E. Everman and C. Richardson. 2018. Phylogeny and taxonomy of flatheads, scorpionfishes, sea robins, and stonefishes (Percomorpha: Scorpaeniformes) and the evolution of the lachrymal saber. Copeia 106: 94–119.

Smith-Vaniz, W.F., G.D. Johnson and J.E. Randall. 1988. Redescription of *Gracila albomarginata* (Fowler and Bean) and *Cephalopholis polleni* (Bleeker) with comments on the generic limits of selected Indo-Pacific groupers (Pisces: Serranidae: Epinephelinae). Proc. Acad. Nat. Sci. Phila. 140: 1–23.

Temminck, C.J. and H. Schlegel. 1843. Pisces. Part 1. pp. 1–20. *In*: von Siebold, P.F. (ed.). Fauna Japonica. Müller, Amsterdam.

Tucker, S.J., E.M. Kurniasih and M.T. Craig. 2016. A new species of grouper (Epinephelus; Epinephelidae) from the Indo-Pacific. Copeia 104: 658–662.

Wu, H., M. Qu, H. Lin, W. Tang and S. Ding. 2020. *Epinephelus tankahkeei*, a new species of grouper (Teleostei, Perciformes, Epinephelidae) from the South China Sea. ZooKeys 933: 125–137. https://doi.org/10.3897/zookeys.933.46406.

Zhuang, X., M. Qu, X. Zhang and S. Ding. 2013. A comprehensive description and evolutionary analysis of 22 grouper (Perciformes, Epinephelidae) mitochondrial genomes with emphasis on two novel genome organizations. PloS One 8: e73561.

APPENDIX I

Proposed classification for the fishes of the subfamily Epinephelinae based on molecular data. Species not included in this or any previous genetic analysis are listed as *insertae sedis* within the most appropriate genus based on a qualitative assessment of their overall morphology, and in accordance with previous classifications, especially Heemstra and Randall (1993).

Family Epinephelidae

Subfamily Epinephelinae **(167 species)**

Genus *Alphestes* Bloch and Schneider **(3 species)**

Type species: *Epinephelus afer* Bloch (by subsequent designation of Jordan and Swain).
Type locality: Gulf of Guinea (Craig et al. 2006).
Included species:

A. afer (Bloch 1793)
A. immaculatus Breder 1936
A. multiguttatus (Günther 1867)

Genus *Cephalopholis* Bloch and Schneider **(28 species)**

Type species: *Cephalopholis argus* Bloch and Schneider.
Type locality: East Indies.
Included species:

C. *aitha* Randall and Heemstra 1991
C. *albomarginatus* Fowler and Bean 1930 (formerly *Gracila*)
C. *argus* Schneider 1801
C. *aurantia* (Valenciennes 1828)
C. *boenak* (Bloch 1790)
C. *colonus* (Valenciennes 1846) (formerly *Paranthias*)
C. *cruentata* (Lacepède 1802)
C. *cyanostigma* (Valenciennes 1828)
C. *formosa* (Shaw 1812)
C. *fulva* (Linnaeus 1758)
C. *furcifer* (Valenciennes 1828) (formerly *Paranthias*)
C. *hemistiktos* (Rüppell 1830)
C. *igarashiensis* Katayama 1957
C. *leopardus* (Lacepède 1801)
C. *microprion* (Bleeker 1852)
C. *miniata* (Forsskål 1775)
C. *nigri* (Günther 1859)
C. *nigripinnis* (Valenciennes 1828)
C. *oligosticta* Randall and Ben-Tuvia 1983
C. *panamensis* (Steindachner 1876)
C. *polleni* (Bleeker 1868)
C. *polyspila* Randall and Satapoomin 2000
C. *rogaa* (Forsskål 1775) (formerly *Aethaloperca*)
C. *sexmaculata* (Rüppell 1830)
C. *sonnerati* (Valenciennes 1828)
C. *spiloparaea* (Valenciennes 1828)
C. *taeniops* (Valenciennes 1828)
C. *urodeta* (Forster 1801)

Genus *Dermatolepis* Gill **(3 species)**

Type species: *Dermatolepis punctatus* Gill (= *Epinephelus dermatolepis* Boulenger; replacement name for *D. punctatus* preoccupied in *Epinephelus* by *Holocentrus punctatus* Bloch).
Type locality: Cabo San Lucas, Baja California, Mexico.
Included species:

D. *dermatolepis* (Boulenger 1895)
D. *inermis* (Valenciennes 1833)
D. *striolata* (Playfair 1867)

Genus *Epinephelus* Bloch **(76 species)**

Type species: *Epinephelus marginalis* Bloch (= *E. fasciatus* (Forsskål) designated under the plenary powers of the IZCN Opinion 93).
Type locality: Red Sea.
Included species:

E. adscensionis (Osbeck 1765)
E. aeneus (Geoffroy Saint-Hilaire 1817)
E. akaara (Temminck and Schlegel 1842)
E. altivelis (Valenciennes 1828) (formerly *Cromileptes*)
E. amblycephalus (Bleeker 1857)
E. analogus Gill 1863
E. areolatus (Forsskål 1775)
E. awoara (Temminck and Schlegel 1842)
E. bilobatus Randall and Allen 1987
E. bleekeri (Vaillant 1878)
E. bontoides (Bleeker 1855)
E. bruneus Bloch 1793
E. chlorostigma (Valenciennes 1828)
E. clippertonensis Allen and Robertson 1999
E. coeruleopunctatus (Bloch 1790)
E. coioides (Hamilton 1822)
E. corallicola (Valenciennes 1828)
E. craigi Frable et al. 2018
E. cyanopodus (Richardson 1846)
E. daemelii (Günther 1876)
E. diacanthus (Valenciennes 1828)
E. erythrurus (Valenciennes 1828)
E. fasciatomaculosus (Peters 1865)
E. fasciatus (Forsskål 1775)
E. faveatus (Valenciennes 1828)
E. flavocaeruleus (Lacepède 1802)
E. fuscoguttatus (Forsskål 1775)
E. fuscomarginatus Johnson and Worthington Wilmer 2019
E. gabriellae Randall and Heemstra 1991
E. geoffroyi (Klunzinger 1870)
E. guttatus (Linnaeus 1758)
E. hexagonatus (Forster 1801)
E. howlandi (Günther 1873)
E. insularis Nakamura and Motomura 2021
E. irroratus (Forster 1801)
E. itajara (Lichtenstein 1822)
E. japonicus (Temminck and Schlegel 1843)
E. labriformis (Jenyns 1840)
E. lanceolatus (Bloch 1790)

E. latifasciatus (Temminck and Schlegel 1842)
E. leucogrammicus (Valenciennes 1828) (formerly *Anyperodon*)
E. longispinis (Kner 1864)
E. macrospilos (Bleeker 1855)
E. maculatus (Bloch 1790)
E. magniscuttis (Postel et al. 1963)
E. malabaricus (Bloch and Schneider 1801)
E. melanostigma Schultz 1953
E. merra Bloch 1793
E. miliaris (Valenciennes 1830)
E. moara (Temminck and Schlegel 1842)
E. morio (Valenciennes 1828)
E. multinotatus (Peters 1876)
E. ongus (Bloch 1790)
E. polylepis Randall and Heemstra 1991
E. polyphekadion (Bleeker 1849)
E. polystigma (Bleeker 1853)
E. quinquefasciatus (Bocourt 1868)
E. quoyanus (Valenciennes 1830)
E. retouti Bleeker 1868
E. rivulatus (Valenciennes 1830)
E. sexfasciatus (Valenciennes 1828)
E. socialis (Günther 1873)
E. spilotoceps Schultz 1953
E. stictus Randall and Allen 1987
E. stoliczkae (Day 1875)
E. striatus (Bloch 1792)
E. summana (Forsskål 1775)
E. tauvina (Forsskål 1775)
E. trimaculatus (Valenciennes 1828)
E. tuamotuensis Fourmanoir 1971
E. tukula Morgans 1959
E. undulatostriatus (Peters 1866)
E. undulosus (Quoy and Gaimard 1824)

Species *incertae sedis*:

*E. chlorocephalus** (Valenciennes 1830)
*E. indistinctus** Randall and Heemstra 1991
E. kupangensis Tucker et al. 2016
*E. lebretonianus** (Hombron and Jacquinot 1853)
*E. suborbitalis** Amaoka and Randall 1990
E. timorensis Randall and Allen 1987
*E. trophis** Randall and Allen 1987
*Known only from type series

Genus *Gonioplectrus* Gill **(1 species)**

Type species: *Plectropoma hispanum* Cuvier.
Type locality: Martinique.
Included species:
G. hispanus (Cuvier 1828)

Genus *Hyporthodus* Gill **(18 species)**

Type species: *Hyporthodus flavicauda* [=*Epinephelus niveatus* (Valenciennes in Cuvier and Valenciennes)]
Type locality: Newport, Rhode Island.
Included species:
H. acanthistius (Gilbert 1892)
H. chabaudi (Castelnau 1861) (formerly *Epinephelus*)
H. cifuentesi (Lavenberg and Grove 1993) (formerly *Epinephelus*)
H. dermopterus (Temminck and Schlegel 1842) (formerly *Triso*)
H. ergastularius (Whitley 1930)
H. exsul (Fowler 1944)
H. flavolimbatus (Poey 1865)
H. haifensis (Ben-Tuvia 1953)
H. mystacinus (Poey 1852)
H. nigritus (Holbrook 1855)
H. niphobles (Gilbert and Starks 1897)
H. niveatus (Valenciennes 1828)
H. octofasciatus (Griffin 1926)
H. quernus (Seale 1901)
H. septemfasciatus (Thunberg 1793)

Species *incertae sedis*:
*H. darwinensis** (Randall and Heemstra 1991)
*H. perplexus** (Randall et al. 1991)
H. drummondhayi (Goode and Bean 1878) (formerly *Epinephelus*)
*Known only from type series

Genus *Mycteroperca* Gill **(27 species)**

Type species: *Serranus olfax* Jenyns (by subsequent designation of Gill 1866).
Type locality: Galapagos Islands.
Included species:
M. acutirostris (Valenciennes 1828)
M. albomarginatus (Boulenger 1903) (formerly *Epinephelus*)
M. andersoni (Boulenger 1903) (formerly *Epinephelus*)
M. bonaci (Poey 1860)
M. cidi Cervigón 1966
M. caninus (Valenciennes 1843) (formerly *Epinephelus*)
M. costae (Steindachner 1878) (formerly *Epinephelus*)

M. epistictus (Temminck and Schlegel 1842) (formerly *Epinephelus*)
M. fusca (Lowe 1838)
M. goreensis (Valenciennes 1830) (formerly *Epinephelus*)
M. heniochus Fowler 1904 (formerly *Epinephelus*)
M. interstitialis (Poey 1860)
M. jordani (Jenkins and Evermann 1889)
M. marginatus (Lowe 1834) (formerly *Epinephelus*)
M. microlepis (Goode and Bean 1879)
M. morrhua (Valenciennes 1833) (formerly *Epinephelus*)
M. olfax (Jenyns 1840)
M. phenax Jordan and Swain 1884
M. poecilonotus (Temminck and Schlegel 1842) (formerly *Epinephelus*)
M. posteli (Fourmanoir and Crosnier 1964) (formerly *Epinephelus*)
M. prionura Rosenblatt and Zahuranec 1967
M. radiatus (Day 1868) (formerly *Epinephelus*)
M. rosacea (Streets 1877)
M. rubra (Bloch 1793)
M. tigris (Valenciennes 1833)
M. venenosa (Linnaeus 1758)
M. xenarcha Jordan 1888

Genus *Plectropomus* Oken **(9 species)**

Type species: *Bodianus maculatus* Bloch (by subsequent designation of Jordan, Tanaka, and Snyder).
Type locality: Reported as Japan but probably an error for Java (Heemstra and Randall 1993).
Included species:
P. areolatus (Rüppell 1830)
P. laevis (Lacepède 1801)
P. leopardus (Lacepéde 1802)
P. maculatus (Bloch 1790)
P. marisrubri (Randall and Hoese 1986)
P. oligacanthus (Bleeker 1855)
P. pessuliferus (Fowler 1904)
P. powelli (Smith 1964) (formerly *Saloptia*)
P. punctatus (Quoy and Gaimard 1824)

Genus *Variola* Swainson **(2 species)**

Type species: *Variola longipinna* Swainson (=*Perca louti* Forsskål by monotypy).
Type locality: Indian Ocean.
Included species:
V. albimarginata Baissac 1953
V. louti (Forsskål 1775)

Chapter 1.2

Early Life Development

Fabiana C. Félix-Hackradt

Introduction

Grouper species are typical top predators of tropical and sub-tropical seas and are of high economic and ecological interest. Some populations have been reduced by more than 50% in the last decades (e.g., *Epinephelus akaara*, *Epinephelus marginatus*, and *Epinephelus striatus*), and 25% of the grouper species are now listed as threatened species on the IUCN red list (Sadovy de Mitcheson et al. 2012). The fate of any population will depend directly on how new individuals, either by birth or immigration, can replenish individual losses to maintain population growth. Reproductive stock reduction can severely impact recruitment (Giménez-Hurtado et al. 2005), and therefore jeopardize population persistence or replenishment (Aguilar-Perera 2006). In addition, about 30% of the grouper species are considered data deficient by IUCN, further hindering direct management actions.

All groupers have a bipartite life cycle with a pelagic larvae and a benthic, and mostly sedentary juvenile and adult phases (Cowen 2001). Given that the dispersal ability of a marine species determines its persistence on our changing world (Mora and Sale 2002), the survival of the larval phase and thus settlement intensity will therefore have great effect on the size and age structure of grouper populations. However, stochastic events (such as oceanographic variation on larval supply) may pose a non-predictable component in the population dynamic puzzle. The lack of data on the early life stages of groupers (i.e., on larval dispersal, settlement and recruitment events, etc.) is therefore particularly worrisome to manage these species efficiently.

Marine Ecology and Conservation Laboratory, Centre of Environmental Sciences, Federal University of Southern Bahia, BR 367 km 10, 45810-000, Porto Seguro, Bahia, Brazil.
Email: fabianacfh@ufsb.edu.br

In this chapter, we will describe the current knowledge on early life history of grouper species, then discuss how these characteristics influence the dynamics of these populations, and to what extent anthropogenic actions are affecting the species' persistence.

Early Life History Traits

Following Pineda et al. (2007), larval dispersal refers to "*the spread of larvae from a spawning ground to a settlement site*", and the final outcome will be a function of life history traits interacting within its environment. These traits include spawning (i.e., timing, location, fertilization rates, etc.), larval transport (i.e., the synergy between hydrodynamics and larval behaviour), survival (i.e., growth, predation and starvation risk) and settlement success (larval supply to suitable habitats). The future of groupers populations therefore relies on a combination of: (1) successful dispersal of the larvae, (2) survival of new settlers, (3) recruitment of juveniles into the adult stock, and (4) reproduction of adults.

Spawning and Surroundings

Since larval dispersal begins with spawning (see Chapter 4 for a complete description), the events occurring at the spawning grounds are decisive for the larval dispersal. Fish aggregation behaviour for reproduction purposes is assumed to be temporally and spatially synchronized with hydrographic features to ensure the survival of the early life stages, being the cause of the high predictability of reproductive events for a number of grouper species (Sedberry et al. 2006). Hjort (1926) first suggested that spawning season is evolutionarily related to optimal conditions for larval development, being species-specific. A temporal match of optimal temperature conditions (Sadovy and Eklund 1999), lunar phase, and tidal regime (Samoilys 1997a) can have an additional effect on groupers' reproductive success, while spatially spawning sites reflect the species' habitat requirements, such as reef profile and location (e.g., reef slope, shelf-edge or break, flat or back reefs, etc.), water depth, current regime, among others (Samoilys 1997a).

Our data compilation shows that grouper species may be classified into two groups regarding their spawning sites requirements: (1) species favouring deep reef environments, such as shelf-edges, continental slopes, and deep channels, versus (2) shallow reef dwellers inhabiting, among others, coral and mid-shelf reefs, seagrass beds, estuarine habitats (Table 1). We find that the genus *Epinephelus* is more frequent in shallow habitats than *Mycteroperca* and *Hyporthodus*, however, there are exceptions to this rule, such as *Hyporthodus septemfasciatus* and *Mycteroperca interstitialis*, more commonly found associated to shallow reefs (see Table 1).

Several studies have found evidences that spawning strategies (benthic or pelagic eggs) and location (in or offshore) can influence the fate of larval dispersal (Macpherson and Raventos 2006). Groupers are pelagic spawners,

Table 1. Compiled information on the spawning season, the site (considering shallow < 50 m and deep > 50 m sites), settlement habitat, settlement size (in mm), minimum and maximum PLD and/or mean value (in days), dispersal distance (in km), and ontogenic movements. Global average size at settlement, PLD, and dispersal distance are also provided for all species.

Species	Winter	Spring	Summer	Fall	Spawning site	Settlement habitat	Settlement size (mm)	Mean PLD (days)	Dispersal distance (km)	Ontogenetic movements	Sources*
Alfestes afer					shallow shelf-edge reefs	seagrass, seaweed, crevices	31	41.52		migrate to shallow reef areas	1, 2, 3, 4
Cephalopholis cruentata					shallow shelf-edge reefs	shallow coral reefs, rubble	24–30	41.52			3, 5, 6, 7
Cephalopholis fulva					deep shelf-edge reefs	shallow coral reefs with high relief, rubble, seagrass	32			migrate to offshore habitats	4, 5, 8, 9
Epinephelus aakara					shallow shelf-edge reefs	hard bottom	30	40			1, 10, 11, 12
Epinephelus adscensionis					shallow shelf-edge reefs	hard bottom	35				4, 5, 13
Epinephelus coioides					deep shelf-edge reefs	estuaries, sand, mud, gravel, mangroves	20–24			migrate to offshore habitats	14, 15, 16
Epinephelus drummondhayi					deep shelf-edge reefs	deep reefs				migrate to offshore habitats	5, 13, 17, 18
Epinephelus fasciatus					shallow reef slopes	hard bottom, estuarine	50	30–35			19, 20, 21

Table 1 Contd. ...

...Table 1 Contd.

Species	Winter	Spring	Summer	Fall	Spawning site	Settlement habitat	Settlement size (mm)	Mean PLD (days)	Dispersal distance (km)	Ontogenetic movements	Sources*
Epinephelus fuscoguttatus					shallow shelf-edge reefs	coral reefs, seagrass	20–24	30			14, 22, 23, 24
Epinephelus guttatus					shallow shelf-edge reefs	patch reefs, limestone rock, corals				migrate to shallow reef habitats	1, 5, 25, 26
Epinephelus itajara					wrecks, creeks, estuaries, shallow reefs (< 50 m)	mangroves, seagrass, swamps, bridges, reefs	15–17	30–80 (mean 50)	110	migrate to lower salinity waters of the tributaries	1, 5, 27, 28, 29
Epinephelus labriformis					shallow rocky bottoms over shelf	hard bottom		60	300		30, 31
Epinephelus marginatus					rocky bottoms over shelf	rocky shores with cavities and scree sites	20–30 (mean 24.7)	20–40 (mean 30)	120	expand the depth range	4, 32, 33, 34, 35, 36, 37, 38
Epinephelus merra					shallow fringing reefs	coral reef flat, mangrove trees, *Acropora* coral	37–43	40–55 (mean 45)			39, 40, 41
Epinephelus morio					deep shelf-edge reefs	inshore and offshore reefs, estuaries, seagrass beds	20–30 (mean 24.7)	30–50 (mean 40)	300–650	migrate to offshore habitats (> 100 mm)	1, 4, 5, 13, 42, 43, 44, 45

Epinephelus striatus													shallow shelf-edge reefs	shallow high relief reefs, offshore reefs, estuaries, seagrass beds, *Porites* sp., tilefish mounds	20–45	30–50 (mean 42)	200	migrate to offshore habitats (> 130 mm)	5, 6, 46, 47, 48
Hyporthodus flavolimbatus													deep shelf-edge rocky reefs and sand mud bottom	hard bottom, shallow reefs	18–38 (mean 24.5)	41.52			5, 13, 49, 50, 51
Hyporthodus mystacinus													deep shelf-edge reef-slopes, rocky pinnacles and ledges	shallow hard bottom	20			migrate to offshore habitats	5, 52, 53
Hyporthodus nigritus													deep shelf-edge reefs	hard bottom, reefs	22				4, 5, 13
Hyporthodus niveatus													deep shelf-edge reefs	shallow rocky reefs, sand or mud bottom, artificial habitats	23–30			migrate to offshore habitats	5, 13, 54
Hyporthodus quernus													hard substrata	hard bottom, deep bank flats	25	30–50	850		55, 55, 57, 58
Hyporthodus septemfasciatus													shallow rocky reefs	shallow reefs	12–20	34–45		expand the depth range	59, 60, 61

Table 1 Contd. ...

...Table 1 Contd.

Species	Winter	Spring	Summer	Fall	Spawning site	Settlement habitat	Settlement size (mm)	Mean PLD (days)	Dispersal distance (km)	Ontogenetic movements	Sources*
Mycteroperca bonaci					shallow shelf-edge reefs	mangroves, seagrass, hard bottom	20	31–57 (mean 41)		migrate to shallow reefs	4, 5, 62, 63
Mycteroperca interstitialis					shallow shelf-edge reefs	hard bottom, mangrove	17–20 (mean 18.5)			migrate to shallow reefs	4, 5, 13, 64
Mycteroperca microlepis					deep shelf-edge reefs	seagrass, mangroves	13–27	26–44 (mean 43)	60–250	migrate to offshore habitats (> 100 mm)	5, 45, 13, 63, 65, 66, 67, 68, 69
Mycteroperca phenax					deep shelf-edge reefs (*Oculina* reefs)	hard bottom, mangroves, reefs				migrate to offshore habitats	5, 13, 43
Mycteroperca rosacea					deep rocky reefs	shallow rocky bottom with *Sargassum* cover	15–25	45–55		migrate to offshore reef habitats	70, 71, 72
Mycteroperca tigris					deep shelf-edge high relief reefs	high relief coastal reefs	20–45				5, 25, 73, 74
Mycteroperca venenosa					deep shelf-edge high relief reefs	*Thalassia* beds, seagrass, hard bottom	20–45 (mean 22)			migrate to offshore habitats	1, 4, 5, 25, 74
Plecotropomus areolatus					shallow coral reefs	shallow hard bottom, coral rubble	25–37	30	33		75, 76

Plecotropomus leopardus													shallow shelf-edge reefs	sand-rubble, live-dead coral reefs	20–25	19–31 (mean 26)	190	migrate to high relief coral reefs	77, 78, 79, 80
Plecotropomus maculatus													shallow shelf-edge reefs	shallow lagoons, reef flat with rubble	37–57 (mean 50)	26	30–110		80, 81
													Average (±SD) values*		27.05 ± 8.75	39.12 ± 9.27	210.30 ± 152		

* References: (1) Thompson and Munro 1978; (2) Reynal et al. 2004; (3) Cherubin et al. 2011; (4) Zavala-Camin 2012; (5) Lindeman et al. 2000; (6) Colin et al. 1997; (7) Nagelkerken 1979; (8) Trott 2006; (9) Claydon and Kroetz 2007; (10) Okumura et al. 2002; (11) Kusaka et al. 2001; (12) Park et al. 2016; (13) Sedberry et al. 2006; (14) Leis et al. 2007; (15) Leis et al. 2009; (16) Grandcourt et al. 2005; (17) Ross 1988; (18) Ziskin 2008; (19) Mishina et al. 2006; (20) Kawabe and Kohno 2009; (21) Kuriiwa et al. 2014; (22) Kohno et al. 1993; (23) Pears et al. 2006; (24) Ch'ng and Senoo 2008; (25) Ojeda-Serrano et al. 2007; (26) Colin et al. 1987; (27) Koenig et al. 2017; (28) Lara et al. 2009; (29) Silva-Oliveira et al. 2008; (30) Craig et al. 2006; (31) Froese and Pauly 2019; (32) La Mesa et al. 2002; (33) Bodilis et al. 2003; (34) Hereu et al. 2006; (35) Macpherson and Raventos 2006; (36) Reñones et al. 2010; (37) Andrello et al. 2013; (38) Faillettaz et al. 2018a; (39) Letourneur et al. 1998; (40) Lee et al. 2002; (41) Jagadis et al. 2006; (42) Sluka et al. 1994; (43) Zatcoff et al. 2004; (44) Gold and Richardson 1998; (45) Coleman et al. 1996; (46) Powell and Tucker 1992; (47) Jackson et al. 2014; (48) Sherman et al. 2017; (49) Jones et al. 1989; (50) Bullock et al. 1996; (51) Marancik et al. 2012; (52) Munro et al. 1973; (53) Brule et al. 2018; (54) Johnson and Keener 1984; (55) Moffitt 2006; (56) Rivera et al. 2004, (57) Rivera et al. 2011; (58) Kobayashi 2006; (59) Kitajima et al. 1991; (60) Shein et al. 2004; (61) Sabate et al. 2009; (62) Keener et al. 1988; (63) Crabtree and Bullock 1998; (64) Scharer et al. 2010; (65) Ross and Moser 1995; (66) Fitzhugh et al. 2005; (67) Koenig and Coleman 1998; (68) Adamski et al. 2012; (69) Weisberg et al. 2014; (70) Martinez-Lagos and Gracia-Lopes 2009; (71) Aburto-Oropeza et al. 2007; (72) Aburto-Oropeza et al. 2010; (73) Heyman and Kjerfve 2008; (74) Tuz-Sulub and Brulé 2015; (75) Hutchinson and Rhodes 2010; (76) Almany et al. 2013; (77) Light and Jones 1997; (78) Samoilys et al. 1997b; (79) Doherty et al. 1994; (80) Williamson et al. 2016; (81) Harrison et al. 2012.

generally in sites located offshore, near continental shelf edges with moderate to high structural relief, or next to promontories with great slope (Coleman et al. 1996, Lindeman et al. 2000, Heyman and Kjerfve 2008, Marancik et al. 2012, Farmer et al. 2017). Grouper larvae have extended pelagic phase (e.g., from 20 days for *E. marginatus* to 80 days for *Epinephelus itajara,* see Table 1), and the strong offshore currents at these locations may transport larvae over long distances, enhancing their dispersal potential (Shanks and Eckert 2005).

However, growing evidence has been shown that larval retention near spawning sites is more common than previously thought (Harrison et al. 2012, Almany et al. 2013, Cinner et al. 2016, Williamson et al. 2016). Numerous studies have demonstrated that topographically-generated coastal eddies and mesoscale currents, as well as other biological (i.e., substrate type, bottom cover, etc.) and oceanographic characteristics (i.e., current direction and intensity), are the main drivers of larval retention and play an important role in structuring the populations (e.g., in *Epinephelus labriformis* (Craig et al. 2006), in *E. striatus* (Heppell et al. 2010, Jackson et al. 2014), or in *E. marginatus* (Andrello et al. 2013). Those gyres are assumed to reduce predation risk by transporting eggs and larvae away from spawning sites, while still bringing them back close to the coast once they are more mobile and able to escape predators (Largier 2003).

For example, *Mycteroperca microlepis* uses bottom currents generated by the Ekman transport to reach coastal habitats, as supported by stable isotopes analysis and the observation of juveniles associated to macroalgae from bottom origin at offshore location (Weisberg et al. 2014). In addition, simulations with Nassau grouper larvae displayed an offshore, followed by downward movement bringing particles toward to the shelf, close to the bottom flow and in contact to chlorophyll maximum (Cherubin et al. 2011). This is consistent with the pelagic survival and predator avoidance hypotheses (Johannes 1978), in which larvae next to adult habitat will suffer more predation pressure from predators (Johannes 1978).

These recent discoveries have important implications for local population adaptation, as low connectivity and high self-recruitment rates can hinder larval supply to neighbouring populations and lead to inbreeding depression (Andrello et al. 2013).

Eggs and Larval Characteristics

The BOFFF (Big Old Fat Fecund Female Fish) hypothesis sustains that bigger, older, fatter, and fecund grouper females contribute more to population dynamics than younger and thinner ones (Hixon et al. 2014). Groupers have high fecundity (e.g., 65×10^5–8×10^6 eggs in *E. marginatus* females, Reñones et al. 2010), which grows exponentially with fish biomass. Egg production of 2×10^6 eggs is found for *Epinephelus fasciatus* and *Epinephelus fuscoguttatus* from captive breeding with high fertilization success (~ 98%)

(Kohno et al. 1993, Kawabe and Kohno 2009), but no *in situ* information is available. Notwithstanding, fertilization of reared *E. akaara* groupers through spontaneous spawning retrieved extremely low successful rates (10–25%) in comparison to assisted fertilization (Okumura et al. 2002), indicating that fertilization success must be highly variable in natural conditions.

Groupers' egg sizes vary with genus: *Cephalopholis* spp. produce the smallest eggs, 0.53–0.61 mm (Nagelkerken 1979, Kawakami et al. 2010, Park et al. 2016), followed by *Mycteroperca* spp. and *Plectropomus* spp. with 0.82–0.92 mm (Masuma et al. 1993, Gracia-López et al. 2004, Carter et al. 2015), while the largest was found among *Epinephelus* genus, with 0.96 mm (*Epinephelus guttatus;* see Colin et al. 1987).

Eggs are buoyant and incubation temperature is around 21–30ºC, depending on the species (Brule et al. 2018). Hatching occurs within the first 24–72 hours, and the yolk-sac larvae absorb its vitellus entirely within the next 3–4 days (Kitajima et al. 1991, Colin 1992, Colin et al. 1997, Martínez-Lagos and Gracia-López 2009). Then, the larvae become planktotrophic and must eventually find their way to suitable settlement habitat. Larval size at hatching varied among species, usually ranging from 1.5 mm in the genus *Mycteroperca, Epinephelus*, and *Plectropomus* (Anderson et al. 2012).

Pelagic Larval Duration

Most fish species, including groupers, depend on the pelagic larval phase to colonize new habitats and maintain population growth. In general, the longer the species' pelagic larval duration (PLD), the larger is its geographic distribution (Kinlan and Gaines 2003). Therefore, PLD was considered as a proxy of the species' dispersal potential and has been studied extensively. However, effective dispersal is not limited to the result of pelagic larval phase, but instead is the combined outcome of many other characteristics, such as (i) environmental/oceanographic forces, (ii) larval swimming abilities, (iii) adult population distribution, (iv) spawning strategies, among others (Shanks et al. 2003, Macpherson and Raventos 2006, Shanks 2009).

As a result of the overfishing of adults, and consequently the rarity of grouper larvae in field samples (Richards 2005), information about early life stages is scarce. Still, here, we collected data from the pelagic larval duration of 18 grouper species (over 166, ~ 10%). Among all groupers, the genus *Epinephelus* and *Mycteroperca* appear as the most studied (Table 1), and the mean pelagic larval duration (PLD) is about 39 ± 9 days (Table 1).

No consistent pattern was observed in PLD between genus; the maximum PLD recorded was found by Lara et al. (2009) in *E. itajara* (PLD = mean 50, range 30 to 80 days) and the lowest at *Plectopomus leopardus* (PLD = mean 25, range 19–31 days) by Doherty et al. (1994). However, the time spent for the larval development is fairly homogeneous among groupers species within genus, still with high potential plasticity in some species (Craig et al. 2006).

Surviving in the open ocean requires finding sufficient food and being able to avoid predators. During the path to settlement habitat, climatic and oceanographic environment may be the first critical steps for grouper larval survival (Colin et al. 2013). High survival rates were observed in Nassau grouper larvae under low turbulence, and high salinity and light levels (Ellis et al. 1997).

Temperature can affect larval development rate by speeding metabolic rate, for example by shortening hatching and yolk consumption (O'Connor et al. 2007). Whereas groupers species are more abundant in warm seas, they are spread over a wide latitudinal range (Sadovy de Mitcheson et al. 2012). It is supposed that tropical species (groupers included) may have fast growing larvae and shorter PLD than sub-tropical or temperate species (Leis et al. 2013), however no direct trend could be observed by data compiled here.

Timing and Size at Settlement

For a number of fish species, settlement periods are consistent among years, but their intensity may vary greatly within and among species (D'Alessandro et al. 2007) to the point that they may not be predicted from larval supply alone (Félix-Hackradt et al. 2013). Larval settlement of the majority of grouper species occurs about two months after spawning, which fluctuates among species and spawning sites. The majority of *Mycteroperca* species for which data is available appear to settle during spring-summer season, while *Epinephelus* species settle later, during the summer-fall months (Table 1).

Among the 32 grouper species evaluated here, the size of grouper larvae at settlement are extremely variable, with a mean settlement size (TL) of 27.05 ± 8.75 mm. Groupers can also delay settlement, which has been observed among *Epinephelus* species. As an example, *E. itajara* can postpone its metamorphosis for about 20 days and actively search for suitable conditions (current direction or velocity, favourable winds or moon phase, etc.) until successfully reaching a settlement habitat (Lara et al. 2009).

Despite clear species plasticity in PLD, the size at settlement appears to be a more consistent parameter than PLD, emphasising that larval growth rates may vary among individuals, while the size at settlement is more likely determined evolutionally (Adamski et al. 2012). In a number of reef species, the faster a larva grows, the less time it spends in the plankton (Wellington and Victor 1992).

Physical Transport of Larvae

Mass settlement events are common in marine reef fishes, and have been reported in some grouper species. Aggregating behaviour during larval phase is supposed to increase survival (Colin 2012). Epinephelid species, such as *Epinephelus oceanicus* in the Indian ocean (Pinault et al. 2015), *E. striatus* in the Bahamas (Shenker et al. 1993), *Epinephelus merra* around the Reunión Islands

(Letourneur et al. 1998), could rely on episodic events, such as hurricane, cyclones, strong frontal systems, storms, etc. to deliver their reproduction effort to coastal nursery habitats. If successful, such strategy (i.e., short but intense settlement pulses) may be sufficient to sustain the population of long-lived species like groupers over long temporal scales (Letourneur et al. 1998), but may impose unpredictable variability for local populations over shorter time scales (Shenker et al. 1993).

The ingress of grouper larvae to settlement sites is often correlated with lunar or oceanographic events. *E. striatus* and *E. merra* larvae use moonless nights to maximize settlement by avoiding diurnal predators (Colin et al. 1997, Chabanet et al. 2005), while *P. leopardus* and *E. itajara* display a more flexible behaviour with settlement around the new moon period but not centred on it (Leis and Carson-Ewart 1999, Lara et al. 2009). Plenty of works aiming at evaluating larval supply of reef fish to coastal/insular habitats through a number of different methods (light traps, reef crest nets, tidal nets, etc.) have demonstrated peaks of larval supply occur during the new moon periods (Dufour and Galzin 1993, Letourneur et al. 1998, Hickford and Schiel 1999, D'Alessandro et al. 2007).

Although correlated, the importance of tidal fluctuations on settlement behaviour seems to be less important than the moon phase (Lyczkowski-shultz et al. 1990, D'Alessandro et al. 2007), probably due to the importance of visual cues next to settlement habitats and/or predator avoidance. Nevertheless, for estuarine-dependent fishes, flood-night tides appear to be commonly used to enhance larval ingress into nursery areas (Churchill et al. 1999), as well as in coastal areas subject to intense tidal regimes (Shenker et al. 1993).

Local climate was the most influencing source of interannual settlement variability (Shenker et al. 1993, Marancik et al. 2012). As an example, settlement pulses in Philippine groupers follow a monsoonal pattern; during northeast monsoon winds, greater settlement peak occurred along eastern sites, and followed an opposite pattern under the influence of southeast blow winds (settlers were common at windward sites) (Mamauag et al. 2001). As meteorological events may be altered by global and/or regional atmospheric regimes, such synchronization with climatic variables, such as El Niño-La Niña events, rainfall, seasonal patterns, etc. may produce variation on young of the year (YOY) class-strength, and consequently on adult population.

Larval Behaviour at Settlement

Grouper larvae possess large dorsal and ventral spines until metamorphosis (Johnson and Keener 1984, Richards 2005), which is assumed to protect them from predation. Contrastingly, those protections may reduce the swimming ability of pre-settlement stages compared to similar-sized larvae of other families (Fuiman and Magurran 1994, Leis 2010). This trade-off situation is only overcome when larvae become able to settle (e.g., mean size of grouper competent larvae ~ 10 mm), once their swimming capacity have increased, as observed in *H. septemfasciatus* post-larval stages (Sabate et al. 2009), one of the fastest reef fish larvae tested (Leis 2010).

Likewise, high swimming capacity and depth control has been observed in other grouper species, like in *P. leopardus* (Leis and Carson-Ewart 1999). They observed that competent coral trout (*P. leopardus*) larvae swam faster in open ocean than near settlement habitats, in which predation rates are assumed to be higher. Such a pattern of decreasing speeds at larger sizes and near settlement habitats has been observed in several other reef species, including *Epinephelus coioides* (Leis et al. 2007, Leis 2010). Yet, once settlement habitat has been detected (or chosen), grouper larvae tend to swim as fast as possible towards the reef, indicating that coral trout larvae moved deliberately, and at a chosen velocity to minimize the risk of predation (Leis and Carson-Ewart 1999). According to a review, mean *in situ* larvae speed can reach 10 cm s^{-1}, considering 8–18 mm reef fish larval size (Leis 2010).

Furthermore, larval endurance, i.e., the time and distance a larvae can swim without feeding, is remarkable in some tropical fish families, reaching 20 km for settlement-sized larvae (considered larvae > 10 mm here) and can reach 50 km for larger specimens (Leis 2010). Although the review done by Leis (2010) does not include any grouper species, assuming that endurance distance increases linearly with fish size enables us to estimate the mean of the endurance swimming around 54 km, considering a mean settlement size of 27 mm (Table 1).

Competent larvae present well-developed sensory and swimming capabilities, and may take advantage of several environmental cues to perceive the surroundings and actively locate their preferential habitat (Milicich and Doherty 1994, Stobutzki and Bellwood 1998). Fish larvae can use magnetic field (Faillettaz et al. 2015, Bottesch et al. 2016), odour (Atema et al. 2002, Wright et al. 2008), sounds (Simpson et al. 2005, Vermeij et al. 2010, Parmentier et al. 2015), and visual stimulus (Lecchini et al. 2005, 2014) as cues to orientate themselves toward settlement habitat, and their perception differs according to spatial scales, i.e., magnetic and olfactory cues for greater distances (10–1000 km and 0–50 km, respectively) than acoustic (0–10 km) and visual (0–100 m) cues (Leis et al. 1996, Staaterman et al. 2012), although sun compass which has been observed for a number of temperate fish species works over unlimited distances (Faillettaz et al. 2015). Other cues, such as the movement of the breaking waves near shore, can be perceived by *M. microlepis* as a settlement cue, producing a behavioural change in the larvae (Weisberg et al. 2014). In the absence of sensory abilities studies in grouper larvae, it is still interesting to notice that Pomacentridae fishes use hearing, olfactory, and visual senses for orientation over the entire larval phase, or even before (Leis 2010). Those skills are especially important for species with pelagic eggs and long PLD as groupers, however more investigation effort is needed to evaluate some missing points.

The behaviours mentioned above involve horizontal displacement of larvae that can directly influence dispersal outcomes, but vertical distribution of larvae may also indirectly alter dispersal patterns (Leis 2010). Irisson et al. (2010) showed that older Serranidae larvae are distributed deeper than pre-flexion and flexion stages, evidencing the inferior swimming abilities of young larvae, limiting them to shallow depths. Such diel vertical migration has also been noticed in serranid larvae, with greater abundance in shallow water during the night. Similar behaviour was addressed in coral trout larvae, which includes directional and depth-stratified swimming preferences, active orientation and reef homing abilities, as potential sources of influence on the observed dispersal patterns (Leis 2007, Leis et al. 2007). By managing vertical distribution and directional swimming, fish larvae are able to control alongshore transport and therefore, fate (Largier 2003, Faillettaz et al. 2018b).

There are a few studies that properly evaluated the diel distribution of grouper larvae, but most suggest that larvae predominantly move into settlement habitat at night. This behaviour was registered for *E. merra* (Letourneur et al. 1998), *E. striatus* (Shenker et al. 1993), *E. itajara* (Koenig et al. 2017), *M. microlepis* (Keener et al. 1988), and other serranid species (Thorrold et al. 1994), and appears as a recurrent pattern for the group. Yet, it does not indicate that diurnal settlement may not occur, e.g., coral trout (Leis and Carson-Ewart 1999).

Biophysical larval dispersal modelling use ocean circulation data coupled with larval behaviour to provide estimate potential connectivity that assist marine management. Biophysical models have incorporated horizontal and vertical larval (such as PLD, spawning site, egg buoyancy, flexion time, etc.) navigation in order to understand the effect of these characteristics in fish settlement success (Staaterman and Paris 2014).

Biophysical models have shown that the sooner the larvae start orienting toward settlement habitats, the greater the survival and self-recruitment rates, and consequently the lower the dispersal distances (Staaterman et al. 2012, Staaterman and Paris 2014). The self-seeding rate fluctuated (20–30%) within taxa traits with PLD, post-flexion timing, swimming behaviour in response to cues, and according to oceanographic data (water turbulence, etc.). Also, it was significantly lower (< 2%) for larvae that lacked horizontal swimming, such as coral larvae, emphasizing that orientation is favorable for larvae that settle close to home (Wolanski and Kingsford 2014). That was the conclusion of the few works addressing grouper larvae (Andrello et al. 2013, 2015), in which high levels of self-recruitment and low connectivity was observed in a Mediterranean MPA network using the dusky grouper, *E. marginatus,* as model (Andrello et al. 2013). They observed that dispersal distance and therefore connectivity increased linearly with PLD, while vertical migration contributed to produce higher retention levels, an effect comparable to reduced PLD length. In addition, the spawning month significantly affected the dispersal distance due to coastal current velocity conditions decreasing dispersal with weakening of coastal current velocities (Andrello et al. 2013).

Habitat Preference

Nursery habitat of most grouper species are localized in coastal and estuarine environments (Eggleston 1995, Fitzhugh et al. 2005), yet with high interspecific variability in microhabitat settlement preferences (Table 1).

Although adult phase is usually found associated with hard bottoms (Lindeman et al. 2000), a number of grouper species are dependent on estuarine region as nursery habitats during the initial life phases. For example, *M. microlepis* usually settle within seagrass beds (Fitzhugh et al. 2005) or oyster reef habitat (Keener et al. 1988), while juveniles of goliath groupers are commonly found near mangrove roots, swamps, and bridges (Sadovy and Eklund 1999, Koenig et al. 2017). Other species, such as black and red grouper can make a facultative use of estuarine region, depending on preferred habitat availability. Nassau grouper (*E. striatus*) settlers were mostly observed on seagrass and macroalgae beds, but can also be found on offshore reefs (Sadovy and Eklund 1999). *Mycteroperca venenosa* make use of *Thalassia* beds as nursery habitats (Thompson and Munro 1978), while *Mycteroperca rosacea* prefers *Sargassum* zones (Martínez-Lagos and Gracia-López 2009) and *Epinephelus oceanicus* settles in recent lava flow areas (Pinault et al. 2015). Others juvenile groupers are highly associated with hard substrata, such as coral reefs for *Cephalopholis cruentata* (Nagelkerken 1979), rocky reefs for *H. septemfasciatus* (Sabate et al. 2009), scree areas for *E. marginatus* (Bodilis et al. 2003), or sand-rubble patches for *P. leopardus* and *Plectropomus maculatus* (Williamson et al. 2016).

In general, groupers do not settle in habitats shallower than 5 m (Lindeman et al. 2000), and remain cryptic during the first months after settlement, inhabiting crevices, holes, and caves, making newly settled juveniles difficult to assess (Roberts 1996).

Dispersal Distance

As suggested by Pineda et al. (2007), dispersal begins with spawning and ends with settlement. The dispersal distance is the result of larval transport from adult spawning habitats to nursery areas at temporal and spatial scales determined by the interaction of species biological traits with environmental conditions.

Following Shanks (2009), groupers belong to one of three functional dispersal groups, consisting of long PLDs (> 1 month), long dispersal distances (> 20 km), and adults living in offshore habitats. These characteristics would imply high connectivity levels among species meta-populations, resulting in high homogeneity. Notwithstanding, a number of epinephelid species, such as *E. marginatus* (Schunter et al. 2011), *E. fasciatus* (Kuriiwa et al. 2014), and *E. striatus* (Jackson et al. 2014) have shown subpopulation structuring at the studied scales (> thousands of km), mainly explained by larval retention events related to local oceanographic patterns. However, other grouper species, such as *Epinephelus morio* and *Mycteroperca phenax*

(Gold and Richardson 1998, Zatcoff et al. 2004), *Epinephelus quernus* (Rivera et al. 2004, 2011), and *E. itajara* (Silva-Oliveira et al. 2008) showed no distinct pattern of geographic heterogeneity across sampled areas. Those differences in larval dispersal patterns reflect the existent variation on larval trajectories and dispersal net distances among species, which can vary within a scale of tens to hundreds of km (Table 1) and is locally influenced by oceanographic conditions as well as larval behaviour (see Section Larval behaviour at settlement).

Just recently, with the aid of molecular tools and connectivity modelling, new information about dispersal distances of fish larvae have been used as fishery management tools. These works highlighted that dispersal is hardly ever passive and linear, but instead highly dependent on a combination of larval and spawning behaviours, local and regional patterns of coastal-offshore currents. Indeed, models considering passive larval behaviour always overestimate dispersal ranges (Paris et al. 2005, Shanks 2009).

The average dispersal distance in groupers was about 210 km (± 150 km), with a great variation between the dispersion range of the grouper species analysed (30–450 km; Table 1). These metrics are relatively shorter than expected by larval passive movement, emphasizing the importance of other factors affecting dispersal. Recently, Harrison et al. (2012), Almany et al. (2013), and Williamson et al. (2016) found that, in coral groupers spawning aggregation sites in Australia, 50% of the larvae settled within 200 km away from their spawning site.

These evidences have important implications on coastal fisheries management, as groupers species are economically and culturally valuable marine resources for traditional communities in all seas. In areas where grouper population are overfished (as *E. marginatus* in the Mediterranean Sea), groupers' remnant populations lie inside protected areas (Hackradt et al. 2014, Di Franco et al. 2018). In such situations, larval dispersal modelling appears as a valuable tool to assist managers in establishing efficient networks of MPAs (see Bode et al. 2006, 2019). Persistence of local sink populations can rely on larval supply from specific source locations (James et al. 2002, Bode et al. 2006, Wolanski and Kingsford 2014). Implementing empirical knowledge about sink-sources regimes, hydrographic patterns, and larval behaviour characteristics in marine conservation policies would enhance the effectiveness of protection measures. A practical example of this application was shown by Almany et al. (2013), when local fishermen initiated the collective management of coral grouper fishery after being aware that sustainability was conditioned by the long-term persistence of larval replenishment source from grouper spawning aggregations.

Post-settlement Survival

Post-settlement mortality may be caused by different factors: food starvation or shelter limitation leading to high predation, disease outbreaks due to high densities, or fishing removal (Letourneur et al. 1998). Moreover, groupers can be highly competitive within settlement habitats. Evidence from captive experiments show that at the beginning of settlement size (15–20 mm), Epinephelids such as *E. septemfasciatus* (Sabate et al. 2009), *Epinephelus lanceolatus* (Hseu et al. 2004), *E. coioides* (Hseu et al. 2003), and *M. rosacea* (Martínez-Lagos and Gracia-López 2009) display aggressive behaviour or cannibalism, therefore increasing conspecific mortality rates. Although the explanation remains speculative, this behaviour may be related to the protection of shelter and resources, as groupers settlers are highly cryptic, closely associated to the bottom, and never stay far from their crevices (Sabate et al. 2009).

In situ observations described high natural mortality on *E. merra* (> 80%) settlers due to cannibalism among settlers and from adults which preyed actively on new settlers (Letourneur et al. 1998). Recent research has shown that cannibalism among predators reduces the ability of top-down control, and the consequences of this effect depend on the competitive ability between the predator and its prey (Hin and Roos 2019). If consumers are superior competitors, then cannibalism will generate a top-down control over consumers and put at risk the predators' persistence; however, if predators are superior, cannibalism will promote the reduction on top-down control on the resource, allowing consumers to coexist within predators (Toscano et al. 2017). In that scenario, as groupers are the commonest top-predators along tropical and subtropical seas, given proper densities, it's plausible that cannibalism among the young of the year (YOY) cohort has evolved to control population dynamics.

Shelter availability is of major importance for reef fish recruitment success (Félix-Hackradt et al. 2014). *Mycteroperca rosacea* juvenile abundance was positively associated with availability of *Sargassum* beds, yet, the persistence and biomass of the brown algae was determined by climatic variation (e.g., El Niño events), affecting annual patterns of leopard grouper recruitment (Aburto-Oropeza et al. 2007, 2010).

Predation usually shapes recruitment patterns among reef species. Top predators like large Nassau groupers can weaken top-down control of intermediate predators as coneys (*Cephalopholis fulva* and *C. cruentata*) over lower trophic-level species by inhibiting coneys' YOY feeding, subsequently increasing recruitment rate of other reef fishes (Stallings 2008). In addition, it is common to observe density-dependent effects over early post-settlement mortality, as shown by Letourneur et al. (1998) for *E. merra*. Limitation of food, shelter, and high incidence of diseases due to high population densities are probable causes of high mortality. In addition, climatic events or environmental factors, might produce great decoupling between larval supply and juvenile abundance (Félix-Hackradt et al. 2013), may act as other sources of density-independent mortality. Nevertheless, the effect of high mortality on juveniles' densities will be directly reflected on the adult population demography (Caley et al. 1996).

Post-settlement Movements

Ontogenetic movements are common amongst groupers. About 20 species of groupers reviewed here undertake post-settlement migration between juvenile and adult habitats. The gag (*M. microlepis*) settles on seagrass meadows, and after a 9-month period within the nursery habitat, the juveniles move to offshore hard-bottom habitats (Koenig and Coleman 1998). Similarly, juveniles of black (*Mycteroperca bonaci*) and red grouper (*E. morio*) which usually settle within estuaries emigrate to offshore reefs when they reach an average size of 100 mm (Ross and Moser 1995). Nassau groupers, on the other hand, remain next to seagrass beds at earlier stages and move into holes as they grow (Beets and Hixon 1994). Narrow settlement habitat preference (e.g., sand-rubble substrata) were also observed for *P. leopardus* settlers, which gradually move to higher relief features (e.g., rubble and live coral) at older juvenile stages (Light and Jones 1997). Other grouper species, such as *E. striatus*, settle in shallow hard bottom areas similar to their adult stages, and move to deep offshore habitats within the first year (Table 1).

Spatiotemporal Patterns

Recruitment is the final step of the early life stages of a fish. Here, it is defined as the entry of benthic juveniles into the resident population (Keough and Downes 1982). Recruitment is usually temporally synchronized with species' reproductive patterns (Robertson 1992) and spatially defined by stochastic variation of larval supply, differential mortality, species-habitat relationship, or post-settlement movements (Félix-Hackradt et al. 2014).

Aburto-Oropeza et al. (2007) showed that recruitment patterns of *M. rosacea* are strongly linked to interannual climate variability of the Multivariate El Niño/Southern Oscillation (ENSO) Index (MEI) due to cascading effects over *Sargassum* beds, the nursery habitat of leopard groupers. Years with high MEI values trigger increasing mortality of *M. rosacea* young fish due to the unfavourable conditions for *Sargassum* beds, in particular strong stratification and poor nutrient waters. Similarly, recruitment of young groupers (e.g., *E. merra, E. fasciatus* and *Epinephelus summana*) at Philippines followed a monsoonal pattern (Mamauag et al. 2001).

In the Caribbean sea, regional differences in recruitment patterns of reef fishes were attributed to habitat changes over the past decade on the Jamaican coast (such as coral and seagrass cover), as well as settlement limitation due to intense fishing effort (Watson and Munro 2004). Recruitment limitation as a result of a decrease in spawning stock size has also been detected for *E. morio*, for which a 25% reduction in adult biomass led to a decrease of over 50% in juveniles of 1–2 years old (Giménez-Hurtado et al. 2005).

Fish recruitment intensity is a critical indicator for predicting future fisheries yields, and could be used to establish management measures in order to prevent overfishing groupers down to no-return thresholds further preventing replenishment through recruitment.

Anthropogenic Effects on Population Dynamics

Overfishing and Population Dynamics

Protogynous species are much more vulnerable to overfishing than gonochoristic ones, since overfished populations will more likely be limited by sperm supply (Sadovy 2001, Huntsman and Schaaf 2004). Although higher fishing pressure toward large sizes classes can lead to growth overfishing, in which size at first maturity is gradually reduced, the prolonged removal of those performant spawners can lead to the reduction of population reproductive potential to a no-return state despite low or even no fishing pressure (Sadovy et al. 2003).

About 60% of spawning aggregations of *E. striatus* have disappeared due to overfishing. Despite the rebuild of the Nassau groupers populations following the establishment of protection measures in the Bahamian archipelago, no effective recovery of previous used spawning grounds has been observed, highlighting the loss of intergenerational transmitting of spawning locations (De Mitcheson et al. 2008). Declines in other aggregating species, such as *Plectropomus areolatus* and *Epinephelus polyphekadion* have been detected in the Pacific (Hamilton et al. 2005), with a decline of 25% of red grouper adult biomass followed by a reduction in recruitment abundance by 50% within a 14-year period (Giménez-Hurtado et al. 2005). Those generalized reductions in spawning stocks can lead to local population extinctions and the decay of the species' evolutionary potential. For the expected increase in the fishing pressure on *C. cruentatus*, the stocks may lose their reproductive capacity and fail to recover even at low levels of fishing mortality (Huntsman and Schaaf 2004).

In the Arabian sea, Grandcourt et al. (2005) estimated that overfishing would lead to a low maximum age (12 of 22 years), an extremely biased sex ration (1:48), and a broad size range in which sex change might occur in *E. coioides*. About 30% of the grouper catches consist of fish under the size of first maturity, and indicate that this Arabian stock suffers growth and recruitment overfishing. Similar patterns have been found in the Indian stocks of *Epinephelus bleekeri* (Richu et al. 2018) and *Hyporthodus flavolimbatus*, in the Gulf of Mexico (Bullock et al. 1996).

Global Warming and Acidification

Temperature has a strong effect on larval duration and so on larval dispersal potential (Leis et al. 2013). High temperatures enhance larval growth and reduce the PLD, therefore reducing dispersal distances (O'Connor et al. 2007, Sswat et al. 2018, Watson et al. 2018). Munday et al. (2009) demonstrated that a reduction of 20% in the PLD led to fewer larvae being dispersed over long distances, which have great effects on recruitment patterns by boosting self-seeding rates. In a not so far warming sea, species distribution range is likely to narrow, as larvae may no longer reach places as distant as before, affecting species' colonization patterns (Andrello et al. 2013). Moreover, with a more restricted connectivity, the population persistence of some isolated populations, such as islands and atolls may rely entirely on self-recruitment, increasing the extinction risk. The weakened gene flow among metapopulations would further lead to allele loss and fixation by drift, with time reduced recruitment rates generating low effective population sizes that would favour inbreeding depression (Waples et al. 2010, Buchholz-Sørensen and Vella 2016).

On the other hand, temperature rise can provoke shifts in the reproductive phenology of species (Green and Fisher 2004, Asch 2015). In groupers (Asch and Erisman 2018), it could result in unsynchronized seasonal patterns of high production and spawning effort in nursery habitats, thus affecting recruitment. Warming seas can also affect ocean circulation, stratification, and turbulence that directly influence larval dispersal outcomes (e.g., the direction of larval supply and larvaes' condition) with species-specific and location-dependent patterns (Munday et al. 2009, Lett et al. 2010).

Recent discoveries show that changes in ocean chemistry will affect larvae growth (Munday et al. 2011) and sensory capabilities (Leis 2018). According to Dixson et al. (2010), in water with high pH levels, reef fish larvae can lose the olfactory ability that help them distinguish predators.

Within an increasingly acidic ocean, coral bleaching will continue to spread and healthy reef habitats will become more and more fragmented, further restricting the nursery habitat availability for a number of fish species (Munday et al. 2009).

Habitat Reduction/Fragmentation

Coastal and estuarine environments are frequently used as nursery habitats for grouper species (see topic *Habitat preference*) and are threatened by pollution, vessel anchoring, unregulated tourism, real estate market, and aquaculture, among others (Halpern et al. 2008).

Inside the Bahamian archipelago, Layman et al. (2004) observed that fish density was sensitive to the degree of mangrove fragmentation, with lower density under highly fragmented habitats. The authors argue that fragmentation decreases tidal exchange, and thus habitat quality, larval and juvenile influx, and halts the movement of transient species to upstream localities.

In addition, Frias-Torres (2006) demonstrated that *E. itajara* juveniles selected highly structurally complex mangrove habitats to settle, which are becoming scarcer worldwide.

MPA Design

Marine protected areas are spatial management zones existing worldwide, and recognized as critical tools for population recovery and sustainability.

Important gaps in ELHS knowledge still hinder the proper implementation of MPAs, however (Sale et al. 2005). Much effort has been done to overcome this caveat, and biophysical modelling has been extensively used to incorporate early life traits into the MPAs' frameworks.

For example, Andrello et al. (2013) suggest that the largest MPA network, gathering 99 marine reserves, showed low connectivity levels for *E. marginatus* species, and left about 20% of the entire Mediterranean shelf without larval supply. Although adult groupers are the stage most affected by area closure, due to their strong site fidelity and heavy fishing pressure (Hackradt et al. 2014), the establishment of no-take areas worldwide does not fully incorporate fish movement patterns, larval dispersal, and the spatial locations of essential fish habitats (Green et al. 2015). Costello et al. (2010) showed that providing enough and accurate spatial information for the design of MPAs would result in increase of fisheries' profits by 10% by focusing the effort on highly productive areas supported by efficient larval supply.

Some species tend to rely essentially on external larval source to maintain viable populations. For example, the populations of *E. merra* in La Réunion are mainly sustained thanks to the larval supply from populations in Mauritius Island, 200 km apart (Crochelet et al. 2013). Preserving the source spawning stocks from the latter island is thus necessary to guarantee the species' connectivity and fisheries in La Réunion.

Recently, Félix-Hackradt et al. (2018) investigated the protection effect of spatial fishing closure over several life stages of reef fish in a Spanish MPA network. They depicted that the "reserve effect" was only perceived in adult benthic fish species, such as groupers, while the high spatial variability observed in juveniles prevented from detecting any positive effect. A "negative" influence of larval abundance was even observed, likely resulting from a combination of geomorphological characteristics (all no-take were localized on promontories) and larval behaviour.

Such work provides relevant information that could be used in MPA sciences, however, only 11% of the newly established MPAs considered connectivity as a major criteria (Balbar and Metaxas 2019).

Conclusions

The combination of new technologies enabled considerable advances in early life history science, where integrated approaches could solve some of the larval dispersal black-box. It is now well established that groupers spawn at specific localities and seasons to maximize the survival of their offspring. Their large eggs produce bigger larvae, which have a higher chance of survival. After hatching, grouper larvae can remain for up to 2 months (~ 40 days) in the water column, during which their sensory abilities may develop as in other fish larvae to eventually locate favourable settlement habitats. Not enough data is available to conclude on the swimming abilities of grouper larvae. Yet, their large larvae and long development provide them with a strong swimming potential. Once settled, the young juveniles become cryptic and aggressive, likely as a strategy to survive this phase of high mortality rates. As they grow, late juveniles or young adults initiate their ontogenetic migration to their distinct, adult habitat. As a consequence of their extensive pelagic larval durations, populations up to hundreds of kilometres apart can still be connected. Then, intense fishing pressure, or even overfishing, may impact the production of larvae, and may ultimately cease recruitment. Synergistic effects of global warming and ocean acidification may further reduce the genes flows, leading sink or isolated populations to a high risk of extinction. Establishing efficient networks of MPAs worldwide would support the recovery of grouper populations and the sustainable exploitation of grouper stocks with moderate fishing levels through spillover. However, information about early life stages of groupers, the fine scale spatial resolution of their habitats, and the local and regional hydrodynamic conditions still need to be further incorporated in order to eventually reach these expected, positive results that would benefit both the biodiversity and the fisheries.

References

Aburto-Oropeza, O., E. Sala, G. Paredes, A. Mendoza and E. Ballesteros. 2007. Predictability of reef fish recruitment. Ecology 88: 2220–2228.

Aburto-Oropeza, O., G. Paredes, I. Mascareñas-Osorio and E. Sala. 2010. Climatic influence on reef fish recruitment and fisheries. Mar. Ecol. Prog. Ser. 410: 283–287.

Adamski, K.M., J.A. Buckel, G.B. Martin, D.W. Ahrenholz and J.A. Hare. 2012. Fertilization dates, pelagic larval durations, and growth in gag (*Mycteroperca microlepis*) from North Carolina, USA. Bul. Mar. Sci. 88(4): 971–986.

Aguilar-Perera, A. 2006. Disappearance of a Nassau grouper spawning aggregation off the southern Mexican Caribbean coast. Mar. Ecol. Prog. Ser. 327: 289–296.

Almany, G.R., R.J. Hamilton, M. Bode, M. Matawai, T. Potuku, P. Saenz-Agudelo, S. Planes, M.L. Berumen, K.L. Rhodes, S.R. Thorrold, G.R. Russ and G.P. Jones. 2013. Dispersal of grouper larvae drives local resource sharing in a coral reef fishery. Curr. Biol. 23: 626–630.

Anderson, S.A., I. Salinas, S.P. Walker, Y. Gublin, S. Pether, Y.Y. Kohn and J.E. Symonds. 2012. Early development of New Zealand hapuku *Polyprion oxygeneios* eggs and larvae. J. Fish Biol. 80: 555–571.

Andrello, M., D. Mouillot, J. Beuvier, C. Albouy, W. Thuiller and S. Manel. 2013. Low connectivity between Mediterranean marine protected areas: a biophysical modeling approach for the dusky grouper *Epinephelus marginatus*. PLoS One 8: e68564.

Andrello, M., D. Mouillot, S. Somot, W. Thuiller and S. Manel. 2015. Additive effects of climate change on connectivity between marine protected areas and larval supply to fished areas. Divers Distrib. 21(2): 139–150.

Asch, R.G. 2015. Climate change and decadal shifts in the phenology of larval fishes in the California Current ecosystem. Proc. Natl. Acad. Sci. 112: E4065–E4074.

Asch, R.G. and B. Erisman. 2018. Spawning aggregations act as a bottleneck influencing climate change impacts on a critically endangered reef fish. Divers Distrib. 24(12): 1712–1728.

Atema, J., M.J. Kingsford and G. Gerlach. 2002. Larval reef fish could use odour for detection, retention and orientation to reefs. Mar. Ecol. Prog. Ser. 241: 151–160.

Balbar, A.C. and A. Metaxas. 2019. The current application of ecological connectivity in the design of marine protected areas. Glob. Ecol. Conserv. e00569.

Beets, J. and M.A. Hixon. 1994. Distribution, persistence and growth of groupers (Pisces: Serranidae) on artificial and natural patch reefs in the Virgin Islands. Bull. Mar. Sci. 55: 470–483.

Bode, M., L. Bode and P.R. Armsworth. 2006. Larval dispersal reveals regional sources and sinks in the Great Barrier Reef. 308: 17–25.

Bode, M., J.M. Leis, L.B. Mason, D.H. Williamson, H.B. Harrison, S. Choukroun and G.P. Jones. 2019. Successful validation of a larval dispersal model using genetic parentage data. PLoS Biol. 17(7): e3000380.

Bodilis, P., A. Ganteaume and P. Francour. 2003. Recruitment of the dusky grouper (*Epinephelus marginatus*) in the North-Western Mediterranean sea. Cybium 27: 123–129.

Bottesch, M., G. Gerlach, M. Halbach, A. Bally, M.J. Kingsford and H. Mouritsen. 2016. A magnetic compass that might help coral reef fish larvae return to their natal reef. Curr. Biol. 26(24): R1266–R1267.

Brule, T., C.M. Teresa, P.D. Esperanza and D. Christian. 2018. Biology, exploitation and management of groupers (Serranidae, Epinephelinae, Epinephelini) and snappers (Lutjanidae, Lutjaninae, Lutjanus) in the Gulf of Mexico.

Buchholz-Sørensen, M. and A. Vella. 2016. Population structure, genetic diversity, effective population size, demographic history and regional connectivity patterns of the endangered dusky grouper, *Epinephelus marginatus* (Teleostei: Serranidae), within Malta's fisheries management zone. PloS One 11(7).

Bullock, L.H., M.F. Godcharles and R.E. Crabtree. 1996. Reproduction of yellowedge grouper, *Epinephelus flavolimbatus*, from the Eastern Gulf of Mexico. Bull. Mar. Sci. 59: 216–224.

Caley, M.J., M.H. Carr, M.A. Hixon, T.P. Hughes, G.P. Jones and B.A. Menge. 1996. Recruitment and the local dynamics of open marine populations. Annu. Rev. Ecol. Syst. 27: 477–500.

Carter, A.B., A.G. Carton, M.I. McCormick, A.J. Tobin and A.J. Williams. 2015. Maternal size, not age, influences egg quality of a wild, protogynous coral reef fish *Plectropomus leopardus*. Mar. Ecol. Prog. Ser. 529: 249–263.

Ch'ng, C.L. and S. Senoo. 2008. Egg and larval development of a new hybrid grouper, tiger grouper *Epinephelus fuscoguttatus x* giant grouper *E. lanceolatus*. Aquacult. Sci. 56(4): 505–512.

Chabanet, P., K. Pothin and M. Moyne-Picard. 2005. Cyclones as mass-settlement vehicles for groupers. Coral Reefs 24: 138.

Cherubin, L.M., R.S. Nemeth and N. Idrisi. 2011. Flow and transport characteristics at an *Epinephelus guttatus* (red hind grouper) spawning aggregation site in St. Thomas (US Virgin Islands). Ecol. Modell. 222: 3132–3148.

Churchill, J.H., R.B. Forward, R.A. Luettich, J.L. Hench, W.F. Hettler, L.B. Crowder and J.O. Blanton. 1999. Circulation and larval fish transport within a tidally dominated estuary. Fish Oceanogr. 8: 173–189.

Cinner, J.E., M.S. Pratchett, N.A.J. Graham, V. Messmer, M.M.P.B. Fuentes, T. Ainsworth, N. Ban, L.K. Bay, J. Blythe, D. Dissard, S. Dunn, L. Evans, M. Fabinyi, P. Fidelman, J. Figueiredo, A.J. Frisch, C.J. Fulton, C.C. Hicks, V. Lukoschek, J. Mallela, A. Moya, L. Penin, J.L. Rummer, S. Walker and D.H. Williamson. 2016. A framework for understanding climate change impacts on coral reef social–ecological systems. Reg. Environ. Chang. 16: 1133–1146.

Claydon, J.A.B. and A.M. Kroetz. 2007. The distribution of early juvenile groupers around South Caicos, Turks and Caicos Islands. In Proc. Gulf Caribb. Fish Inst. 60: 345–350.

Coleman, F.C., C.C. Koenig and L.A. Collins. 1996. Reproductive styles of shallow-water groupers (Pisces: Serranidae) in the eastern Gulf of Mexico and the consequences of fishing spawning aggregations. Environ. Biol. Fishes 47: 129–141.

Colin, P.L., D.Y. Shapiro and D. Weiler. 1987. Aspects of the reproduction of two groupers, *Epinephelus guttatus* and *E. striatus* in the West Indies. Bull. Mar. Sci. 40: 220–230.

Colin, P.L. 1992. Reproduction of the Nassau grouper, *Epinephelus striatus* (Pisces: Serranidae) and its relationship to environmental conditions. Env. Biol. Fish 34(4): 357–377.

Colin, P.L., W.A. Laroche and E.B. Brothers. 1997. Ingress and settlement in the Nassau grouper, *Epinephelus striatus* (Pisces: Serranidae), with relationship to spawning occurrence. Bull. Mar. Sci. 60: 656–667.

Colin, P.L. 2012. Aggregation spawning: Biological aspects of the early life history. pp. 191–224. *In*: Mitcheson Y.S. de and P. Colin (eds.). Reef Fish Spawning Aggregations: Biology, Research and Management, Fish & Fis. Springer Science+Business Media.

Colin, P.L., Y. Sadovy de Mitcheson and T.J. Donaldson. 2013. Grouper spawning aggregations: Be careful what you measure and how you measure it: A rebuttal of Golbuu and Friedlander (2011). Estuar. Coast Shelf Sci. 123: 1–6.

Costello, C., A. Rassweiler, D. Siegel, G. De Leo, F. Micheli and A. Rosenberg. 2010. The value of spatial information in MPA network design. Proc. Natl. Acad. Sci. 107: 18294–18299.

Cowen, R.K. 2001. Oceanographic influences on larval dispersal and retention and their consequences for population connectivity. pp. 171–199. *In*: Sale, P.F. (ed.). Coral Reef Fishes: Dynamics and Diversity in a Complex Ecosystem. Academic Press, San Diego.

Crabtree, R.E. and L.H. Bullock. 1998. Age, growth, and reproduction of black grouper, *Mycteroperca bonaci*. In Florida waters. Fish Bull. 96(4): 735–753.

Craig, M.T., P.A. Hastings, D. Ross and J.A. Rosales-casia. 2006. Phylogeography of the flag cabrilla *Epinephelus labriformis* (Serranidae): implications for the biogeography of the Tropical Eastern Pacific and the early stages of speciation in a marine shore fish. J. Biogeogr. 33: 969–979.

Crochelet, E., P. Chabanet, K. Pothin, E. Lagabrielle, J. Roberts, G. Pennober, R. Lecomte-Finiger and M. Petit. 2013. Validation of a fish larvae dispersal model with otolith data in the Western Indian Ocean and implications for marine spatial planning in data-poor regions. Ocean Coast Manag. 86: 13–21.

D'Alessandro, E., S. Sponaugle and T. Lee. 2007. Patterns and processes of larval fish supply to the coral reefs of the upper Florida Keys. Mar. Ecol. Prog. Ser. 331: 85–100.

Di Franco, A., J.G. Plass-Johnson, M. Di Lorenzo, B. Meola, J. Claudet, S.D. Gaines, J.A. García-Charton, S. Giakoumi, K. Grorud-Colvert, C.W. Hackradt, F. Micheli and P. Guidetti. 2018. Linking home ranges to protected area size: The case study of the Mediterranean Sea. Biol. Conserv. 221: 175–181.

Dixson, D.L., P.L. Munday and G.P. Jones. 2010. Ocean acidification disrupts the innate ability of fish to detect predator olfactory cues. Ecol. Lett. 13: 68–75.

Doherty, P.J., A.J. Fowler, M.A. Samoilys and D.A. Harris. 1994. Monitoring the replenishment of coral trout (Pisces: Serranidae) populations. Bull. Mar. Sci. 54(1): 343–355.

Dufour, V. and R. Galzin. 1993. Colonization patterns of reef fish larvae to the lagoon at Moorea Island, French Polynesia. Mar. Ecol. Prog. Ser. 102: 143–152.

Eggleston, D.B. 1995. Recruitment in Nassau grouper *Epinephelus striatus*: post-settlement abundance, microhabitat features, and ontogenetic habitat shifts. 124: 9–22.

Ellis, S.C., W.O. Watanabe and E.P. Ellis. 1997. Temperature effects on feed utilization and growth of post-settlement stage nassau grouper. Trans Am. Fish Soc. 126: 309–315.

Faillettaz, R., A. Blandin, C.B. Paris, P. Koubbi and J.O. Irisson. 2015. Sun-compass orientation in Mediterranean fish larvae. PloS One 10(8): e0135213.

Faillettaz, R., L. Gilletta, F. Petit, P. Francour and J.O. Irisson. 2018a. First records of dusky grouper *Epinephelus marginatus* settlement-stage larvae in the Ligurian Sea. J. Oceanogr. Res. Data 10(1): 1–6.

Faillettaz, R., C.B. Paris and J.O. Irisson. 2018b. Larval fish swimming behavior alters dispersal patterns from marine protected areas in the north-western Mediterranean Sea. Front. Mar. Sci. 5: 97.

Farmer, N.A., W.D. Heyman, M. Karnauskas, S. Kobara, T.I. Smart, J.C. Ballenger, M.J.M. Reichert, D.M. Wyanski, M.S. Tishler, K.C. Lindeman, S.K. Lowerre-barbieri, T.S. Switzer, J.J. Solomon, K. Mccain, M. Marhefka and G.R. Sedberry. 2017. Timing and locations of reef fish spawning off the southeastern United States.

Félix-Hackradt, F., C. Hackradt, J. Treviño-Otón, A. Pérez-Ruzafa and J.A. García-charton. 2018. Effect of marine protected areas on distinct fish life-history stages. Mar. Environ. Res. 140: 200–209.

Félix-Hackradt, F.C., C.W. Hackradt, J. Treviño-Otón, A. Pérez-Ruzafa and J.A. García-Charton. 2013. Temporal patterns of settlement, recruitment and post-settlement losses in a rocky reef fish assemblage in the South-Western Mediterranean Sea. Mar. Biol. 160: 2337–2352.

Félix-Hackradt, F.C., C.W. Hackradt, J. Treviño-Otón, A. Pérez-Ruzafa and J.A. García-Charton. 2014. Habitat use and ontogenetic shifts of fish life stages at rocky reefs in South-western Mediterranean Sea. J. Sea Res. 88: 67–77.

Fitzhugh, G.R., C.C. Koenig, F.C. Coleman and C.B. Grimes. 2005. Spatial and temporal patterns in fertilization and settlement of young gag (*Mycteroperca microlepis*) along the west Florida Shelf. Bull. Mar. Sci. 77: 377–396.

Frias-Torres, S. 2006. Habitat use of juvenile goliath grouper. Endanger Species Res. 2: 1–6.

Froese, R. and D. Pauly (eds.). 2019. FishBase. World Wide Web Electronic Publication. www.fishbase.org (12/2019).

Fuiman, L.A. and A.E. Magurran. 1994. Development of predator defences in fishes. Rev. Fish Biol. Fish 4: 145–183.

Giménez-Hurtado, E., R. Coyula-Pérez-Puelles, S.E. Lluch-Cota, A.A. González-Yañez, V. Moreno-García and R. Burgos-De-La-Rosa. 2005. Historical biomass, fishing mortality, and recruitment trends of the Campeche Bank red grouper (*Epinephelus morio*). Fish Res. 71: 267–277.

Gold, J.R. and L.R. Richardson. 1998. Genetic homogeneity among geographic samples of snappers and groupers: evidence of continuous gene flow? pp. 709–726. *In*: Proceedings of the 50th Gulf and Caribbean Fisheries Institute. Gulf and Caribbean Fisheries Institute.

Gracia-López, V., M. Kiewek-Martínez and M. Maldonado-García. 2004. Effects of temperature and salinity on artificially reproduced eggs and larvae of the leopard grouper *Mycteroperca rosacea*. Aquaculture 237: 485–498.

Grandcourt, E.M., T.Z. Al Abdessalaam, F. Francis and A.T. Al Shamsi. 2005. Population biology and assessment of the orange-spotted grouper, *Epinephelus coioides* (Hamilton, 1822), in the southern Arabian Gulf. Fish Res. 74: 55–68.

Green, A.L., A.P. Maypa, G.R. Almany, K.L. Rhodes, R. Weeks, R.A. Abesamis, M.G. Gleason, P.J. Mumby and A.T. White. 2015. Larval dispersal and movement patterns of coral reef fishes, and implications for marine reserve network design. Biol. Rev. 90: 1215–1247.

Green, B.S. and R. Fisher. 2004. Temperature influences swimming speed, growth and larval duration in coral reef fish larvae. J. Exp. Mar. Bio. Ecol. 299: 115–132.

Hackradt, C.W., J.A. García-Charton, M. Harmelin-Vivien, Á. Pérez-Ruzafa, L. Le Diréach, J. Bayle-Sempere, E. Charbonnel, D. Ody, O. Reñones, P. Sanchez-Jerez and C. Valle. 2014. Response of rocky reef top predators (Serranidae: Epinephelinae) in and around marine protected areas in the Western Mediterranean Sea. PLoS One 9.

Halpern, B.S., K.L. McLeod, A.A. Rosenberg and L.B. Crowder. 2008. Managing for cumulative impacts in ecosystem-based management through ocean zoning. Ocean Coast Manag. 51: 203–211.

Hamilton, R.J., M. Matawai, W. Kama, P. Lahui, J. Warku and A.J. Smith. 2005. Applying local knowledge and science to the management of grouper aggregation sites in Melanesia. SPC Live Reef Fish Inf. Bull. 14: 7–19.

Harrison, H.B., D.H. Williamson, R.D. Evans, G.R. Almany, S.R. Thorrold, G.R. Russ, K.A. Feldheim, L. Van Herwerden, S. Planes, M. Srinivasan, M.L. Berumen and G.P. Jones. 2012. Larval export from marine reserves and the recruitment benefit for fish and fisheries. Curr. Biol. 22: 1023–1028.

Heppell, S., B.X. Semmens, C.V. Pattengill-Semmens, P.G. Bush, B.C. Johnson, C.M. Mccoy, J. Gibb and S.S. Heppell. 2010. Oceanographic patterns associated with nassau grouper aggregation spawn timing: shifts in surface currents on the nights of peak spawning. pp. 2010–2012. *In*: Gulf and Caribbean Fisheries Institute Proceedings.

Hereu, B., D. Diaz, J. Pasqual, M. Zabala and E. Sala. 2006. Temporal patterns of spawning of the dusky grouper *Epinephelus marginatus* in relation to environmental factors. Mari. Ecol. Progr. Ser. 325: 187–194.
Heyman, W.D. and B. Kjerfve. 2008. Characterization of transient multi-species reef fish spawning aggregations at Gladden Spit, Belize. Bull. Mar. Sci. 83: 531–551.
Hickford, M.J.H. and D.R. Schiel. 1999. Evaluation of the performance of light traps for sampling fish larvae in inshore temperate waters. Mar. Ecol. Progr. Ser. 186: 293–302.
Hin, V. and A.M. Roos. 2019. Cannibalism prevents evolutionary suicide of ontogenetic omnivores in life-history intraguild predation systems. Ecol. Evol. 1–16.
Hixon, M.A., D.W. Johnson and S.M. Sogard. 2014. Structure in fishery populations. ICES J. Mar. Sci. 71: 2171–2185.
Hjort, J. 1926. Fluctuations in the year classes of important food fishes. J. Cons Int. Explor. Mer. 1: 5–38.
Hseu, J.R., H.F. Chang and Y.Y. Ting. 2003. Morphometric prediction of cannibalism in larviculture of orange-spotted grouper, *Epinephelus coioides*. Aquaculture 218: 203–207.
Hseu, J.R., P.P. Hwang and Y.Y. Ting. 2004. Morphometric model and laboratory analysis of intracohort cannibalism in giant grouper *Epinephelus lanceolatus* fry. Fish Sci. 70: 482–486.
Huntsman, G.R. and W.E. Schaaf. 2004. Simulation of the impact of fishing on reproduction of a protogynous grouper, the graysby. North Am. J. Fish Manag. 14: 41–52.
Hutchinson, N. and K.L. Rhodes. 2010. Home range estimates for squaretail coralgrouper, *Plectropomus areolatus* (Rüppell 1830). Coral Reefs 29(2): 511–519.
Irisson, J.O., C.B. Paris, C. Guigand and S. Planes. 2010. Vertical distribution and ontogenetic "migration" in coral reef fish larvae. Limnol. Oceanogr. 55: 909–919.
Jackson, A.M., B.X. Semmens, Y.S. De Mitcheson, R.S. Nemeth, S.A. Heppell, P.G. Bush, A. Aguilar-Perera, J.A.B. Claydon, M.C. Calosso, K.S. Sealey, M.T. Schärer and G. Bernardi. 2014. Population structure and phylogeography in Nassau grouper (*Epinephelus striatus*), a mass-aggregating marine fish. PLoS One 9.
Jagadis, I., B. Ignatius, D. Kandasami and M.A. Khan. 2006. Embryonic and larval development of honeycomb grouper *Epinephelus merra* Bloch. Aquacult. Res. 37(11): 1140–1145.
James, M.K., P.R. Armsworth, L.B. Mason and L. Bode. 2002. The structure of reef fish metapopulations: Modelling larval dispersal and retention patterns. Proc. R Soc. Biol. Sci. B 269: 2079–2086.
Johannes, R.E. 1978. Reproductive strategies of coastal marine fishes in the tropics. Environ. Biol. Fishes 3: 65–84.
Johnson, G.D. and P. Keener. 1984. Aid to identification of American grouper larvae. Bull. Mar. Sci. 34: 106–134.
Jones, R.S., E.J. Gutherz, W.R. Nelson and G.C. Matlock. 1989. Burrow utilization by yellowedge grouper, *Epinephelus flavolimbatus*, in the northwestern Gulf of Mexico. Environ. Biol. Fish 26(4): 277–284.
Kawabe, K. and H. Kohno. 2009. Morphological development of larval and juvenile blacktip grouper, *Epinephelus fasciatus*. Fish Sci. 75: 1239–1251.
Kawakami, T., J. Aoyama and K. Tsukamoto. 2010. Morphology of pelagic fish eggs identified using mitochondrial DNA and their distribution in waters west of the Mariana Islands. Environ. Biol. Fishes 87: 221–235.
Keener, P., G.D. Johnson, B.W. Stender, E.B. Brothers and H.R. Beatty. 1988. Ingress of postlarval gag, *Mycetoperca microlepis* (Pisces: Serranidae), through a South Carolina barrier Island inlet. Bull. Mar. Sci. 42: 376–396.
Keough, M.J. and B.J. Downes. 1982. International association for ecology recruitment of marine invertebrates: the role of active larval choices and early mortality. Oecologia 54: 348–352.
Kinlan, B.P. and S.D. Gaines. 2003. Propagule dispersal in marine and terrestrial environments: a community perspective. Ecology 84: 2007–2020.
Kitajima, C., M. Takaya, Y. Tsukashima and T. Arakawa. 1991. Development of eggs, larvae and juveniles of the grouper, *Epinephelus septemfasciatus*, reared in the laboratory. Japanese J. Ichthyol. 38: 47–55.

Kobayashi, D.R. 2006. Colonization of the Hawaiian Archipelago via Johnston Atoll: a characterization of oceanographic transport corridors for pelagic larvae using computer simulation. Coral Reefs 25(3): 407–417.

Koenig, C., L. Bueno, F. Coleman, J. Cusick, R. Ellis, K. Kingon, J. Locascio, C. Malinowski, D. Murie and C. Stallings. 2017. Diel, lunar, and seasonal spawning patterns of the Atlantic goliath grouper, *Epinephelus itajara*, off Florida, United States. Bull. Mar. Sci. 93: 391–406.

Koenig, C.C. and F.C. Coleman. 1998. Absolute abundance and survival of juvenile gags in sea grass beds of the Northeastern Gulf of Mexico. Trans Am. Fish Soc. 127: 44–55.

Kohno, H., S. Diani and A. Supriatna. 1993. Morphological development of larval and juvenile grouper, *Epinephelus fuscoguttatus*. Japan J. Ichthyol. 40: 3017–316.

Kuriiwa, K., S.N. Chiba, H. Motomura and K. Matsuura. 2014. Phylogeography of blacktip grouper, *Epinephelus fasciatus* (Perciformes: Serranidae), and influence of the Kuroshio current on cryptic lineages and genetic population structure. Ichthyol. Res. 61: 361–374.

Kusaka, A., K. Yamaoka, T. Yamada, M. Abe and I. Kinoshita. 2001. Early development of dorsal and pelvic fins and their supports in hatchery-reared red-spotted grouper, *Epinephelus akaara* (Perciformes: Serranidae). Ichthyol. Res. 48(4): 355–360.

La Mesa, G., P. Louisy and M. Vacchi. 2002. Assessment of microhabitat preferences in juvenile dusky grouper *(Epinephelus marginatus*) by visual sampling. Mar. Biol. 140(1): 175–185.

Lara, M., J. Schull, D. Jones and R. Allman. 2009. Early life history stages of goliath grouper *Epinephelus itajara* (Pisces: Epinephelidae) from Ten Thousand Islands, Florida. Endanger. Species Res. 7: 221–228.

Largier, J.L. 2003. Considerations in estimating larval dispersal distances from oceanographic data. Ecol. Appl. 13: S71–S89.

Layman, C.A., D.A. Arrington, R.B. Langerhans and B.R. Silliman. 2004. Degree of fragmentation affects fish assemblage structure in Andros Island (Bahamas) estuaries. Caribb. J. Sci. 40: 232–244.

Lecchini, D., S. Planes and R. Galzin. 2005. Experimental assessment of sensory modalities of coral-reef fish larvae in the recognition of their settlement habitat. Behav. Ecol. Sociobiol. 58(1): 18–26.

Lecchini, D., K. Peyrusse, R.G. Lanyon and G. Lecellier. 2014. Importance of visual cues of conspecifics and predators during the habitat selection of coral reef fish larvae. CR Biol. 337(5): 345–351.

Lee, Y.D., S.H. Park, A. Takemura and K. Takano. 2002. Histological observations of seasonal reproductive and lunar related spawning cycles in the female honeycomb grouper *Epinephelus merra* in Okinawan waters. Fish Sci. 68(4): 872–877.

Leis, J., K. Wright and R. Johnson. 2007. Behaviour that influences dispersal and connectivity in the small, young larvae of a reef fish. Mar. Biol. 153: 103–117.

Leis, J.M., H.P.A. Sweatman and S.E. Reader. 1996. What the pelagic stages of coral reef fishes are doing out in blue water: Daytime field observations of larval behavioural capabilities. Mar. Freshw. Res. 47: 401–411.

Leis, J.M. and B.M. Carson-Ewart. 1999. *In situ* swimming and settlement behaviour of larvae of an Indo-Pacific coral-reef fish, the coral trout *Plectropomus leopardus* (Pisces: Serranidae). Mar. Biol. 134: 51–64.

Leis, J.M. 2007. Behaviour as input for modelling dispersal of fish larvae: behaviour, biogeography, hydrodynamics, ontogeny, physiology and phylogeny meet hydrography. Mar. Ecol. Prog. Ser. 347: 185–193.

Leis, J.M., A.C. Hay, M.M. Lockett, J.P. Chen and L.S. Fang. 2007. Ontogeny of swimming speed in larvae of pelagic-spawning, tropical, marine fishes. Mar. Ecol. Prog. Ser. 349: 255–267.

Leis, J.M., A.C. Hay and G.J. Howarth. 2009. Ontogeny of *in situ* behaviours relevant to dispersal and population connectivity in larvae of coral-reef fishes. Mar. Ecol. Progr. Ser. 379: 163–179.

Leis, J.M. 2010. Pacific coral-reef fishes: the implications of behaviour and ecology of larvae for biodiversity and conservation, and a reassessment of the open population paradigm. Env. Biol. Fish 65(2): 199–208.

Leis, J.M., J.E. Caselle, I.R. Bradbury, T. Kristiansen, J.K. Llopiz, M.J. Miller, M.I. O'Connor, C.B. Paris, A.L. Shanks, S.M. Sogard, S.E. Swearer, E.A. Treml, R.D. Vetter and R.R. Warner. 2013. Does fish larval dispersal differ between high and low latitudes? Proc. Biol. Sci. 280: 20130327.

Leis, J.M. 2018. Paradigm lost: ocean acidification will overturn the concept of larval-fish biophysical dispersal. Front Mar. Sci. 5: 1–9.

Letourneur, Y., P. Chabanet, L. Vigliola and M. Harmelin-Vivien. 1998. Mass settlement and post-settlement mortality of *Epinephelus merra* (Pisces: Serranidae) on Réunion coral reefs. Mar. Biol. Assoc. United Kingdom 78: 307–319.

Lett, C., S.D. Ayata, M. Huret and J.O. Irisson. 2010. Biophysical modelling to investigate the effects of climate change on marine population dispersal and connectivity. Prog. Oceanogr. 87: 106–113.

Light, P.R. and G.P. Jones. 1997. Habitat preference in newly settled coral trout (*Plectropomus leopardus*, Serranidae). Coral Reefs 16: 117–126.

Lindeman, K.C., R. Pugliese, G.T. Waugh and J.S. Ault. 2000. Developmental patterns within a multispecies reef fishery: Management applications for essentials fish habitats and protected areas. Bull. Mar. Sci. 66: 929–956.

Lyczkowski-shultz, J., D.L. Ruple, S.L. Richardson and J.H. Cowan. 1990. Distribution of fish larvae relative to time and tide in a Gulf of Mexico Barrier island pass. Bull. Mar. Sci. 46: 563–577.

Macpherson, E. and N. Raventos. 2006. Relationship between pelagic larval duration and geographic distribution of Mediterranean littoral fishes. Mar. Ecol. Prog. Ser. 327: 257–265.

Mamauag, S., L. Penolio and P. Aliño. 2001. Deriving recruitment and spawning patterns from a survey of juvenile grouper (Pisces: Serranidae) occurrences in the Philippines. Fish 54–65.

Marancik, K.E., D.E. Richardson, J. Lyczkowski-Shultz, R.K. Cowen and M. Konieczna. 2012. Spatial and temporal distribution of grouper larvae (Serranidae: Epinephelinae: Epinephelini) in the Gulf of Mexico and Straits of Florida. Fish Bull. 110: 1–20.

Martínez-Lagos, R. and V. Gracia-López. 2009. Morphological development and growth patterns of the leopard grouper *Mycteroperca rosacea* during larval development. Aquac. Res. 41: 120–128.

Masuma, S., T. Nobuhiro and T. Kazuhisa. 1993. Embryonic and morphological development of larval and juvenile coral trout, *Plectropomus leopardus*. Japan J. Ichthyol. 40: 333–342.

Milicich, M.J. and P.J. Doherty. 1994. Larval supply of coral reef fish populations: magnitude and synchrony of replenishment to Lizard Island, Great Barrier Reef. Mar. Ecol. Prog. Ser. 110: 121–134.

Mishina, H., B. Gonzales, H. Pagaliawan, M. Moteki and H. Kohno. 2006. Reproductive biology of blacktip grouper, *Epinephelus fasciatus*, in Sulu Sea, Philippines. La mer 44(23): 31.

Mitcheson, Y.S. De, A. Cornish, M. Domeier, P.L. Colin, M. Russell and K.C. Lindeman. 2008. A global baseline for spawning aggregations of reef fishes. Conserv. Biol. 22: 1233–1244.

Moffitt, R.B. 2006. Biological data and stock assessment methodologies for deep-slope bottomfish resources in the Hawaiian archipelago. pp. 301–308. *In*: Deep Sea 2003: Conference on the Governance and Management of Deep-sea Fisheries.

Mora, C. and P.F. Sale. 2002. Are populations of coral reef fish open or closed? Trends Ecol. Evol. 17: 422–428.

Munday, P.L., D.L. Dixson, J.M. Donelson, G.P. Jones, M.S. Pratchett, G.V. Devitsina and K.B. Døving. 2009. Ocean acidification impairs olfactory discrimination and homing ability of a marine fish. Proc. Natl. Acad. Sci. 106: 1848–1852.

Munday, P.L., M. Gagliano, J.M. Donelson, D.L. Dixson and S.R. Thorrold. 2011. Ocean acidification does not affect the early life history development of a tropical marine fish. Mar. Ecol. Prog. Ser. 423: 211–221.

Munro, J.L., V.C. Gaut, R. Thompson and P.H. Reeson. 1973. The spawning seasons of Caribbean reef fishes. J. Fish Biol. 5(1): 69–84.

Nagelkerken, W.P. 1979. Biology of the Graysby, *Epinephelus cruentatus,* on the coral reef of Curacao. Studies on the fauna of Curaçao and other Caribbean Islands 60: 1–118.

O'Connor, M.I., J.F. Bruno, S.D. Gaines, B.S. Halpern, S.E. Lester, B.P. Kinlan and J.M. Weiss. 2007. Temperature control of larval dispersal and the implications for marine ecology, evolution, and conservation. Proc. Natl. Acad. Sci. 104: 1266–1271.

Ojeda-Serrano, E., R.S. Appeldoorn and I. Ruiz-Valentin. 2007. Reef fish spawning aggregations of the Puerto Rican Shelf. 59th Gulf and Caribbean Fisheries Institute 59: 401–408.

Okumura, S., K. Okamoto, R. Oomori and A. Nakazono. 2002. Spawning behavior and artificial fertilization in captive reared red spotted grouper, *Epinephelus akaara*. Aquaculture 206: 165–173.

Paris, C.B., R.K. Cowen, R. Claro and K.C. Lindeman. 2005. Larval transport pathways from Cuban snapper (Lutjanidae) spawning aggregations based on biophysical modeling. Mar. Ecol. Prog. Ser. 296: 93–106.

Park, J.Y., J.K. Cho, M.H. Son, K.M. Kim, K.H. Han and J.M. Park. 2016. Artificial spawning behavior and development of eggs, larvae and juveniles of the red spotted grouper, *Epinephelus akaara* in Korea. Dev. Reprod. 20: 31–40.

Parmentier, E., L. Berten, P. Rigo, F. Aubrun, S.L. Nedelec, S.D. Simpson and D. Lecchini. 2015. The influence of various reef sounds on coral-fish larvae behaviour. J. Fish Biol. 86(5): 1507–1518.

Pears, R.J., J.H. Choat, B.D. Mapstone and G.A. Begg. 2006. Demography of a large grouper, *Epinephelus fuscoguttatus*, from Australia's Great Barrier Reef: implications for fishery management. Mar. Ecol. Progr. Ser. 307: 259–272.

Pinault, M., J.P. Quod and R. Galzin. 2015. Mass-settlement of the Indian ocean black-tip grouper *Epinephelus oceanicus* (Lacepède, 1802) in a shallow volcanic habitat following a tropical storm. Environ. Biol. Fishes 98: 705–711.

Pineda, J., J.A. Hare and S. Sponaugle. 2007. Larval transport and dispersal in the coastal ocean and consequences for population connectivity. Oceanography 20: 22–39.

Powell, A.B. and J.W. Tucker. 1992. Egg and larval development of laboratory-reared Nassau grouper, *Epinephelus striatus* (Pisces, Serranidae). Bull. Mar. Sci. 50(1): 171–185.

Reñones, O., A. Grau, X. Mas, F. Riera and F. Saborido-Rey. 2010. Reproductive pattern of an exploited dusky grouper *Epinephelus marginatus* (Lowe 1834) (Pisces: Serranidae) population in the western Mediterranean. Sci. Mar. 74(3): 523–537.

Reynal, L., V. Druault-Aubin, A. Lagin and J.J. Rivoalen. 2004. Fishing on spawning aggregation sites of groupers in Martinique (Mutton hamlet–*Alphestes afer*). Proc. Gulf Caribb. Fish. Inst. 55: 608–613.

Richards, W.J. 2005. Early Stages of Atlantic Fishes (WJ Richards, Ed.). Taylor & Francis, Boca Raton.

Richu, A., N. Dahanukar, Ali, K. AnvarRanjeet and R. Raghavan. 2018. Population dynamics of a poorly known serranid, the duskytail grouper *Epinephelus bleekeri* in the Arabian Sea. J. Fish Biol. 93: 741–744.

Rivera, M.A.J., K.R. Andrews, D.R. Kobayashi, J.L.K. Wren, C. Kelley, G.K. Roderick and R.J. Toonen. 2011. Genetic analyses and simulations of larval dispersal reveal distinct populations and directional connectivity across the range of the Hawaiian grouper (*Epinephelus quernus*). J. Mar. Biol. 2011: 1–11.

Rivera, M.A.N.A.J., C.D. Kelley and G.K. Roderick. 2004. Subtle population genetic structure in the Hawaiian grouper, *Epinephelus quernus* (Serranidae) as revealed by mitochondrial DNA analyses. Biol. J. Linn. Soc. 81: 449–468.

Roberts, C.M. 1996. Settlement and beyond: population regulation and community structure of reef fishes. *In*: Polunin, N.V.C. and C.M. Roberts (eds.). Reef Fisheries. Springer Science Business Media, Dordrecht.

Robertson, D.R. 1992. Patterns of lunar settlement and early recruitment in Caribbean reef fishes at Panamá. Mar. Biol. 114: 527–537.

Ross, S.W. 1988. Xanthic coloration as the normal color pattern of juvenile speckled hind, *Epinephelus drummondhayi* (Pisces: Serranidae). Copeia 3: 780–784.

Ross, S.W. and M.L. Moser. 1995. Life history of juvenile gag, in North Carolina estuaries. Bull. Mar. Sci. 56: 222–237.
Sabate, F. de la S., Y. Sakakura, M. Shiozaki and A. Hagiwara. 2009. Onset and development of aggressive behavior in the early life stages of the seven-band grouper *Epinephelus septemfasciatus*. Aquaculture 290: 97–103.
Sadovy de Mitcheson, Y., M.T. Craig, A.A. Bertoncini, K.E. Carpenter, W.W. Cheung, J.H. Choat, A.S. Cornish, S.T. Fennessy, B.P. Ferreira, P.C. Heemstra, M. Liu, R.F. Myers, D.A. Pollard, K.L. Rhodes, L.A. Rocha, B.C. Russell, M.A. Samoilys and J. Sanciangco. 2012. Fishing groupers towards extinction: a global assessment of threats and extinction risks in a billion dollar fishery. Fish Fish 14: 119–136.
Sadovy, Y. and A.-M. Eklund. 1999. Synopsis of Biological Data on the Nassau Grouper, *Epinephelus striatus*. Seattle, Washington.
Sadovy, Y. 2001. The threat of fishing to highly fecund fishes. J. Fish Biol. 59: 90–108.
Sadovy, Y.J., T.J. Donaldson, T.R. Graham, F. McGilvray, G.J. Muldoon, M. Philipps, M.A. Rimmer, A. Smith and B. Yeeting. 2003. While Stocks Last: The Live Reef Food Fish Trade. Asian Development Bank, 169p.
Sale, P.F., R.K. Cowen, B.S. Danilowicz, G.P. Jones, J.P. Kritzer, K.C. Lindeman, S. Planes, N.V.C. Polunin, G.R. Russ, Y.J. Sadovy and R.S. Steneck. 2005. Critical science gaps impede use of no-take fishery reserves. Trends Ecol. Evol. 20: 74–80.
Samoilys, M. 1997a. Movement in a large predatory fish: Coral trout, *Plectropomus leopardus* (Pisces: Serranidae), on Heron reef. Aust. Coral Reefs 16: 151–158.
Samoilys, M.A. 1997b. Periodicity of spawning aggregations of coral trout *Plectropomus leopardus* (Pisces: Serranidae) on the northern Great Barrier Reef. Mar. Ecol. Prog. Ser. 160: 149–159.
Scharer, M.T., M.I. Nemeth and R.S. Appeldoorn. 2010. Protecting a multi-species spawning aggregation at Mona Island, Puerto Rico. In Proc. Gulf Caribb. Fish. Inst. 62: 252–259.
Schunter, C., J. Carreras-Carbonell, S. Planes, E. Sala, E. Ballesteros, M. Zabala, J.-G. Harmelin, M. Harmelin-Vivien, E. Macpherson and M. Pascual. 2011. Genetic connectivity patterns in an endangered species: The dusky grouper (*Epinephelus marginatus*). J. Exp. Mar. Bio. Ecol. 401: 126–133.
Sedberry, G.R., O. Pashuk, D.M. Wyanski, J.A. Stephen and P. Weinbach. 2006. Spawning locations for Atlantic reef fishes off the Southeastern U.S. pp. 463–514. *In*: 57th Gulf and Caribbean Fisheries Institute.
Shanks, A.L., B.A. Grantham and M.H. Carr. 2003. Propagule dispersal distance and the size and spacing of marine reserves. Ecol. Appl. 13: S159–S169.
Shanks, A.L. and G.L. Eckert. 2005. Population persistence of california current fishes and benthic crustaceans: a marine drift paradox. Ecol. Monogr. 75: 505–524.
Shanks, A.L. 2009. Pelagic larval duration and dispersal distance revisited. Biol. Bull. 216: 373–385.
Shein, N.L., H. Chuda, T. Arakawa, K. Mizuno and K. Soyano. 2004. Ovarian development and final oocyte maturation in cultured sevenband grouper *Epinephelus septemfasciatus*. Fish Sci. 70(3): 360–365.
Shenker, J.M., E.D. Maddox, E. Wishinski, A. Pearl, S. Thorrold and N. Smith. 1993. Onshore transport of settlement-stage Nassau grouper *Epinephelus striatus* and other fishes. Exuma Sound, Bahamas Mar. Ecol. Prog. Ser. 98: 31–43.
Sherman, K.D., R.A. King, C.P. Dahlgren, S.D. Simpson, J.R. Stevens and C.R. Tyler. 2017. Historical processes and contemporary anthropogenic activities influence genetic population dynamics of nassau grouper (*Epinephelus striatus*) within the Bahamas. Front. Mar. Sci. 4: 393.
Silva-Oliveira, G.C., P.S. Do Rêgo, H. Schneider, I. Sampaio and M. Vallinoto. 2008. Genetic characterisation of populations of the critically endangered Goliath grouper (*Epinephelus itajara*, Serranidae) from the Northern Brazilian coast through analyses of mtDNA. Genet. Mol. Biol. 31: 988–994.

Simpson, S.D., M. Meekan, J. Montgomery, R. McCauley and A. Jeffs. 2005. Homeward sound. Science 308(5719): 221–221.

Sluka, R., M. Chiappone and K.M. Sullivan. 1994. Comparison of juvenile grouper populations in southern Florida and the central Bahamas. Bul. Mar. Sci. 54(3): 871–880.

Sswat, M., M.H. Stiasny, F. Jutfelt, U. Riebesell and C. Clemmesen. 2018. Growth performance and survival of larval Atlantic herring, under the combined effects of elevated temperatures and CO_2. PLoS One 13: e0191947.

Staaterman, E., C.B. Paris and J. Helgers. 2012. Orientation behavior in fish larvae A missing piece to Hjort's critical period hypothesis. J. Theor. Biol. 1–9.

Staaterman, E. and C.B. Paris. 2014. Modelling larval fish navigation: The way forward. ICES J. Mar. Sci. 71: 918–924.

Stallings, C. 2008. Indirect effects of an exploited predator on recruitment of coral-reef fishes. Ecology 89: 2090–2095.

Stobutzki, I.C. and D.R. Bellwood. 1998. Nocturnal orientation to reefs by late pelagic stage coral reef fishes. Coral Reefs 17: 103–110.

Thompson, R. and J.L. Munro. 1978. The biology, ecology and bionomics of the Caribbean reef fishes: Serranidae (hinds and groupers). J. Fish Biol. 12: 115–146.

Thorrold, S.R., J.M. Shenker, E. Wishinski, R. Mojica and E.D. Maddox. 1994. Larval supply of shorefishes to nursery habitats around Lee Stocking Island, Bahamas. I. Small-scale distribution patterns. Mar. Biol. 118: 555–566.

Toscano, B.J., V. Hin and V.H.W. Rudolf. 2017. Cannibalism and intraguild predation community dynamics: coexistence, competitive exclusion, and the loss of alternative stable states. Am. Nat. 190: 617–630.

Trott, T.M. 2006. Preliminary analysis of age, growth, and reproduction of coney (Cephalopholis fulva) at Bermuda. GCFI 57: 385–399.

Tuz-Sulub, A. and T. Brulé. 2015. Spawning aggregations of three protogynous groupers in the southern Gulf of Mexico. J. Fish Biol. 86(1): 162–185.

Vermeij, M.J., K.L. Marhaver, C.M. Huijbers, I. Nagelkerken and S.D. Simpson. 2010. Coral larvae move toward reef sounds. PloS One 5(5).

Waples, R.S., D.W. Jensen and M. McClure. 2010. Eco-evolutionary dynamics: fluctuations in population growth rate reduce effective population size in chinook salmon. Ecology 91(3): 902–914.

Watson, M. and J.L. Munro. 2004. Settlement and recruitment of coral reef fishes in moderately exploited and overexploited Caribbean ecosystems: Implications for marine protected areas. Fish Res. 69: 415–425.

Watson, S., B.J.M. Allan, D.E. Mcqueen, S. Nicol, D.M. Parsons, S.M.J. Pether, S. Pope, A.N. Setiawan, N. Smith, C. Wilson and P.L. Munday. 2018. Ocean warming has a greater effect than acidification on the early life history development and swimming performance of a large circumglobal pelagic fish. Glob. Chang. Biol. 1–18.

Weisberg, R.H., L. Zheng and E. Peebles. 2014. Gag grouper larvae pathways on the West Florida Shelf. Cont. Shelf Res. 88: 11–23.

Wellington, G.M. and B.C. Victor. 1992. Regional differences in duration of the planktonic larval stage of reef fishes in the eastern Pacific Ocean. Mar. Biol. 113: 491–498.

Williamson, D.H., H.B. Harrison, G.R. Almany, M.L. Berumen, M. Bode, M.C. Bonin, S. Choukroun, P.J. Doherty, A.J. Frisch, P. Saenz-Agudelo and G.P. Jones. 2016. Large-scale, multidirectional larval connectivity among coral reef fish populations in the Great Barrier Reef Marine Park. Mol. Ecol. 25: 6039–6054.

Wolanski, E. and M. Kingsford. 2014. Oceanographic and behavioural assumptions in models of coral reef larval fish oceanography. J. R Soc. Interface 11: 1–19.

Wright, K.J., D.M. Higgs, A.J. Belanger and J.M. Leis. 2008. Auditory and olfactory abilities of larvae of the Indo-Pacific coral trout *Plectropomus leopardus* (Lacepède) at settlement. J. Fish Bio. 72(10): 2543–2556.

Zatcoff, M.S., A.O. Ball and G.R. Sedberry. 2004. Population genetic analysis of red grouper, *Epinephelus morio*, and scamp, *Mycteroperca phenax*, from the southeastern U.S. Atlantic and Gulf of Mexico. Mar. Biol. 144: 769–777.

Zavala-Camin, L.A. 2012. Ocorrência de juvenis de Serranidae nas regiões Sudeste e Sul do Brasil. Bioikos 22(2): 63–79.

Ziskin, G.L. 2008. Age, growth and reproduction of speckled hind, *Epinephelus drummondhayi*, off the Atlantic coast of the southeast United States. Master thesis. College of Charleston. 130 p.

Chapter 1.3

Sexual Patterns and Reproductive Behaviours in Groupers

Beatrice Padovani Ferreira,[1,*] *Simone Marques,*[1]
Mario Vinicius Condini[2] and *Min Liu*[3]

Introduction

The reproductive biology of teleosts shows astonishing diversity, surpassing most vertebrate groups (Helfman et al. 2009). They exhibit different adaptations regarding sexual patterns (gonochorism and hermaphroditism, simultaneously or sequentially), mating systems (promiscuity, polygamy, and monogamy), modes of fertilization (external, internal, or oral), spawning behaviours (ranging from a couple spawning into an elaborate nest to hundreds of individuals spawning in large aggregations), among many other outstanding traits (Jakobsen et al. 2009, Wootton and Smith 2015). Functional hermaphroditism is a natural but uncommon feature in most teleosts; it is only found in seven out of 40 orders, predominantly in marine tropical perciforms (Nelson 2006, Sadovy de Mitcheson and Liu 2008).

Epinephelidae groupers (previously known as the subfamily Epinephelinae of the family Serranidae) comprise 16 genera and more than 160 species, including some with high commercial and ecological values, distributed worldwide in subtropical and tropical seas (Craig et al. 2011, Ma and Craig 2018). Species in this family exhibit great interspecific variability in sexual pattern, reproductive behaviour, and gonadal development. Studies on these topics have been conducted in the last five decades using several

[1] Department of Oceanography, Federal University of Pernambuco, Brazil.
[2] Department of Agricultural and Biological Sciences, Federal University of Espirito Santo, Brazil.
[3] State Key Laboratory of Marine Environmental Science, College of Ocean and Earth Sciences, Xiamen University, Xiamen, Fujian, China.
* Corresponding author: beatrice.ferreira@ufpe.br

tools and techniques. Most of these studies sought to provide information to assist management and conservation of endangered or overfished groupers species. As with most marine fishes, species-specific data is required, because reproductive life-history traits are variable both within and among grouper species (Sadovy 1996). Thus, in such a speciose group, it is expected that substantial knowledge gaps still remain on reproductive biology of groupers. This chapter summarizes information on sexual pattern, gonad development, and spawning periodicity, including sexual behavior and reproductive aggregation of groupers, providing an up-to-date overview of these emblematic teleosts.

Sexual Patterns

Most marine fishes are gonochorists, i.e., genders are separate and no sex change occurs along the lifetime of the individual, which reproduces solely as female or male during its lifetime. In contrast, the hermaphroditic species function as both females and males. Hermaphrodites can be simultaneous, which can reproduce both as male and female at the same time, or be sequential, in which the functional female phase develops first and transitions to male (protogynous), or has a previous functional male development and then transitions to a female phase (protandrous), or changes sex in both ways (bi-directional) (Sadovy and Shapiro 1987, Sadovy de Mitcheson and Liu 2008).

The features of functional hermaphroditism are described for 27 teleost families (Sadovy de Mitcheson and Liu 2008). In fact, only approximately one-third of grouper species have their sexual patterns examined, with protogynous hermaphroditism being the most commonly found, present in 54 species of seven genera (Table 1).

Protogynous hermaphroditism was described in groupers since early investigations on their reproductive biology (e.g., Van Oordt 1933, Smith 1965), and this led to the general belief that this pattern extended to all species. However, more recent investigations have revealed that sexuality among groupers can be more complex and diverse, including gonochorism, protogyny with one or two different pathways among male individuals, and bi-directional sex change with males changing or reversing to females in a couple of diagnosed protogynous species (Sadovy de Mitcheson and Liu 2008). Figure 1 summarizes those reproductive patterns.

Reinboth (1962, 1970) recognized two forms of protogyny based on male developmental pathways and coined the terms monandry, in which all males are derived from females through sex change, and diandry, in which males can be born either as males or derived from females. While the distinction between primary and secondary males was clear for some species of Labridae species, in which morphological differences in testicular structure clearly reflected whether a male was originated by sex change (secondary male) or born as a male (primary male), it was not the case for groupers, in which all males, sex change or not, had a secondary testes, i.e., with ovarian structure

Table 1. Bibliographical review of the known sexual patterns of Epinephelidae, showing the number (n) and percentage (%) of species by genus.

Genus	Total Species	Gonochoristic n (%)	Bi-directional n (%)	Protogynous n (%)	Not Determined n (%)	References
Aethaloperca	1				1 (100.0)	1
Alphestes	3			1 (33.3)	2 (66.7)	2
Anyperodon	1				1 (100.0)	1
Cephalopholis	24		1 (4.2)	8 (33.3)	15 (62.5)	1, 3, 4
Cromileptes	1			1 (100.0)	0 (0.0)	5
Dermatolepis	3				3 (100.0)	1
Epinephelus	86	3 (3.5)	1 (1.2)	24 (27.9)	58 (67.4)	1, 6, 7, 8, 9, 10, 11, 12, 13
Gonioplectrus	1				1 (100.0)	1
Gracila	1				1 (100.0)	1
Hyporthodus	14			7 (50.0)	7 (50.0)	1
Mycteroperca	15	1 (6.7)		9 (60.0)	5 (33.3)	1
Paranthias	2				2 (100.0)	1
Plectropomus	7			3 (42.9)	4 (57.1)	1, 14
Saloptia	1				1 (100.0)	1
Triso	1				1 (100.0)	1
Variola	2				2 (100.0)	1
Total species	163	4	2	53	106	

References: (1) Craig et al. 2011; (2) Marques and Ferreira 2011; (3) Erisman et al. 2010; (4) Liu and Sadovy 2004; (5) Liu and Sadovy de Mitcheson 2009; (6) Nakamura and Motomura 2021; (7) Bulanin et al. 2017; (8) Liu et al. 2016; (9) Bright et al. 2016; (10) Tucker et al. 2016; (11) Frable et al. 2019; (12) Johnson et al. 1998; (13) Wu et al. 2020; (14) Adams 2003.

* Protogynous species: *Alphestes afer; Cephalopholis cruentata; C. cyanostigma; C. fulva; C. hemistiktos; C. miniata; C. paramensis; C. taeniops; C. urodeta; Cromileptes altivelis; Epinephelus adscensionis; E. aeneus; E. albomarginatus; E. andersoni; E. awoara; E. chlorostigma; E. coioides; E. daemelii; E. diacanthus; E. drummondhayi; E. fasciato maculosus; E. fuscoguttatus; E. guttatus; E. labriformes; E. malabaricus; E. marginatus; E. merra; E. morio; E. ongus; E. polylepis; E. quoyanus; E. rivulatus; E. coeruleopunctatus; E. akaara; E. lanceolatus; H. ergastularius; H. flavolimbatus; H. nigritus; H. niveatus; H. octofasciatus; H. quernus; H. septemfasciatus; Mycteroperca bonaci; M. fusca; M. interstitialis; M. microlepis; M. olfax; M. phenax; M. rubra; M. tigris; M. venenosa; Plectropomus laevis; P. leopardos; P. maculatus.*

and a lumen. Development of males directly from immature female phase (i.e., juvenile phase) was later described for several grouper species and often referred as pre-maturational sex change (Warner and Robertson 1978, Sadovy and Shapiro 1987, Ferreira and Russ 1995). The increase in the number of studies in the last decades led to a multitude of terms that were used in the literature, and often not clearly related to sexual function or gonad ontogeny/morphology. Sadovy de Mitcheson and Liu (2008) proposed a redefinition of terms to represent the pathways observed for protogynous species, with monandry being termed for species in which all males are derived from

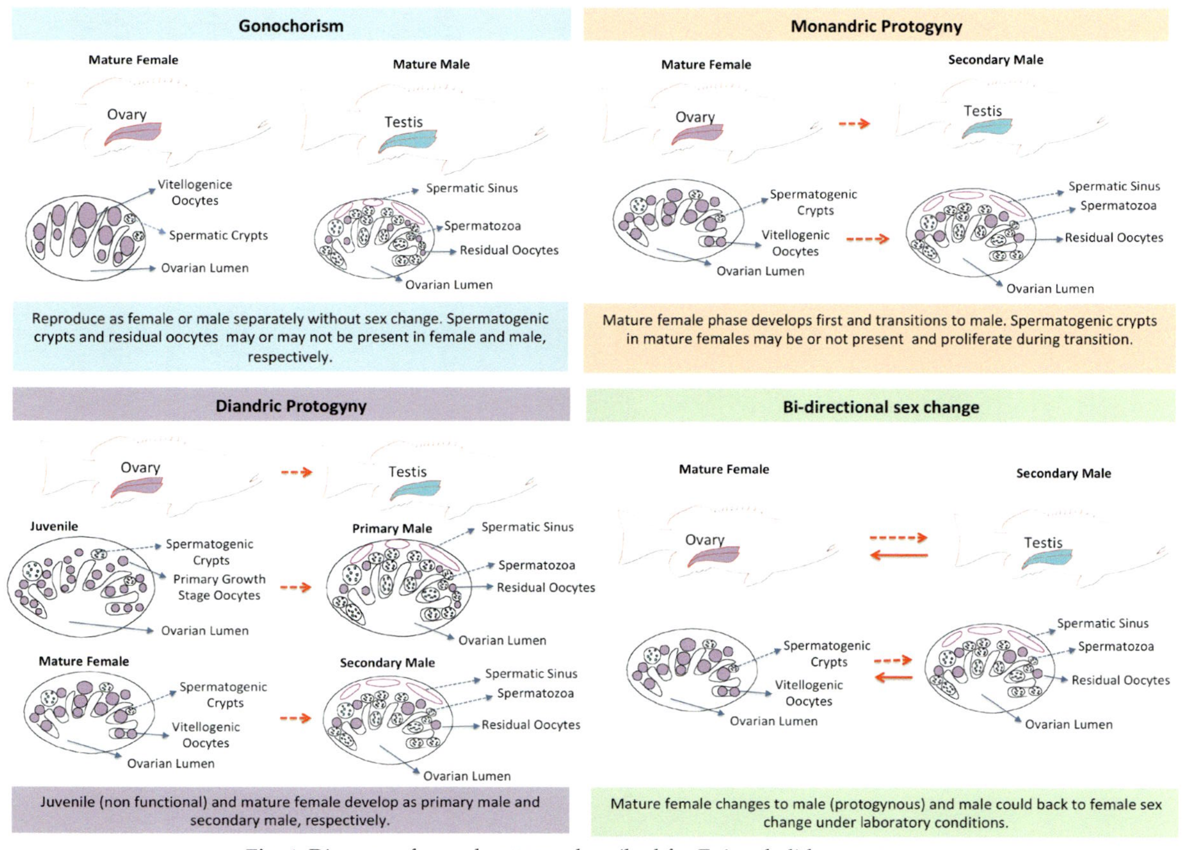

Fig. 1 Diagram of sexual patterns described for Epinephelidae groupers.

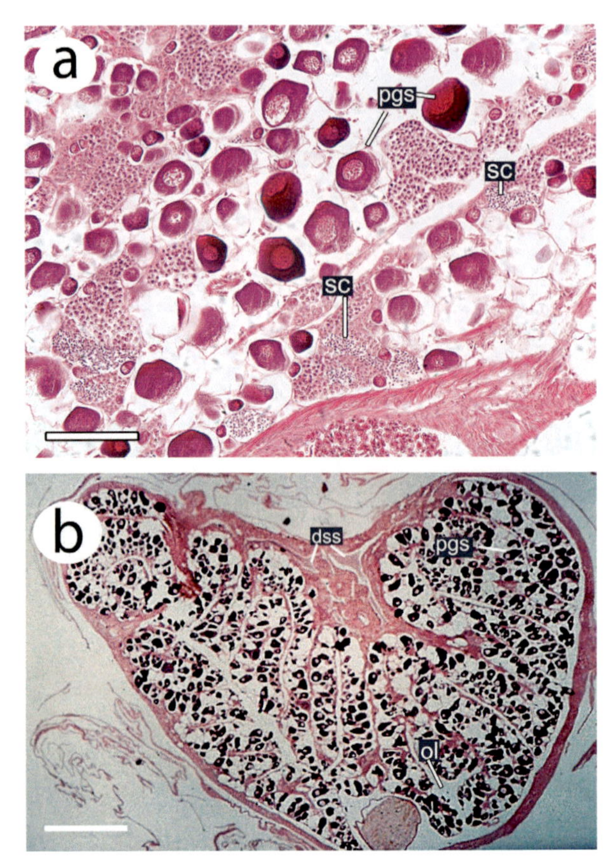

Fig. 2 (a) Gonad of a primary male from *Epinephelus akaara* (14.6 cm TL), presence of primary growth stage oocytes (pgs) and sc – spermatogenic crypts. (b) Gonad of a primary male from *Plectropomus leopardus* (32 cm TL); presence of dorsal sperm sinus (dss) filled with spermatozoa and an ovarian lumen (ol). Scale bars: 100 µm (a) and 500 µm (b).

functional females, thus defined as secondary males, and diandry for species in which there are two types of males: primary males, that are males at the first sexual maturation, and secondary males, arising from a sex change in functional females, irrespective of their similar testicular structure (Fig. 1). Figure 2 shows two examples of primary male grouper gonads.

While protogyny is the prevalent type of sexual development among groupers, functional gonochorism has been revealed so far in four (2.5%) species, including *Epinephelus striatus* (Sadovy and Colin 1995), *Mycteroperca rosacea*, *E. polyphekadion* (Rhodes et al. 2011), and *E. latifasciatus* (Sujatha and Shrikanya 2013) (Table 1; Fig. 1). These groupers mature as either males or females, but may retain the potential for protogynous sex change. The predominance of primary males in a *Alphestes* species in combination with other features has been considered as a precursor to gonochorism (Marques and Ferreira 2011).

As for bi-directional sex change, only two grouper species have been observed under laboratory conditions, *Cephalopholis boenak* (Liu and Sadovy 2004a) and *E. akaara* (Okumura 2001), both previously determined as protogynous hermaphroditism, with female to male and male to female sex change observed.

Diagnosis of Sexual Pattern

Assessing sexual patterns in groupers typically requires assessing specific indicators, some being more reliable than others. After identifying some drawbacks of earlier attempts, Sadovy and Shapiro (1987) proposed rigorous criteria to determine hermaphroditism, based on a combination of direct and indirect characteristics.

Indirect indicators include demographic features, such as bimodal size-frequency distributions derived from significantly larger males, typical of protogynous succession. However, population structure on its own cannot be taken as a reliable indicator of sexual patterns, because other biological attributes may also influence this process, such as size dimorphism, differential rates of growth, maturation, mortality by sex, and selective removal by fisheries. Gonadal structure is also considered an indirect indicator in groupers, as a testis containing a lumen, a female feature, retained by all males (Smith 1965, Sadovy and Shapiro 1987). Thus, the appearance of lumen and primary growth stage (not yolked) oocytes cannot be used to identify a transitional gonad, since these features are known to appear in both gonochoric and primary males, as well as in juveniles and immature females (Liu and Sadovy 2004b, Liu and Sadovy de Mitcheson 2009). Another confounding factor in groupers' sexual pattern diagnosis is the juvenile bisexual phase development (i.e., male and female germinal cells occurring in the same gonad) displayed by both gonochoric and hermaphroditic species.

Functional sex change should be identified through direct evidence of previous female function, such as the occurrence of atretic vitellogenic oocytes and the concurrent to proliferation of testicular tissue with small crypts (see Fig. 3).

Difficulties of diagnosis increase when functional males are the same size as females (Reinboth 1970), or even some species can present functional males smaller than females with possibilities of diandric protogynous or gonochoric patterns (Liu and Sadovy 2004b). Erisman et al. (2009) proposed the sperm rank competition index (SR) to evaluate the intensity of sperm competition in groupers, and suggested that higher indexes were observed for species with higher sperm competition, where gonochorism was progressively favoured against protogyny. The mutton hamlet, *Alphestes afer* is a possible example, as the males in this species are smaller than females, and presented a high SR index, related to species that were protogynous or unconfirmed gonochorist (Marques and Ferreira 2011) (Fig. 4).

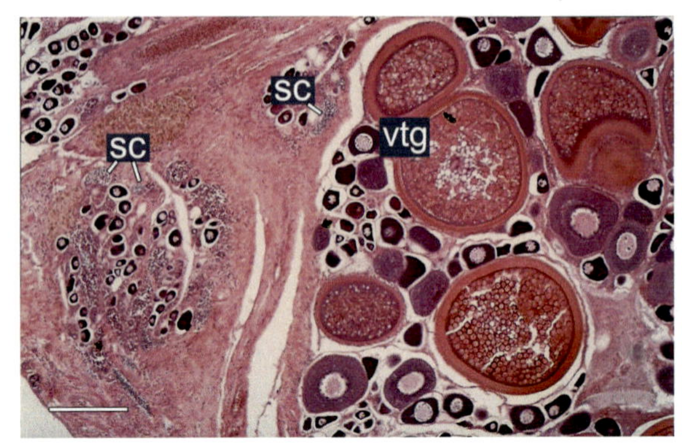

Fig. 3 The proliferation of spermatogenic crypts in the ovary of a mature female of *Plectropomus leopardus*. sc, spermatogenic crypts; ytg, vitellogenic stage oocytes. Scale bar: 500 μm.

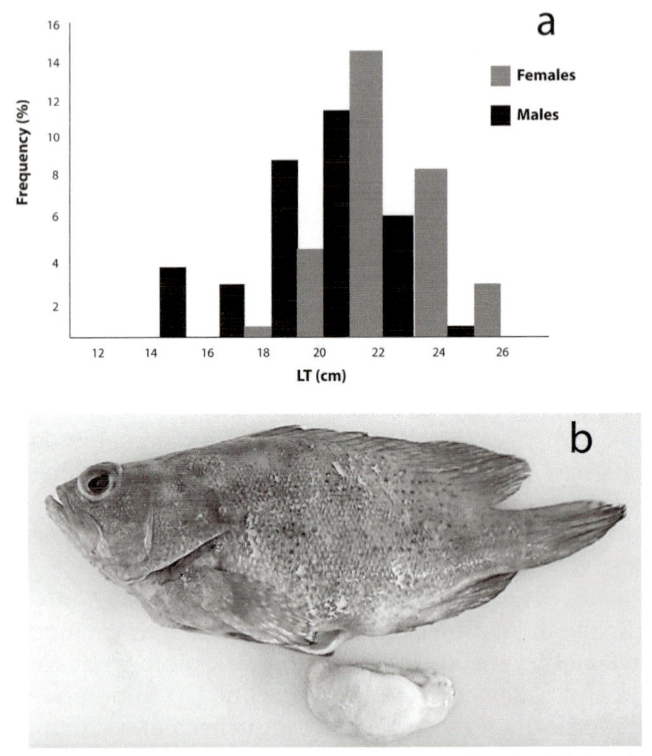

Fig. 4 (a) Size distribution pattern of *Alphestes afer* males and females. (b) A male was smaller than females and presented large testes (18.5 cm TL). Source: Marques and Ferreira (2011).

In some cases, experimental manipulations have been pointed out as necessary for determining sexual patterns (Liu and Sadovy 2004a, Liu and Sadovy de Mitcheson 2011). Indeed, with the crescent increase in rearing and farming of groupers in Asia, this seems to be one of the fastest growing fields of study, although limited to a few species. However, due to the apparent sexual plasticity of groupers, sex changes observed under laboratory conditions may not occur in wild populations (Sadovy and Domeier 2005).

Although strict criteria have been established on determining functional hermaphroditism in teleosts (Sadovy and Shapiro 1987, Sadovy de Mitcheson and Liu 2008), sexual patterns are not fully resolved in many species of groupers. The complexity of grouper reproductive patterns and pragmatic difficulties in studying transitional individuals in the wild are challenges to be overcome in order to improve the current scenario. Indeed, sex change may be extremely rapid in some grouper species (e.g., three weeks in *E. rivulatus*—Mackie 2003), and specimen collections with a wide body size range from juveniles to adults are needed to observe the ephemerous moments of sex differentiating and sex changing and to obtain more precise diagnoses (e.g., Siau 1994, Marino et al. 2001, Fennessy and Sadovy 2002, Liu and Sadovy 2004b, Liu and Sadovy de Mitcheson 2009, Liu et al. 2016).

The Role of Social Factors

The size advantage model (Ghiselin 1969) provides the most widely accepted explanation for the adaptive significance of protogyny in groupers, as well as in functional hermaphrodites. According to this theory, protogyny has evolved in social systems where large males that exclude smaller ones monopolize mating. This is indeed the case of most groupers, where large males actively control haremic systems over discrete territories (e.g., Colin et al. 1987, Zabala et al. 1997a, Mackie 2000). Mating system in this case typically consists of pair spawning, with females smaller than males (but see Section 1.4.3 for other spawning modes). In some protogynous species, however, females may approach or even overcome the size of the largest males (e.g., Marino et al. 2001, Reñones et al. 2010). The retention of the initial sex in these large individuals can be due a persisting suppression of sex change either by social or physical factors (Sadovy and Shapiro 1987).

Like many other hermaphroditic teleosts, sex change in groupers may be triggered or influenced by social factors. Based on a male removal experiment, Mackie (2003) demonstrated that female to male transition in *E. rivulatus* is controlled by a suppressive dominance of males and a threshold sex ratio. Liu and Sadovy (2004a) also demonstrated that female to male sex change in *C. boenak* occurred after the removal of the larger male under controlled experiments. Meanwhile, sexual differentiation of groupers can also be influenced by social factors, and the percentage of primary males can be significantly elevated under laboratory conditions in *C. boenak* and *E. coioides* (Liu and Sadovy 2004a, Liu and Sadovy de Mitcheson 2011).

Larger individuals are the main targets of some gears, such as hook and line and the first ones to be taken out of the fish stock (Gulland 1983). Therefore, higher withdrawals of males by fishery may occur, causing a considerable population imbalance in face of fishing pressure (Ferreira and Russ 1995, Russ et al. 1998). Ferreira and Russ (1995) observed that the coral trout *Plectropomus leopardus* in unprotected reefs in the Great Barrier Reef (GBR) had a smaller proportion of larger males compared to closed reefs, but that seemed to be compensated for by a larger proportion of transitional-stage fish and young males. They suggested that sex change of *P. leopardus* results from a combination of developmental and behavioural processes. Adams et al. (2000) investigated the same species in two geographic regions of the GBR, and observed a greater proportion of males above the minimum size and age of harvest on reefs closed to fishing for 8–10 years than on reefs open to fishing, but also noted that the effect varied with geographic region, independent of reef closure status, suggesting regional variation may also be a factor altering reproductive strategies. Later, Carter et al. (2014) also compared sex ratio, proportion of vitellogenic females, and the spawning fraction of local populations of *P. leopardus* populations in areas open or closed to fishing in four geographic regions of the GBR. Their results indicated that while variation in sex ratios and spawning frequency was primarily driven by water temperature, no-take management zones influenced spawning frequency, and concluded that variations in reproductive traits indicated the need for a regional approach to management.

Long-term observations may help to elucidate fishing effects for some protogynous populations. For the dusky grouper *E. marginatus*, earlier studies in Mediterranean Sea showed that sexual transition in this species occurred in individuals of 9 to 16 years old and with body sizes ranging from 68 to 90 cm total length (TL) (Chauvet 1988). In contrast, recent observations indicated that sexual transition is occurring in younger and smaller individuals (7 years old and 52.1 cm TL, respectively) (Reñones et al. 2010).

Gonad Development and Periodicity of Spawning

The morphological description of gonads and of their developmental stages is one of the key aspects of reproductive studies in teleosts. Gonad maturity scales, containing the description of these stages, are available for a number of groupers at a variable level of detail. In protogynous species, this number of development stages increases to encompass the transitional stages in addition to female and male stages. For females, the same pattern of ovarian development is observed as in other teleost species, with asynchronous development and multiple spawning being the predominant type. As for males, the testis has a typical characteristic of retaining a central lumen and dorsal sperm sinuses as mentioned above (Sadovy and Shapiro 1987).

The number of described ovarian and testicular stages in protogynous groupers may vary according to the species and to the authors' interpretation of morphological traits. For examples, Carter et al. (1994) mentioned 11 different stages of gonad development for *E. striatus* (5 ovarian stages, 1 transitional stage, 5 testicular stages); Marino et al. (2001) cited 13 stages

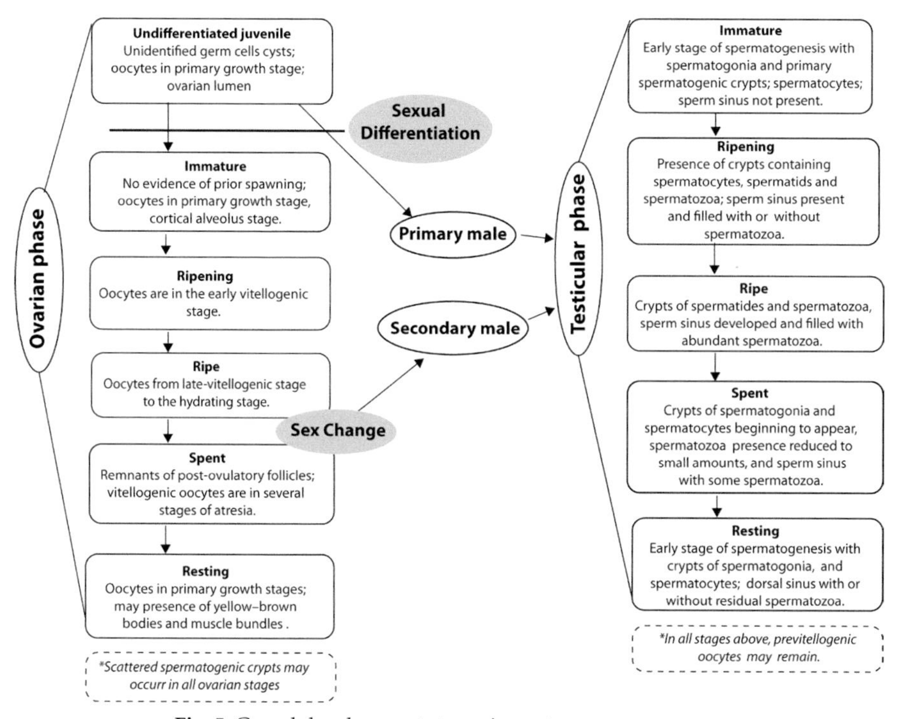

Fig. 5 Gonad development stages in protogynous groupers.

for *E. marginatus* (7 ovarian stages, 3 testicular stages, 3 intersexual stages); DeMartini et al. (2011) cited 12 stages for *Hyporthudus quernus* (7 ovarian stages, 3 testicular stages, 1 bisexual stage and 1 transitional stage); 9 stages by Frisch et al. (2007) to *P. leopardus* (5 ovarian stages, 2 testicular stages, and 2 transitional stages); Marques and Ferreira (2011) cited 12 stages for *A. afer* (5 ovarian stages, 5 testicular stages, and 2 transitional stages); and 12 stages for *C. panamensis* (5 ovarian stages, 5 testicular stages, 1 bisexual stage, and 1 transitional stage) by Erisman et al. (2010). Figure 5 summarizes ovarian and testicular stages in protogynous groupers.

Periodicity and Timing of Spawning

Spawning activity in teleosts including groupers are associated with several factors, such as temperature, photoperiod, lunar and tidal cycles, water's flow rate, and chemical composition (Jakobsen et al. 2009, Wootton and Smith 2015). The temperature is one of the determinant environmental factors in the seasonality of spawning, with most grouper species in tropical regions seeming to reproduce in temperatures around 25–30°C (Colin 1992, Mackie 2000, Rhodes et al. 2012, 2016, Sanchez et al. 2017, Marques and Ferreira 2018), which overlap with seasonal lows in sea surface water temperatures. While most Atlantic species seem to spawn during winter to early spring months, e.g., *C. fulva* (Freitas et al. 2011), *A. afer* (Marques and Ferreira 2011), *M. bonaci* (Teixeira et al. 2004, Freitas et al. 2011, Brule et al. 2003) and *E. morio* (Freitas et al. 2011), *E. itajara* has been reported to spawn during summer months in both Southern (Freitas et al. 2015) and Northern hemispheres (Bullock et al. 1992, Mann et al. 2009, Koenig et al. 2017).

Most groupers studied present asynchronous oocyte development, and are multiple spawners. Females can spawn repeatedly for many days and even months with new batches of oocytes developing each time, while males produce sperm continuously. Overall, the majority of groupers exhibit reproductive patterns that parallel lunar cycles, with spawning occurring during the full moon and near dusk (Colin et al. 1987, Shapiro et al. 1993, Aguilar-Perera and Aguilar-Dávila 1996, Rhodes and Sadovy 2002, Grandcourt et al. 2009, Sadovy de Mitcheson and Colin 2011, Ohta and Ebisawa 2015, Koenig et al. 2017).

Reproductive Behaviour

Groupers have diverse reproductive modes, encompassing different migration behaviours, mating systems, and sexual patterns. Hermaphroditism can influence both the structure of a mating group and sperm competition (Erisman et al. 2009, 2013). These authors used a comparative phylogenetic approach to reconstruct the evolutionary history of reproduction in groupers and to evaluate the association between mating system characteristics and sexual patterns. Based on this approach, they evaluated whether evolutionary shifts in sexual patterns and, in particular, the loss of sex change were influenced by variations from paired to group spawning and associated increases in sperm competition among males. Erisman et al. (2009) determined mating group separated into three categories: (i) paired-spawning, in which adults are known to release their gametes in male-female pairs, (ii) group spawning, where species release gametes within groups containing several males and one or more females, and (iii) mixed spawning, where species exhibiting both behaviours. These results showed a remarkable congruence between the mating systems and sexual patterns of groupers. They found that the majority of grouper species exhibited a polygynous mating system, with large males often monopolizing mating with several females. These males have relatively small testes, indicative of low levels of sperm competition and the absence of group mating, once they are paired spawning. On the other hand, spawning in gonochoric species occurs in groups with several males and one or more females, where male territorial behaviour is rare or absent, and males have relatively large testes (a high gonadosomatic index, expressed as the percentage of body weight devoted to gonad) indicative of sperm competition.

Such diversity in reproductive patterns in groupers have been summarised by Mackie (2007) in four types: (i) non-migratory, polygamous, and pair-spawning strategy of smaller species such as *C. spiloparaea* and *C. urodeta* (Donaldson 1995), (ii) migratory, aggregating, polygamous, and pair-spawning strategy of larger species, such as *E. guttatus* (Colin et al. 1987), *M. tigris* (Sadovy et al. 1994), and *P. leopardus* (Samoilys and Squire 1994), (iii) non-migratory, monogamous, and pair-spawning mode of reproduction *C. boenak* (Donaldson 1989) and *C. hemistiktos* (Shpigel and Fishelson 1991) being considered a less common type, and (iv) migratory, aggregating, and group-spawning as *E. striatus* (Colin 1992), *E. itajara* (Koenig et al. 2017), and *M. rosacea* (Erisman and Buckhorn 2007).

Spawning Aggregations

A fish spawning aggregation (FSA) is a repeated concentration of conspecific fish species, which is gathered for the purpose of spawning in a predictable fashion in time and space. The locations and cycles associated with FSA are dictated by the adaptation of various species in relation to several factors, such as geomorphology, habitat features, and ocean dynamics. These interactions produce complex, localized, and ephemeral linkages through ocean food webs and attract top predators and megaplanktivores (Heyman et al. 2001, Pittman and McAlpine 2003, Petitgas et al. 2010, Ezer et al. 2011). FSA results in a mass point source of offspring. These events usually involve a great number of females and males, being at least four times compared with densities outside the aggregation (Domeier 2012, Russell et al. 2014).

Regarding their formation, spawning aggregations follow basically two main patterns: migratory or transient and non-migratory or resident. Some groupers travel long distances until reaching spawning aggregation sites, for instance, the larger grouper species, *M. tigris*, *P. leopardus*, and *E. striatus* (Sadovy et al. 1994, Zeller 1998, Bolden 2000). In contrast, non-migratory groupers aggregate at their area of residence without migrating extensive distances. This is usually the case for smaller groupers, such as *C. boenak* and *C. miniata* (Donaldson 1989, Shpigel and Fishelson 1991). Once in the reproduction site, males and females can approach each other in different ways during spawning according to their mating system. Monogamous pairs usually exhibit some degree of bonding (e.g., *C. hemistiktos*, Shpigel and Fishlson 1991). In polygynous mating systems, large males form harems with several females, but courtship behaviour may target a single female and precede pair spawning (Rowell et al. 2019). These males establish mating territories and usually are aggressive with other males (e.g., *E. marginatus*, Zabala et al. 1997b). Males use acoustic communication, producing sounds as part of the courtship repertoire (Rowell et al. 2019, Sanchez et al. 2017). In contrast, in polyandrous mating systems, a female-led spawns with multiple males (e.g., *M. rosacea*, Erisman and Buckhorn 2007). In this case, sperm competition is the key to fertilization success (Erisman and Buckhorn 2007).

According to Domeier (2012) in the resident type, individuals are drawn to a site within or nearby their adult home range. They usually (i) occur at a specific time of day over numerous days, (ii) last only a few hours or less, (iii) occur daily over an often-lengthy reproductive period of the year, and (iv) may occur year-round. A single day of spawning for an individual participating in a resident spawning aggregation represents a small fraction of that individual's annual reproductive effort. In contrast, in the transient type, individuals are drawn to a site well outside their typical adult home range, where they often (i) occur during a very specific portion of one or two months of the year, (ii) persist for a period of days or at most a few weeks, and (iii) do not occur year-round. A single transient spawning aggregation may represent the total reproductive effort for participating individuals. Moreover, species associated with the transient aggregations are typically large, whereas those characterised as resident aggregations are usually smaller.

Colin and collaborators (2003) proposed a suite of direct and indirect indicators of fish spawning aggregations. The direct indicators are: (i) undisputed spawning observations, (ii) females with hydrated eggs, and (iii) presence of post ovulatory follicles in the ovaries of aggregating females. Another criterion proposed by Domeier (2012) is the identification of very early-stage eggs and larvae that can be positively associated with the aggregating species. Among indirect indications of spawning aggregations, it can be particular behaviours or colour patterns, if these are demonstrably known to be associated only with spawning, as well as gonadosomatic index data or the presence of swollen abdomens (indicating the presence of late vitellogenic oocytes or hydrated oocytes) in a large percentage of the aggregated individuals.

Fisheries

Transient FSA events gather large fish biomass and attract fisheries' attention, and thus have clear management implications. Harvest of fish spawning aggregations has been a major contributor to extinction risk in numerous species of groupers (Robinson et al. 2015). Dramatic declines of species, such as the nassau grouper *E. striatus* have been well documented (Sala et al. 2001, Aguilar-Perera 2006), as well as of other species, raising concern over the consequences of fisheries heavily targeting those high densities events, and prompting global conservation actions. A global web-based database is maintained and managed by the Science and Conservation of Fish Aggregations (SCRFA) that provides current information on the occurrence of FSA around the world. SCRFA have reported 977 episodes of fish spawning aggregations across all five oceans and encompassing 51 families, 110 genera, and 243 species of fishes (SCRFA 2018). Out of all aggregations reported by SCRFA, 517 reports were of spawning aggregations of 48 species of Epinephelidae, belonging to five genera (*Cephalopholis*, *Dermatolepis*, *Epinephelus*, *Mycteroperca*, and *Plectropomus*) and occurred in 34 countries.

Management is crucial, as intensive fishery in FSAs areas may cause sudden stock colapse, due to hyperstability of catches caused by the aggregation behaviour. Many aggregations have been eradicated by intense fishing pressure (e.g., Sala et al. 2001, Sadovy and Domeier 2005), and regions of particularly heavy fishing pressure tend to be devoid of spawning aggregations (Sadovy de Mitcheson et al. 2008). Therefore, it is a crucial step to identify FSAs for conservation and fishery management purposes. Recent evidence suggests that even extirpated FSAs in the Caribbean and Western Atlantic have recovered, following implementation of strict management measures (Chollett et al. 2020).

Implications for Conservation

There are no studies on sexual patterns of nearly 65% of groupers (Table 1). This gap causes concern not only because the knowledge of the sexual pattern is important for a comprehensive portrayal of their reproductive biology and ecology, but also for being crucial information to guide conservation and sustainable fishery management actions of these species. Gonochorist as well as protogynous hermaphrodite populations tend to decrease their reproductive potential as fishing increases due to decreasing in stock biomass and, consequently, in egg production as larger individuals are more fecund (Huntsman and Schaaf 1994).

Selective removal through fisheries of the big females out of the population can decrease population resilience according to the BOFFFF theory (Big Old Fat Fecund Female Fish), in which older and biggest female fish produce larger numbers of eggs/larvae than younger and smaller females (Hixon et al. 2014, Peck 2015). BOFFFF produces larger eggs/larvae that would represent a larger reserve of fat and culminate in a higher growth and survival rates of these eggs/larvae during the period of low exogenous feeding. Therefore, this theory predicts that BOFFFFs will have higher fecundity and more healthy and resistant offspring (Hixon et al. 2014, Peck 2015).

For protogynous species, size-selective fisheries would lead to a female dominated sex ratio with possible sperm limitation. Once fisheries preferentially remove the larger males, numerical reduction of older age-classes in the population would lead to dramatic male: female sex ratio imbalance, leading to sperm limitation that is reducing the capability of male fertilization during spawning aggregations (Reinboth 1980, Alonzo and Mangel 2004, Condini et al. 2018). Exogenous controlled populations may in turn regulate sex ratios through transitioning of larger females to males, but this can also lead to a reduced fecundity through the loss of the most productive females (Coleman et al. 1996).

Final Remarks

Groupers show considerable phenotypic plasticity of sexual expression with functional sexual patterns, which may differ over time and vary considerably according to species and to social and environmental conditions. The complex and variable nature of this phenomena challenges our general understanding of teleost sexual patterns. Still, further studies are needed for several species of protogynous hermaphrodites, to unravel not only patterns of sex change but also their reproductive strategies. In addition, is always important to review reliable sex change indicators in order to confirm the sexual pattern of a grouper species.

Protogynous hermaphrodites like groupers are generally considered more susceptible to fishing pressure than gonochoristic species (Bannerot et al. 1987, Alonzo and Mangel 2004). Long term studies, such as the ones carried on the GBR, Australia, for the common coral trout *P. leopardus* (Ferreira and Russ 1995, Adams et al. 2000, Carter et al. 2014) and on the Mediterranean for the dusky grouper *E. striatus* (Chauvet 1988, Reñones et al. 2010) are crucial to reveal effects of fishing and management measures on populations.

Recent assessments of the status of conservation of groupers according to IUCN red list criteria revealed that species that are under greater risk are large bodied, longer-lived groupers that form large spawning aggregations (Erisman et al. 2015, Sadovy de Mitchelson et al. 2013, Sadovy de Mitchelson et al. 2020). Of special concern are species that are poorly known, and for which fisheries are unmanaged. Understanding the driving forces controlling reproduction and sexual development are particularly important for the management and conservation of groupers, particularly in face of new threats, such as rising new trade markets and climate change.

References

Adams, S., B.D. Mapstone, G.R. Russ and C.R. Davies. 2000. Geographic variation in the sex ratio, sex specific size, and age structure of *Plectropomus leopardus* (Serranidae) between reefs open and closed to fishing on the Great Barrier Reef. Canadian Journal of Fisheries and Aquatic Sciences 57(7): 1448–1458.

Adams, S. 2003. Morphological ontogeny of the gonad of three Plectropomid species through sex differentiation and transition. Journal of Fish Biology 63: 22–36.

Aguilar-Perera, A. and W. Aguilar-Dávila. 1996. A spawning aggregation of Nassau grouper *Epinephelus striatus* (Pisces: Serranidae) in the Mexican Caribbean. Environmental Biology of Fishes 45: 351–361.

Aguilar-Perera, A. 2006. Disappearance of a Nassau grouper spawning aggregation off the southern Mexican Caribbean coast. Marine Ecology Progress Series 327: 289–96.

Alonzo, S.H. and M. Mangel. 2004. The effects of size-selective fisheries on the stock dynamics of and sperm limitation in sex-changing fish. Fishery Bulletin 102: 1–13.

Bannerot, S., W.W. Fox Jr. and J.E. Powers. 1987. Reproductive strategies and the management of snappers and groupers in the Gulf of Mexico and Caribbean. pp. 561–603. *In*: Poots, G.W. and R.J. Wooton (eds.). Fish Reproduction: Strategies and Tactics. Academic Press, Orlando.

Bolden, S.K. 2000. Long-distance movement of a Nassau grouper (*Epinephelus striatus*) to a spawning aggregation in the central Bahamas. Fishery Bulletin 96: 642–645.

Bright, D., A. Reynolds, N.H. Nguyen, R. Knuckey, W. Knibb and A. Elizur. 2016. A study into parental assignment of the communal spawning protogynous hermaphrodite, giant grouper (*Epinephelus lanceolatus*). Aquaculture 459: 19–25.

Brulé, T., X. Renán, T. Colás-Marrufo, Y. Hauyon, A.N. Tuz-Sulub and C. Déniel. 2003. Reproduction in the protogynous black grouper Mexico (Mycteroperca bonaci (Poey)) from the southern Gulf of Mexico. Fishery Bulletin 101(3): 463–475.

Bulanin, U., M. Masrizal and Z.A. Muchlisin. 2017. Hermaphroditism in the white spot grouper *Epinephelus coeruleopunctatus* (Pisces: Serranidae) harvested from Padang City waters, Indonesia. F1000Research 6.

Bullock, L.H., M.D. Murphy, M.F. Godcharles and M.E. Mitchell. 1992. Age, growth, and reproduction of jewfish *Epinephelus itajara* in the eastern Gulf of Mexico. Fishery Bulletin 90: 243–249.

Carter, A.B., G.R. Russ, A.J. Tobin, A.J. Williams, C.R. Davies and B.D. Mapstone. 2014. Spatial variation in the effects of size and age on reproductive dynamics of common coral trout *Plectropomus leopardus*. Journal of Fish Biology 84(4): 1074–1098.

Carter, J., G.J. Marrow and V. Pryor. 1994. Aspects of the ecology and reproduction of Nassau grouper, *Epinephelus striatus*, off the coast of Belize, Central America. Proceedings of the 43rd Gulf and Caribian Fisheries Institute 48.

Chauvet, C. 1988. Étude de la croissance du mérou *Epinephelus Guaza* (Linné, 1758) des côtes Tunisiennes. Aquatic Living Resources 1: 277–288.

Chollett, I., M. Priest, S. Fulton and W.D. Heyman. 2020. Should we protect extirpated fish spawning aggregation sites? Biological Conservation 241: 108395.

Coleman, F.C., C.C. Koenig and A. Collins. 1996. Reproductive styles of shallow-water groupers (Pisces: Serranidae) in the eastern Gulf of Mexico and the consequences of fishing spawning aggregations. Environmental Biology of Fishes 47: 129–141.

Colin, P.L., D.Y. Shapiro and D. Weiler. 1987. Aspects of the reproduction of two groupers, *Epinephelus guttatus* and *E. striatus* in the West Indies. Bulletin of Marine Science 40: 220–230.

Colin, P.L. 1992. Reproduction of the Nassau grouper, *Epinephelus striatus* (Pisces: Serranidae) and its relationship to environmental conditions. Environmental Biology of Fishes 34: 357–377.

Colin, P.L., Y.J. Sadovy and M.L. Domeier. 2003. Manual for the study and conservation of reef fish spawning aggregations. Society for the Conservation of Reef Fish Aggregations Special Publication No. 1: 1–98.

Condini, M.V., J.A. García-Charton and A.M. Garcia. 2018. A review of the biology, ecology, behavior and conservation status of the dusky grouper, *Epinephelus marginatus* (Lowe 1834). Reviews in Fish Biology and Fisheries 28: 301–330.

Craig, M.T., Y.J. Sadovy de Mitcheson and P.C. Heemstra. 2011. Groupers of the world. A field and market guide. NISC (Pty) Ltd. Grahamstown, South Africa, 1–47.

DeMartini, E.E., A.R. Everson and R.S. Nichols. 2011. Estimates of body sizes at maturation and at sex change, and the spawning seasonality and sex ratio of the endemic Hawaiian grouper (*Hyporthodus quernus*, F. Epinephelidae). Fishery Bulletin 109: 123–134.

Domeier, M.L. 2012. Revisiting spawning aggregations: definitions and challenges. *In*: Sadovy de Mitcheson, Y. and P.L. Colin (eds.). Reef Fish Spawning Aggregations: Biology, Research and Management. Fish & Fisheries Series, Springer.

Donaldson, T.J. 1989. Pair spawning of *Cephalopholis boenak* (Serranidae). Japanese Journal of Ichthyology 35: 497–500.

Donaldson, T.J. 1995. Courtship and spawning behavior of the pygmy grouper, *Cephalopholis spiloparaea* (Serranidae: Epinephelinae), with notes on *C. argus* and *C. urodeta*. Environmental Biology of Fishes 43: 363–370.

Erisman, B.E. and M.L. Buckhorn. 2007. Spawning patterns in the leopard grouper, *Mycteroperca rosacea*, in comparison with other aggregating groupers. Marine Biology 151: 1849–1861.

Erisman, B.E., M.T. Craig and P.A. Hastings. 2009. A phylogenetic test of the size-advantage model: evolutionary changes in mating behavior influence the loss of sex change in a fish lineage. The American Naturalist 174: E83–E99.

Erisman, B.E., M.T. Craig and P.A. Hastings. 2010. Reproductive biology of the Panama graysby *Cephalopholis panamensis* (Teleostei: Epinephelidae). Journal of Fish Biology 76: 1312–1328.

Erisman, B.E., C.W. Petersen, P.A. Hastings and R.R. Warner. 2013. Phylogenetic perspectives on the evolution of functional hermaphroditism in teleost fishes. Integrative and Comparative Biology 53: 736–754.

Erisman, B.E., W. Heyman, S. Kobara, T. Ezer, S. Pittman, O. Aburto-Oropeza and R.S. Nemeth. 2015. Fish spawning aggregations: where well-placed management actions can yield big benefits for fisheries and conservation. Fish and Fisheries 18: 128–144.

Ezer, T., W.D. Heyman, C. Houser and B. Kjerfve. 2011. Modelling and observations of high-frequency flow variability and internal waves at a Caribbean reef spawning aggregation site. Ocean Dynamics 61: 581–598.

Fennessy, S.T. and Y. Sadovy. 2002. Reproductive biology of a diandric protogynous hermaphrodite, the serranid *Epinephelus andersoni*. Marine and Freshwater Research 53: 147–158.

Ferreira, B.P. and G.R. Russ. 1995. Population structure of the leopard coralgrouper, *Plectropomus leopardus*, on fished and unfished reefs off Townsville, Central Great Barrier Reef, Australia. Fishery Bulletin 93(4): 629–642.

Frable, B.W., S.J. Tucker and H.J. Walker. 2019. A new species of grouper, Epinephelus craigi (Perciformes: Epinephelidae), from the South China Sea. Ichthyological Research 66(2): 215–224.

Freitas, M.O., R.L. De Moura, R.B. Francini-Filho and C.V. Minte-Vera. 2011. Spawning patterns of commercially important reef fish (Lutjanidae and Serranidae) in the tropical western South Atlantic. Scientia Marina 75: 135–146.

Freitas, M.O., V. Abilhoa, V.J. Giglio, M. Hostim-Silva, R.L. de Moura, R.B. Francini-Filho and C.V. Minte-Vera. 2015. Diet and reproduction of the goliath grouper, *Epinephelus itajara* (Actinopterygii: Perciformes: Serranidae), in eastern Brazil. Acta Ichthyologica et Piscatoria 1: 45.

Frisch, A.J., M.I. McCormick and N.W. Pankhurst. 2007. Reproductive periodicity and steroid hormone profiles in the sex-changing coral-reef fish, *Plectropomus leopardus*. Coral Reefs 26: 189–197.

Ghiselin, M.T. 1969. The evolution of hermaphroditism among animals. Quarterly Review Biology 44: 189–208.

Grandcourt, E.M., T.Z. Al Abdessalaam, F. Francis, A.T. Al Shamsi and S.A. Hartmann. 2009. Reproductive biology and implications for management of the orange-spotted grouper *Epinephelus coioides* in the southern Arabian Gulf. Journal of Fish Biology 74: 820–841.

Gulland, J.A. 1983. Fish Stock Assessment: A Manual of Basic Methods. Wiley, New York.

Helfman, G.S., B.B. Collet, D.E. Facey and B.W. Bowen. 2009. The Diversity of Fishes: Biology, Evolution, and Ecology. Second Edition. Wiley-Blackweel.

Heyman, W., R. Graham, B. Kjerfve and R. Johannes. 2001. Whale sharks *Rhincodon typus* aggregate to feed on fish spawn in Belize. Marine Ecology Progress Series 215: 275–282.

Hixon, M.A., D.W. Johnson and S.M. Sogard. 2014. BOFFFFs: on the importance of conserving old-growth age structure in fishery populations. ICES Journal of Marine Science 71: 2171–2185.

Huntsman, G.R. and W.E. Schaaf. 1994. Simulation of the impact of fishing on reproduction of a protogynous grouper, the Graysby. North American Journal of Fisheries Management 4: 41–52.

Johnson, K., P. Thomas and R.R. Wilson Jr. 1998. Seasonal cycles of gonadal development and plasma sex steroid levels in Epinephelus morio, a protogynous grouper in the eastern Gulf of Mexico. Journal of Fish Biology 52(3): 502–518.

Jakobsen, T., M.J. Fogarty, B.A. Megrey and E. Moksness. 2009. Fish Reproductive Biology: Implications for Assessment & Management. Blackwell Publishing Ltd.

Koenig, C.C., L.S. Bueno, F.C. Coleman, J.A. Cusick, R.D. Ellis, K. Kingon and C.D. Stallings. 2017. Diel, lunar, and seasonal spawning patterns of the Atlantic goliath grouper, *Epinephelus itajara*, off Florida, United States. Bulletin of Marine Science 93(2): 391–406.

Liu, M. and Y. Sadovy. 2004a. The influence of social factors on adult sex change and juvenile sexual differentiation in a diandric, protogynous epinepheline, *Cephalopholis boenak* (Pisces, Serranidae). Journal of Zoology 264: 239–248.

Liu, M. and Y. Sadovy. 2004b. Early gonadal development and primary males in the protogynous epinepheline, *Cephalopholis boenak*. Journal of Fish Biology 65: 987–1002.

Liu, M. and Y. Sadovy de Mitcheson. 2009. Gonad development during sexual differentiation in hatchery-produced orange-spotted grouper (*Epinephelus coioides*) and humpback grouper (*Cromileptes altivelis*) (Pisces: Serranidae, Epinephelinae). Aquaculture 287: 191–202.

Liu, M. and Y. Sadovy de Mitcheson. 2011. The influence of social factors on sexual differentiation in a diandric, protogynous grouper *Epinephelus coioides*. Ichthyological Research 58(1): 84–89.

Liu, M., Y.Y. Wang, X.J. Shan, B. Kang and S.X. Ding. 2016. Primary male development of two sequentially hermaphroditic groupers, *Epinephelus akaara* and *Epinephelus awoara* (Perciformes: Epinephelidae). Journal of Fish Biology 88(4): 1598–1613.

Ma, K.Y. and M.T. Craig. 2018. An inconvenient monophyly: an update on the taxonomy of the groupers (Epinephelidae). Copeia 106(3): 443–456.

Mackie, M.C. 2000. Reproductive biology of the halfmoon grouper, *Epinephelus rivulatus*, at Ningaloo Reef, Western Australia. Environmental Biology of Fishes 57: 363–376.

Mackie, M.C. 2003. Socially controlled sex-change in the half-moon grouper, *Epinephelus rivulatus*, at Ningaloo Reef, Western Australia. Coral Reefs 22: 133–142.

Mackie, M.C. 2007. Reproductive behavior of the halfmoon grouper, *Epinephelus rivulatus*, at Ningaloo Reef, Western Australia. Ichthyological Research 54: 213–220.

Mann, D.A., J.V. Locascio, F.C. Coleman and C.C. Koenig. 2009. Goliath grouper *Epinephelus itajara* sound production and movement patterns on aggregation sites. Endangered Species Research 7: 229–236.

Marino, G., E. Azzurro, A. Massari, M.G. Finoia and A. Mandich. 2001. Reproduction in the dusky grouper from the southern Mediterranean. Journal of Fish Biology 58: 909–927.

Marques, S. and B.P. Ferreira. 2011. Sexual development and reproductive pattern of the Mutton hamlet, *Alphestes afer* (Teleostei: Epinephelidae): a diandric, hermaphroditic reef fish. Neotropical Ichthyology 9: 547–558.

Marques, S. and B.P. Ferreira. 2018. Sexual development and demography of the rock hind *Epinephelus adscensionis*, a protogynous grouper, in the south-west Atlantic. Marine and Freshwater Research 69: 300–312.

Nakamura, J. and H. Motomura. 2021. Epinephelus insularis, a new species of grouper from the western Pacific Ocean, and validity of E. japonicus (Temminck and Schlegel 1843), a senior synonym of Serranus reevesii Richardson 1846 and E. tankahkeei Wu et al. 2020 (Perciformes: Epinephelidae). Ichthyological Research 68(2): 263–276.

Nelson, J.S. 2006. Fishes of the World. John Wiley and Sons. Inc., Hoboken, New Jersey.

Ohta, I. and A. Ebisawa. 2015. Reproductive biology and spawning aggregation fishing of the white-streaked grouper, *Epinephelus ongus*, associated with seasonal and lunar cycles. Environmental Biology of Fishes 98: 1555–1570.

Okumura, S. 2001. Evidence of sex reversal towards both directions in reared red spotted grouper *Epinephelus akaara*. Fisheries Science 67: 535–537.

Peck, M.A. 2015. Bigger mothers = better chances: the first test of a central hypothesis in marine fish ecology—editorial comment on the feature article by Saenz-Agudelo et al. Marine Biology 162: 1–2.

Petitgas, P., D.H. Secor, I. McQuinn and G. Huse. 2010. Stock collapses and their recovery: mechanisms that establish and maintain life-cycle closure in space and time. ICES Journal of Marine Science 67: 1841–1848.

Pittman, S.J. and C.A. McAlpine. 2003. Movements of marine fish and decapod crustaceans: process, theory and application. Advances in Marine Biology 44: 205–294.

Reinboth, R. 1962. Morphologische und funktionelle Zweigeschlechtlichkeit marinen Teleostiern (Serranidae. *Sparidae, Centracanth*).

Reinboth, R. 1970. Intersexuality in fishes. Society Member of Endocrinology 515: 515–543.

Reinboth, R. 1980. Can sex inversion be environmentally induced? Biology of Reproduction 22: 49–59.

Reñones, O., A. Grau, X. Mas, F. Riera and F. Saborido-Rey. 2010. Reproductive pattern of an exploited dusky grouper *Epinephelus marginatus* (Lowe 1834) (Pisces: Serranidae) population in the western Mediterranean. Marine Science 74: 523–537.

Rhodes, K.L. and Y. Sadovy. 2002. Temporal and spatial trends in spawning aggregations of camouflage grouper, *Epinephelus polyphekadion*, in Pohnpei, Micronesia. Environmental Biology of Fish 63: 27–39.

Rhodes, K.L., B.M. Taylor and J.L. Mcllwain. 2011. Detailed demographic analysis of an *Epinephelus polyphekadion* spawning aggregation and fishery. Marine Ecology Progress Series 421: 183–198.

Rhodes, K.L., J.L. McIlwain and E. Joseph. 2012. Reproductive movement, residency and fisheries vulnerability of brown-marbled grouper, *Epinephelus fuscoguttatus* (Forsskål, 1775). Coral Reefs 31: 443–453.

Rhodes, K.L., B.M. Taylor, D. Hernandez-Ortiz and J. Cuetos-Bueno. 2016. Growth and reproduction of the highfin grouper *Epinephelus maculatus*. Journal of Fish Biology 88(5): 1856–1869.

Robinson, J., N.A. Graham, J.E. Cinner, G.R. Almany and P. Waldie. 2015. Fish and fisher behaviour influence the vulnerability of groupers (Epinephelidae) to fishing at a multispecies spawning aggregation site. Coral Reefs 34(2): 371–82.

Rowell, T.J., O. Aburto-Oropeza, J.J. Cota-Nieto, M.A. Steele and B.E. Erisman. 2019. Reproductive behaviour and concurrent sound production of Gulf grouper *Mycteroperca jordani* (Epinephelidae) at a spawning aggregation site. Journal of Fish Biology 94(2): 277–96.

Russ, G.R., D.C. Lou, J.B. Higgs and B.P. Ferreira. 1998. Mortality rate of a cohort of the coral trout, *Plectropomus leopardus*, in zones of the Great Barrier Reef Marine Park closed to fishing. Marine and Freshwater Research 49: 507–11.

Russell, M.W., Y. Sadovy, B.E. Erisman, R.J. Hamilton, B.E. Luckhurst and R.S. Nemeth. 2014. Status Report World's Fish Aggregations 2014. Science and Conservation of Fish Aggregations. Science and Conservation of Fish Aggregations, California, USA. International Coral Reef Initiative.

Sadovy, Y.J. 1996. Reproduction of reef fishery species. pp. 15–59. *In*: Polunin, N.V.C. and C.M. Roberts (eds.). Reef Fisheries. Chapman & Hall, London.

Sadovy, Y. and D.Y. Shapiro. 1987. Criteria for the diagnosis of hermaphroditism in fishes. Copeia 1987(1): 136–156.

Sadovy, Y., A. Rosario and A. Román. 1994. Reproduction in an aggregating grouper, the red hind, *Epinephelus guttatus*. Environmental Biology of Fishes 41: 269–286.

Sadovy, Y. and P.L. Colin. 1995. Sexual development and sexuality in the Nassau grouper. Journal of Fish Biology 46: 961–976.

Sadovy, Y. and M. Domeier. 2005. Are aggregation-fisheries sustainable? Reef fish fisheries as a case study. Coral Reefs 24: 254–262.

Sadovy de Mitcheson, Y. and M. Liu. 2008. Functional hermaphroditism in teleosts. Fish and Fisheries 9: 1–43.

Sadovy de Mitcheson, Y., A. Cornish, M. Domeier, P.L. Colin, M. Russel and K.C. Lindeman. 2008. A global baseline for spawning aggregations of reef fishes. Conservation Biology 22: 1233–1244.

Sadovy de Mitcheson, Y. and P.L. Colin. 2011. Reef Fish Spawning Aggregations: Biology, Research and Management. Fish and Fisheries Series 35. Springer.

Sadovy de Mitcheson, Y., M.T. Craig, A.A. Bertoncini, K.E. Carpenter, W.W. Cheung, J.H. Choat, A.S. Cornish, S.T. Fennessy, B.P. Ferreira, P.C. Heemstra, M. Liu, R.F. Myers, D.A. Pallard, K.L. Rhodes, L.A. Rocha, B.C. Russell, M.A. Samoilys and J. Sanciangco. 2013. Fishing groupers towards extinction: a global assessment of threats and extinction risks in a billion dollar fishery. Fish and Fisheries 14(2): 119–36.

Sadovy de Mitcheson, Y.J., P.L. Colin, S.J. Lindfield and A. Bukurrou. 2020. A decade of monitoring an Indo-Pacific grouper spawning aggregation: Benefits of protection and importance of survey design. Frontiers in Marine Science 7: 853.

Sala, E., E. Ballesteros and R.M. Starr. 2001. Rapid decline of Nassau grouper spawning aggregations in Belize: fishery management and conservation needs. Fisheries 26: 23–30.

Samoilys, M.A. and L.C. Squire. 1994. Preliminary observations on the spawning behavior of coral trout, *Plectropomus leopardus* (Pisces: Serranidae), on the Great Barrier Reef. Bulletin of Marine Science 54: 332–342.

Sanchez, P.J., R.S. Appeldoorn, M.T. Schärer-Umpierre and J.V. Locascio. 2017. Patterns of courtship acoustics and geophysical features at spawning sites of black grouper (*Mycteroperca bonaci*). Fishery Bulletin 115(2): 186–195.

SCRFA, Science and Conservation of Fish Aggregations, 2018. SCRFA database. Web online database. https://www.scrfa.org/database/: SCRFA. [2013-onwards].

Shapiro, D.Y., Y. Sadovy and M.A. McGehee. 1993. Size, composition, and spatial structure of the annual spawning aggregation of the red hind, *Epinephelus guttatus* (Pisces: Serranidae). Copeia 1993(2): 399–406.

Shpigel, M. and L. Fishelson. 1991. Territoriality and associated behaviour in three species of the genus *Cephalopholis* (Pisces: Serranidae) in the Gulf of Aqaba, Red Sea. Journal of Fish Biology 38: 887–896.

Siau, Y. 1994. Population structure, reproduction and sex-change in a tropical East Atlantic grouper. Journal of Fish Biology 44: 205–211.

Smith, C.L. 1965. The patterns of sexuality and the classification of serranid fishes. American Museum Novitates 2207.

Sujatha, K. and K.V.L. Shrikanya. 2013. Reproductive biology of striped grouper *Epinephelus latifasciatus* (Temminck and Schelegel, 1842), off Visakhapatnam, middle east coast of India. Indian Journal of Geo-Marine Sciences 42: 183–190.

Teixeira, S.F., B.P. Ferreira and I.P. Padovan. 2004. Aspects of fishing and reproduction of the black grouper *Mycteroperca bonaci* (Poey, 1860) (Serranidae: Epinephelinae) in the Northeastern Brazil. Neotropical Ichthyology 2: 19–30.

Tucker, S.J., E.M. Kurniasih and M.T. Craig. 2016. A new species of grouper (Epinephelus; Epinephelidae) from the Indo-Pacific. Copeia 104(3): 658–662.

Van Oordt, G.J. 1933. Zur sexualität der gattung Epinephelus (Serranidae, Teleostei). Z. Mikr. Anat. Forsch. 33: 525–533.

Warner, R.R. and D.R. Robertson. 1978. Sexual patterns in the labroid fishes of the western Caribbean, I the wrasses (Labridae). Smithsonian Contributions to Zoology.

Wootton, R.J. and C. Smith. 2015. Reproductive Biology of Teleost Fishes. Wiley Blackwell.

Wu, X., Y. Yang, C. Zhong, Y. Guo, S. Li, H. Lin and X. Liu. 2020. Transcriptome profiling of laser-captured germ cells and functional characterization of zbtb40 during 17alpha-methyltestosterone-induced spermatogenesis in orange-spotted grouper (Epinephelus coioides). BMC Genomics 21(1): 1–13.

Zabala, M., A. Garcia-Rubies, P. Louisy and E. Sala. 1997a. Spawning behaviour of the Mediterranean dusky grouper *Epinephelus marginatus* (Lowe, 1834) (Pisces, Serranidae) in the Medes Islands Marine Reserve (NW Mediterranean, Spain). Scientia Marina 61: 65–77.

Zabala, M., P. Louisy, A. Garcia-Rubies and V. Gracia. 1997b. Socio-behavioural context of reproduction in the Mediterranean dusky grouper *Epinephelus marginatus* (Lowe, 1834) (Pisces, Serranidae) in the Medes Islands Marine Reserve (Nw Medtierranean, Spain). Scientia Marina 61: 79–89.

Zeller, D.C. 1998. Spawning aggregations: patterns of movement of the coral trout *Plectropomus leopardus* (Serranidae) as determined by ultrasonic telemetry. Marine Ecology Progress Series 162: 253–263.

CHAPTER 1.4

Feeding Biology of Groupers

M. Harmelin-Vivien[1,2,*] and *J.G. Harmelin*[1,2]

Introduction

To survive, fish, as most animals, need to feed frequently, if not daily. Feeding is thus a strong evolving force for the speciation of Teleost fish (e.g., Harmelin-Vivien 2002, Lobato et al. 2014), and the link between morphology and feeding behaviour has been largely studied and discussed (Wainwright and Richard 1995, Wainwright and Bellwood 2002). Groupers (Serranidae, Epinephelinae) are common predators that occur in warm-temperate and tropical coastal ecosystems. Among the 163 epinephelid species recorded in the world (Froese and Pauly 2015), most feed on fish, crustaceans, and cephalopods in various proportions throughout the majority of their life history (Parrish 1987). Their functional morphology, particularly patterns of dentition and jaw morphology, is related to their feeding ecology and the main prey consumed (Weaver 1996). Two types of predation strategies are encountered among groupers for catching prey: active hunting tactic or ambush strategy (Hobson 1974, Shpigel and Fishelson 1989). Prey type, predation technics, and feeding apparatus morphology are often linked. For example, *Mycteroperca* spp. generally swim in mid-water, actively pursue and capture their prey (mostly fish), and they present large canine tooth size and slender mandibular bone (Weaver 1996). In contrast, *Epinephelus* spp. are demersal species, and mainly ambush predators that rely on suction feeding for prey capture, and they present smaller but more numerous teeth, along with higher and thicker jaws. Only the genus *Paranthias* departs from this general feeding pattern of epinephelids with differing diet and morphology

[1] Aix Marseille Université, CNRS/INSU, Université de Toulon, IRD, Mediterranean Institute of Oceanography (MIO) UM 110, 13288, Marseille, France.

[2] GIS Posidonie, MIO - Case 901, Campus universitaire de Luminy, 13 288 Marseille cedex 9 France.

* Corresponding author: mireille.harmelin@mio.osupytheas.fr

(Wainwright and Bellwood 2002). *Paranthias* spp. are specialized to feed on zooplankton all through their life, and display specialized gracile feeding structures and elongated body shape (Randall 1967, Wainwright and Bellwood 2002).

Fish diet is generally studied through the analysis of gut contents using various indices of prey importance (Hyslop 1980, Liao et al. 2001, Baker et al. 2014). *In situ* observation of fish behaviour is also a useful way to get information on the type of prey consumed by fish, particularly in coral reefs, although it is less commonly used (Shpigel and Fishelson 1989, Bruggeman et al. 1994). More recently, the analysis of stable isotopes of carbon and nitrogen allows the assessment of the feeding habits and trophic level of fish, and to determine their role in the food webs of ecosystems (e.g., Vander Zanden and Rasmussen 2001, Post 2002, Fry 2006). While gut contents represent snapshots in time, stable isotope analyses give a more integrated information on the food assimilated a few weeks or months before sampling, but have to be combined with a knowledge of prey composition to be correctly interpreted (Letourneur et al. 2013, Condini et al. 2015).

In epinephelids, a large range of body size is observed between species that are similar in shape (Wainwright and Richards 1995). How does the maximum body size reached by adult groupers influence their diets?

Patterns of prey use generally change ontogenetically from small juveniles to large adults as the mechanical properties of their feeding apparatus change when fish are growing (Wainwright and Bellwood 2002). So, how do grouper diets vary with individual size within a species, and do patterns of ontogenetic feeding differ among species?

Difference in diets can be related also to difference in feeding activity and prey availability during the day/night cycle or according to season (Harmelin-Vivien and Bouchon 1976, Shpigel and Fishelson 1989). How these temporal factors affect the perception of grouper diets can be obtained from studies of gut contents or *in situ* fish behaviour.

Independently of their size, groupers are often qualified as apex or top-predators (Page et al. 2013, Hackradt et al. 2014). But one can wonder if data on their feeding behaviour and trophic level estimation justify such a trophic position in food webs. Recent data provided by stable isotope analysis give a contrasted perception of this problem.

Finally, what do we still have to learn about grouper feeding biology to fully understand their role in coastal ecosystem functioning?

These different questions are explored in this chapter, with emphasis on some epinephelid species, for which detailed data are available in the literature. However, this chapter does not pretend to present exhaustive information due to the vast amount of literature available on grouper diets and their feeding behaviours.

Patterns of Diet with Adult Size: Small versus Large Species

As emphasized by Wainwright and Bellwood (2002), and except for the zooplanktivorous genus *Paranthias*, the major morphological axis of the epinephelid radiation is body size from a similar shape. Epinephelid species may be less than 30 centimeters to more than 2.5 m long when adults, and this is particularly remarkable in the genus *Epinephelus* (Table 1). Whatever the geographic distribution, the habitat, and the depth range of species, three main types of prey are found in groupers' diets: crustaceans, fish (mostly teleosts), and cephalopods. An increase in the size of the dominant prey types with the size of the predator species is generally observed. Small benthic crustaceans like shrimps and crabs dominate in the diet of small species (< 50 cm) like *Alphestes* spp., while large crustaceans (e.g., stomatopods, spinny lobsters, sleepy lobsters), cephalopods (octopi, squids and cuttlefish), and teleost fish constitute the main prey of groupers larger than 1 m long (Table 1, Fig. 1). Above 2 m body size, giant groupers feed essentially on large elusive prey like fish and cephalopods, and may also catch marine turtles. This trend in feeding behaviour was observed by Wainwright and Bellwood (2002) for Caribbean groupers, but is general for all epinephelids (Fig. 1). However, if there is a general increase in individual prey size with the size of the predator species, some prey types are consumed by both small and large species. Crustaceans are the dominant prey in the diet of the small-sized *Alphestes* spp. (30 cm) and the larger-sized *Hyporthodus* spp. (115–230 cm). Both small and large epinephelid species may be exclusive (or quite exclusive) piscivores. Piscivory is observed in the small *Gonioplectrus hispanus* (30 cm) and *Gracila albomarginata* (40 cm), in the medium-sized *Aniperodon leucogrammicus* and *Variola* spp. (65–83 cm), and in several species of the large-sized groupers like *Mycteroperca* spp. and *Plectropomus* spp. (85–200 cm) (Table 1). Piscivorous groupers may swim in midwater above the bottom like *Gracila, Mycteroperca*, and *Variola* actively pursuing their prey, or stay ambushed on or close to the bottom like *Aniperodon* and *Plectropomus*.

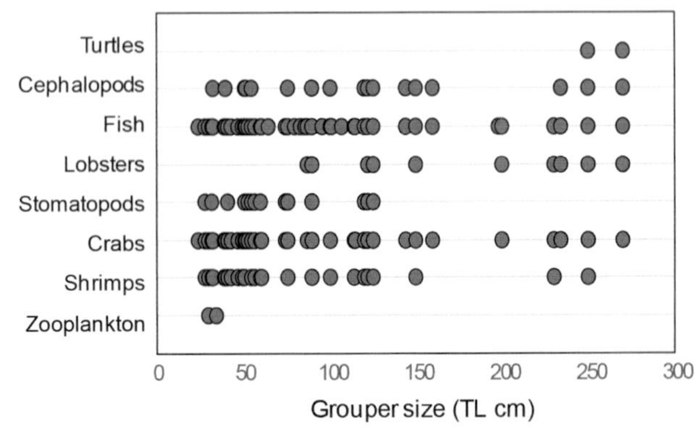

Fig. 1 Main types of prey ingested by adult groupers according to the maximum size reached by the species (TL: total length in cm).

Table 1. Main prey types consumed by groupers in decreasing order of importance in adult fish diet. The first number into brackets following genus name indicates the total number of species in the genus and the second the number of species for which information on diet was recorded. Max size: maximum size reached by the species according to FishBase (Froese and Pauly 2015); TL: total length in cm; References: numbers refer to the main references found in the literature by alphabetical order.

Genus species (Tot nb spp. - with diet)	**Max size (TL cm)**	**Prey 1**	**Prey 2**	**Prey 3**	**References**
***Aethaloperca* (1-1)**					
Ae. rogaa	60	Fish	Crustaceans (stomatopods, crabs)		38, 57
***Alphestes* (3-3)**					
Al. afer	33	Crustaceans (crabs, shrimps)	Fish	Cephalopods (octopi)	66
Al. immaculatus	30	Crustaceans	Fish		35, 40
Al. multiguttatus	30	Crustaceans	Fish		40
***Anyperodon* (1-1)**					
An. leucogrammicus	65	Fish			41
***Cephalopholis* (24-16)**					
C. argus	60	Fish	Crustaceans (crabs, shrimps, stomatopods)	Polychaetes	26, 38, 41, 42, 67, 73
C. aurantia	60	Fish	Crustaceans		53
C. boenak	30	Fish	Crustaceans		8, 9, 47
C. cruentata	43	Fish	Crustaceans (shrimps)		62, 63, 66
C. cyanostigma	40	Fish	Crustaceans		8, 40, 53, 59
C. fulva	41	Fish	Crustaceans (crabs, shrimps, stomatopods)		29, 62, 66
C. hemistiktos	35	Fish	Crustaceans		53, 73
C. igarashiensis	43	Fish	Crustaceans		21, 40
C. leopardus	24	Fish	Crustaceans		40
C. micropion	25	Fish	Crustaceans		21
C. miniata	50	Fish	Crustaceans (crabs, shrimps)		9, 41, 47, 67, 73
C. nigripinnis	28	Fish	Crustaceans		21
C. panamensis	39	Fish	Crustaceans		21
C. sexmaculata	50	Fish	Crustaceans		3, 40, 53, 59

Table 1 Contd. ...

...Table 1 Contd.

Genus species (Tot nb spp. - with diet)	**Max size (TL cm)**	**Prey 1**	**Prey 2**	**Prey 3**	**References**
C. sonnerati	57	Fish	Crustaceans (crabs, shrimps, stomatopods)		9, 38, 40, 53
C. urodeta	28	Fish	Crustaceans (crabs, shrimps, hermit crabs)		41, 53, 61, 67
Cromileptes **(1-1)**					
Cr. altivelis	70	Fish	Crustaceans		59
Dermatolepis **(3-3)**					
D. dermatolepis	100	Fish	Crustaceans		35
D. inermis	91	Fish			62
D. striolata	85	Fish	Crustaceans		53
Epinephelus **(87-54)**					
E. adscensionis	61	Fish	Crustaceans (crabs, shrimps)	Gastropods	62, 66
E. aeneus	120	Fish	Crustaceans (crabs, shrimps)	Cephalopods (coctopi, cuttlefish, squids)	22, 25
E. akaara	58	Fish	Crustaceans		53
E. albomarginatus	100	Fish	Crustaceans	Cephalopods	53
E. analogus	114	Fish	Crustaceans (crabs, shrimps)		40
E. andersoni	87	Fish	Crustaceans (crabs, lobsters)		40
E. areolatus	47	Crustaceans (crabs, shrimps)	Fish		45, 47, 53
E. awoara	60	Crustaceans (crabs, shrimps)	Fish	Sea urchins, Bivalv	51, 53
E. bruneus	136	Fish	Crustaceans		53
E. caninus	164	Fish	Crustaceans	Molluscs	48, 76
E. chlorostigma	75	Fish	Crustaceans (crabs, stomatopods)		40
E. coeruleopunctatus	76	Fish	Crustaceans		47, 53, 70

Table 1 Contd. ...

...Table 1 Contd.

Genus species (Tot nb spp. - with diet)	**Max size (TL cm)**	**Prey 1**	**Prey 2**	**Prey 3**	**References**
E. coioides	120	Fish	Crustaceans (crabs, shrimps)		45, 47, 53
E. costae	140	Fish	Crustaceans	Molluscs	24, 25, 33, 34, 78
E. cyanopodus	122	Fish	Crustaceans (crabs, shrimps)		41, 47, 53
E. daemelii	200	Fish	Crustaceans		40
E. epistictus	80	Fish	Crustaceans		53
E. fasciatus	40	Fish	Crustaceans (crabs, shrimps)	Cephalopods (octopi, cuttlefish), Sea urchins	38, 40, 47, 48, 57
E. fasciomaculosus	30	Crustaceans (crabs, shrimps)	Fish		40
E. flavocaeruleus	90	Fish	Crustaceans (crabs, shrimps)	Cephalopods (cuttlefish)	40
E. fuscoguttatus	120	Fish	Crustaceans (crabs, shrimps)		41, 53
E. goreensis	140	Fish	Crustaceans	Cephalopods	22
E. guttatus	76	Fish	Crustaceans (crabs, shrimps, stomatopods)	Cephalopods	55, 62, 66
E. hexagonatus	28	Fish	Crustaceans (crabs, stomatopods)	Polychaetes	38, 41, 53
E. howlandi	55	Fish	Crustaceans (crabs, shrimps)	Cephalopods (octopi)	41, 53
E. itajara	250	Fish	Crustaceans (crabs, shrimps, lobsters)	Cephalopods, Turtles	15, 30
E. labriformis	60	Fish	Crustaceans		40
E. lanceolatus	270	Fish	Crustaceans (crabs, lobsters)	Cephalopods, Turtles	49, 53
E. latifasciatus	137	Fish	Crustaceans		53

Table 1 Contd. ...

...Table 1 Contd.

Genus species (Tot nb spp. - with diet)	**Max size (TL cm)**	**Prey 1**	**Prey 2**	**Prey 3**	**References**
E. longispinnis	55	Fish	Crustaceans (crabs, shrimps, stomatopods)	Cephalopods (cuttlefish)	40
E. macrospilos	51	Fish	Crustaceans (crabs, shrimps, stomatopods)	Cephalopods (octopi, cuttlefish), Polychaetes	38, 40, 41, 47
E. maculatus	61	Fish	Crustaceans	Cephalopods (cuttlefish)	40, 47, 53
E. malabaricus	234	Fish	Crustaceans	Cephalopods (octopi)	40, 47, 53
E. marginatus	150	Fish	Crustaceans (crabs, shrimps, lobsters)	Cephalopods (octopi)	7, 18, 25, 34, 37, 50, 52, 64,
E. melanostigma	35	Fish	Crustaceans		53
E. merra	32	Fish	Crustaceans (crabs, shrimps, stomatopods)	Cephalopods (octopi)	38, 41, 47, 67
E. miliaris	53	Fish	Crustaceans (crabs, stomatopods)	Gastropods	53, 57
E. morrhua	90	Fish	Crustaceans (crabs, shrimps, lobsters)		53
E. morrio	125	Fish	Crustaceans (crabs, shrimps, lobsters, stomatopods)	Cephalopods (octopi), Bivalves	12, 17, 66
E. multinotatus	100	Fish	Crustaceans (crabs, shrimps)		45
E. ongus	40	Crustaceans	Fish		53
E. poecilonotus	65	Fish	Crustaceans		53
E. polyphekadion	90	Fish	Crustaceans (crabs, shrimps)	Cephalopods (octopi, cuttlefish), Gastropods	3, 40, 47, 53
E. quoyanus	40	Crustaceans (crabs)	Fish	Bivalves	77

Table 1 Contd. ...

...Table 1 Contd.

Genus species (Tot nb spp. - with diet)	**Max size (TL cm)**	**Prey 1**	**Prey 2**	**Prey 3**	**References**
E. radiatus	70	Fish	Crustaceans		53
E. retouti	50	Fish	Crustaceans		53
E. rivulatus	45	Fish	Crustaceans	Bivalves	53, 77
E. sexfasciatus	40	Crustaceans (crabs, shrimps)			69
E. socialis	52	Fish	Crustaceans (crabs)	Cephalopods (octopi)	40
E. striatus	122	Fish	Crustaceans (crabs, shrimps, lobsters, stomatopods)	Cephalopods (cuttlefish)	28, 36, 65, 66, 72
E. tauvina	90	Fish	Crustaceans (crabs, lobsters)		40, 44
E. trimaculatus	50	Fish	Crustaceans		53
E. tukula	200	Fish	Crustaceans (crabs, lobsters)		40
E. undulosus	120	Fish	Crustaceans (shrimps, stomatopods)	Ascidians	40
***Gonioplectrus* (1-1)**					
Go. hispanus	30	Fish			21
***Gracila* (1-1)**					
Gr. albomarginata	40	Fish			49
***Hyporthodus* (13-6)**					
H. flavolimbatus	115	Crustaceans	Fish (crabs)		27, 40
H. mystacinus	160	Crustaceans	Fish	Cephalopods (cuttlefish, squids)	10, 40
H. nigritus	230	Crustaceans (crabs, shrimps, lobsters)	Fish		40
H. niveatus	122	Crustaceans	Fish (crabs)	Cephalopods, Gastrop	40
H. quernus	122	Fish	Crustaceans		40, 43
H. septemfasciatus	155	Crustaceans	Fish		53

Table 1 Contd. ...

...*Table 1 Contd.*

Genus species (Tot nb spp. - with diet)	**Max size (TL cm)**	**Prey 1**	**Prey 2**	**Prey 3**	**References**
***Mycteroperca* (15-11)**					
M. acutirostris	80	Fish	Crustaceans		31, 40
M. bonaci	150	Fish	Crustaceans (crabs)		32, 62, 66
M. interstitialis	80	Fish			16, 17, 62, 66
M. jordani	198	Fish			39
M. microlepis	145	Fish	Crustaceans		1, 13, 14, 58
M. olfax	120	Fish			68
M. phenax	107	Fish			54, 62
M. rosacea	86	Fish			40
M. rubra	144	Fish	Crustaceans	Cephalopods (cuttlefish, squids)	4, 25, 40, 71
M. tigris	101	Fish			62, 66
M. venenosa	100	Fish	Cephalopods (squids)		66
***Paranthias* (2-2)**					
Pa. colonus	35	Zooplankton	Fish		35
Pa. furcifer	30	Zooplankton			62, 66
***Plectropomus* (7-6)**					
Pl. areolatus	80	Fish			41, 59
Pl. laevis	125	Fish		Crustaceans	49
Pl. leopardus	120	Fish			46, 47, 74, 75
Pl. maculatus	125	Fish	Cephalopods (cuttlefish, squids)		46, 53
Pl. oligacanthus	75	Fish	Crustaceans		49
Pl. punctatus	96	Fish			57
***Saloptia* (1-0)**	39				
***Triso* (1-0)**	68				
***Variola* (2-2)**					
V. albimarginata	65	Fish			40
V. louti	83	Fish			2, 38, 41, 47

(1) Adams 1976; (2) Allen 2004; (3) Anderson and Hafiz 1989; (4) Aronov and Goren 2008; (5) Artero et al. 2014; (6) Artero 2014; (7) Barreiros and Santos 1998; (8) Beukers-Stewart and Jones 2004; (9) Blaber et al. 1990; (10) Bowman et al. 2000; (11) Brulé et al. 1993; (12) Brulé et al. 1999; (13) Brulé et al. 2005; (14) Brulé et al. 2011; (15) Bullock et al. 1992; (16) Bullock et al. 1994; (17) Bullock and Smith 1991; (18) Condini et al. 2011; (19) Connell 1998; (20) Craig 2007;

Table 1 Contd. ...

...Table 1 Contd.

(21) Craig et al. 2011; (22) Da Silva Monteiro 1998; (23) Derbal and Kara 1996; (24) Derbal and Kara 2007; (25) Diatta et al. 2003; (26) Dierking et al. 2009; (27) Dodrill et al. 1993; (28) Eggleston et al. 1998; (29) Gathaz et al. 2013; (30) Gerhardinger et al. 2006; (31) Gibran 2007; (32) Gómez-Canchong et al. 2004; (33) Göthel 1992; (34) Gracia López and Castelló i Orvay 2005; (35) Grove and Lavenberg 1997; (36) Grover et al. 1998; (37) Harmelin and Harmelin-Vivien 1999; (38) Harmelin-Vivien and Bouchon 1976; (39) Heemstra 1995; (40) Heemstra and Randall 1993; (41) Hiatt and Strasburg 1960; (42) Hobson 1974; (43) Honebrink 1990; (44) Jeyaseelan 1998; (45) Kailola et al. 1993; (46) Kingsford 1992; (47) Kulbicki et al. 2005; (48) Kyrtatos 1982; (49) Lieske and Myers 1994; (50) Linde et al. 2004; (51) Liu 1971; (52) Machado et al. 2008; (53) Masuda and Allen 1993; (54) Matheson et al. 1986; (55) Menzel 1960; (56) Meyer and Dierking 2011; (57) Morgans 1982; (58) Mullaney and Gale 1996; (59) Myers 1999; (60) Nagelkerken 1979; (61) Nakai et al. 2001; (62) Pattengill et al. 1997; (63) Popple and Hunte 2005; (64) Reñones et al. 2002; (65) Randall 1965; (66) Randall 1967; (67) Randall and Brock 1960; (68) Rodriguez 1984; (69) Salini et al. 1994; (70) Sano et al. 1984; (71) Sazima 1986; (72) Silva Lee 1974; (73) Shpigel and Fishelson 1989; (74) St John 1999; (75) St John et al. 2000; (76) Tortonese 1986; (77) van der Elst and Adkin 1991; (78) Zaidi et al. 2013; (79) Zeller 1997.

However, most grouper species (*Aethaloperca, Cephalopholis, Cromileptes, Dermatolepis*, and *Epinephelus*, along with some *Mycteroperca* and *Plectropomus*) feed on fish, crustaceans, and cephalopods, ingested in different proportions according to individual size, habitat, and period. Thus, size is not necessarily a good proxy for inferring the diet of an epinephelid species.

Ontogenetic Changes in Grouper Diet

Patterns of prey use by groupers change ontogenetically when fish are growing, as in most fish species (Wootton 1999). However, changes in feeding behaviour with size occurred more or less early depending on the species, its diet, and size. Fish dominated early in the diet of the mostly piscivorous species belonging to the *Plectropomus, Mycteroperca*, and *Cephalopholis* genera, whatever the maximum size attained by these species when adults. Small *Cephalopholis argus* mainly prey on fish at a length < 5 cm TL (Harmelin-Vivien and Bouchon 1976). *Plectropomus leopardus* already consume more fish than crustaceans at 10 cm TL (St John 1999) (Fig. 2) and *Mycteroperca microlepis* at 15 cm TL (Mullaney 1994, Brulé et al. 2011) (Fig. 3). In these piscivorous groupers, the main crustaceans consumed by the smallest individuals are shrimps (decapods Natantia). In less piscivorous groupers which feed on crustaceans and fish when adults (plus a few other prey groups like mollusks), a longer transition period is observed from a diet dominated by crustaceans to a diet dominated by fish (and/or cephalopods) when individuals are growing. The size at which fish become the dominant prey increases generally with the species-specific maximum size attained by the species: 10 cm in *Epinephelus merra* (Harmelin-Vivien and Bouchon 1976) (Fig. 4), 30 cm TL in *Epinephelus striatus* (Eggleston et al. 1998) (Fig. 5), around 60 cm TL in *E. marginatus* (Azevedo et al. 1995, Barreiros and Santos 1998, Reñones et al. 2002) (Figs. 6a and b), and 140 cm TL in the giant grouper

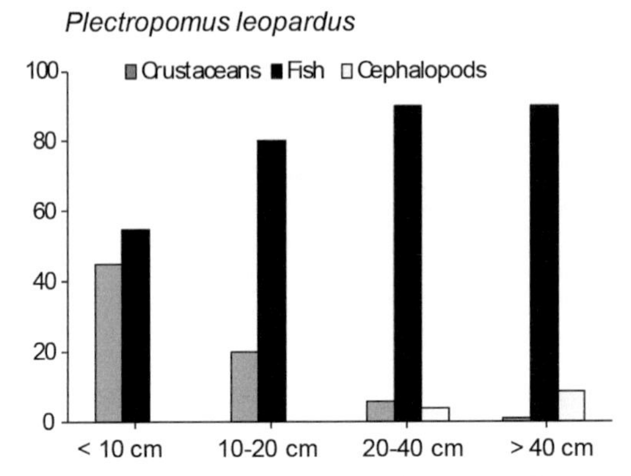

Fig. 2 Ontogenetic changes in the diet of leopard coral grouper, *Plectropomus leopardus*, on the Great Barrier Reef, Australia (redrawn from St John 1999).

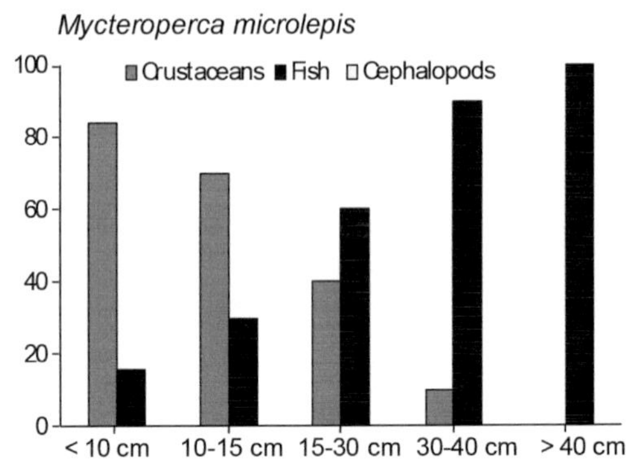

Fig. 3 Ontogenetic changes in the diet of gag, *Mycteroperca microlepis*, in South Carolina, USA (redrawn from Mullaney 1994).

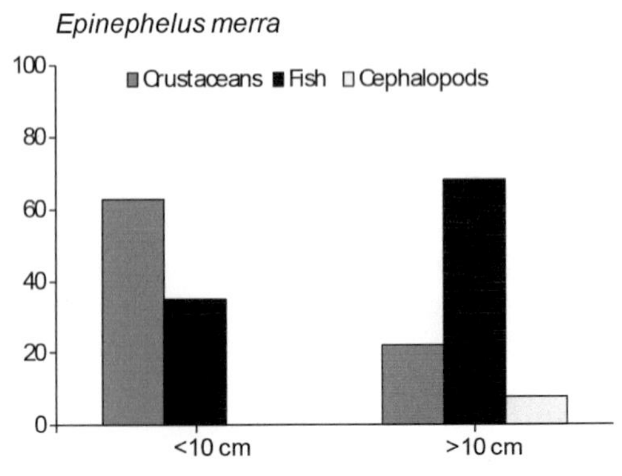

Fig. 4 Ontogenetic changes in the diet of honeycomb grouper, *Epinephelus merra*, in Toliara, SW Madagascar (drawn from Harmelin-Vivien and Bouchon 1976).

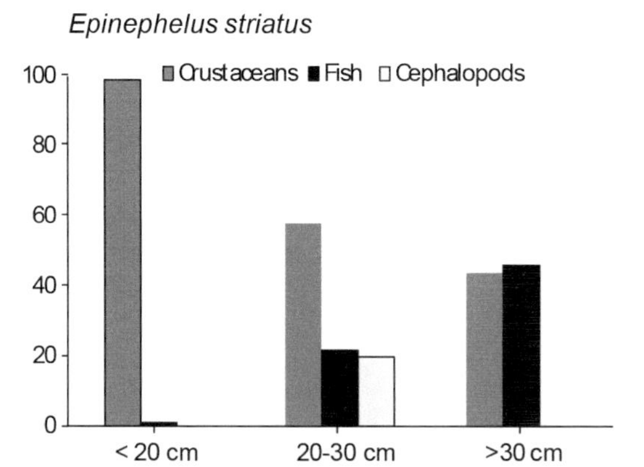

Fig. 5 Ontogenetic changes in the diet of Nassau grouper, *Epinephelus striatus,* near Lee Stocking Island, Bahamas (drawn from Eggleston et al. 1998).

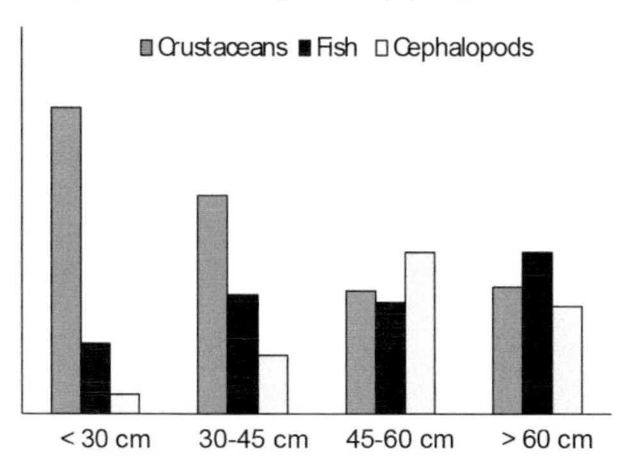

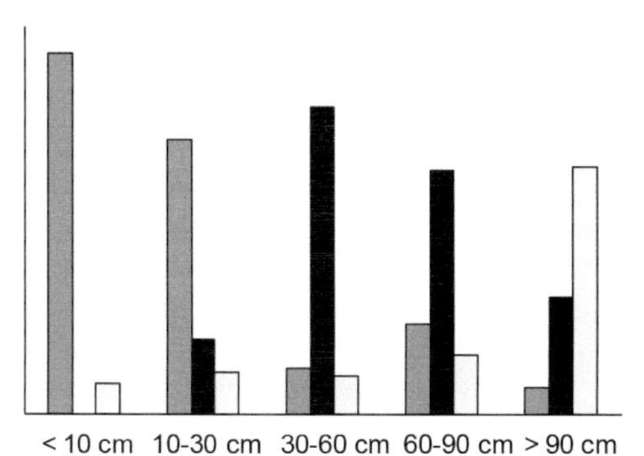

Fig. 6 Ontogenetic changes in the diet of brown grouper, *Epinephelus marginatus.* (a) in Balearic Islands, Mediterranean (drawn from Reñones et al. 2002); (b) in the Azores, Atlantic (drawn from Azevedo et al. 1995, and Barreiros and Santos 1998).

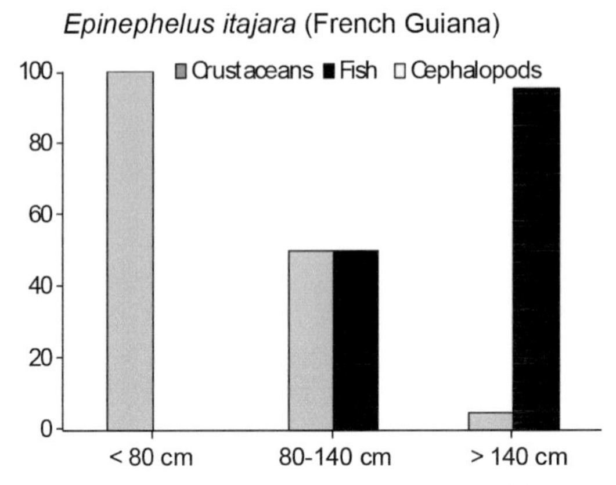

Fig. 7 Ontogenetic changes in the diet of Goliath grouper, *Epinephelus itajara*, in French Guiana (drawn from Artero et al. 2014).

E. itajara (Artero 2014, Artero et al. 2014) (Fig. 7). Crabs (brachyurans) constitute the main crustacean prey consumed by *E. merra*, *E. striatus*, and *E. itajara*. In the dusky grouper *E. marginatus*, the main crustaceans ingested by the smallest individuals (< 10 cm TL) are amphipods, crabs, and shrimps. Afterwards, crabs dominated in the diet of fish from 10 to 45 cm TL, and crabs, spinny lobsters, and sleepy lobsters in individuals larger than 45 cm TL (Azevedo et al. 1995, Derbal and Kara 1996, Barreiros and Santos 1998, Reñones et al. 2002, Gracia López and Castelló y Orvay 2005). The largest dusky groupers (> 90 cm TL) seem to preferentially consume cephalopods, particularly octopi, when possible (Smale 1986, Barreiros and Santos 1998, Linde et al. 2004) (Fig. 6b). Ontogenetic variations of diet are often associated with changes in habitat from shallow coastal waters for juveniles to deeper offshore waters for adults, as in *E. marginatus* (Reñones et al. 2002, Machado et al. 2008), from estuaries to offshore reefs, as in *Mycteroperca microlepis* (Ross and Moser 1995), or from coastal mangroves to offshore rocky reefs, as in *E. itajara* (Artero et al. 2014).

Along with changes in diet composition, an increase in mean and maximum prey sizes with the size of the predator was observed in all grouper species analysed: *Epinephelus marginatus* (Reñones et al. 2002), *Cephalopholis boenak* and *C. cyanostigma* (Beukers-Stewart and Jones 2004), *C. argus* (Meyer 2008), and *E. itajara* (Artero et al. 2014), among others. The size range of prey also increases with predator size, as larger fish continue to feed on small-sized prey. For example, the mean prey size of *C. boenak* on the Great Barrier Reef is 7 ± 8 mm (range 5–20 mm) for 10 cm SL individuals, and 33 ± 18 mm (range 13–65 mm) for 18 cm SL individuals (Beukers-Stewart and Jones 2004). These authors considered that the maximum size of prey consumed is limited to approximately one-third of the body length of the predator. Increase in prey size with the size of the predator is a general trend in predator-prey relationships that has been widely explained in terms of energy optimization and resource partitioning (Wootton 1999). However, the persistence of the ingestion of small-sized prey indicated that groupers are able to continue to feed on small crustaceans (e.g., swarms of shrimps) and fish (settling juveniles, small pelagic fish) if they are abundant in the environment (St John 1999, Beukers-Stewart and Jones 2004).

Temporal Variations in Grouper Feeding Behavior

Diel feeding periodicity in groupers differs with species, and the types of study performed, such as *in situ* observations, stomach content analysis, or acoustic tracking provide complementary information (Parrish 1987). Generally, groupers which forage more in the water column tend to be mainly active by day, while species preying more on benthic crustaceans tend to feed more actively during the night, like *Alphestes* spp. (Hobson 1974, Parrish 1987). However, many species of groupers which feed on benthic prey are nevertheless reported to be diurnally active, like *C. boenak* (Liu and Sadovy 2005), *C. cruentata* (Nagelkerken 1979), *C. hemistiktos* (Shpigel and Fishelson 1989), *P. leopardus* (Zeller 1997), and *E. marginatus* (Gibran 2007, Azzurro et al. 2013, Koeck et al. 2014). Peaks of feeding activity at crepuscular periods are also observed in different species like *E. striatus* and *E. fulvus* in the Bahamas (Sluka and Sullivan 1996) or *C. argus* and *C. miniata* in the Red Sea (Shpigel and Fishelson 1989). However, it seems that many species of groupers are able to feed by both day and night. In these species, the dominant prey ingested by groupers can change as an adaptation to prey behaviour and accessibility. In these cases, groupers consume more fish during the day, and more crustaceans and cephalopods at night when these preys emerge from their shelters (Hobson 1974, Harmelin-Vivien and Bouchon 1976).

Feeding periodicity seems also to differ among regions within a species. If *C. argus* feeds mainly during crepuscular periods in the Red Sea (Shpigel and Fishelson 1989), it is active by both day and night in Madagascar (Harmelin-Vivien and Bouchon 1976). In Hawaii, on the contrary, Meyer (2008) finds that *C. argus* is active during the day and quiescent at night, as observed when following fish implanted with transmitters. The question remains to be answered if such changes in feeding periodicity in a species are due to differences in prey composition, abundance, or availability in the environment, or to interspecific competition between closely related predators.

In a few grouper species, variations of feeding composition and intensity with seasons were investigated. On the Great Barrier Reef, *P. leopardus* mainly prey on fish in summer and winter, but the main families caught differ between seasons, with small pelagic fish (clupeids and engraulids) being more consumed in summer (Kingsford 1992). However, on the Senegalese coasts, little difference in fish feeding is observed between winter and summer in *E. aeneus, E. marginatus*, and *M. rubra* (Diatta et al. 2003).

Spatial Variations of Feeding Behavior

The feeding ecology of groupers may differ between regions because of differences in resource availability and the effects of competition. However, the way of quantifying prey importance in stomach contents (% in number: %N, volume: %V or weight: %W) make comparisons sometimes difficult. Some species seem to present a rather constant diet composition in all sites, like *C. argus* which feeds mainly on fish in most of its geographical range. It consumes 96%W of fish in Madagascar, Reunion, and Mauritius (Harmelin-Vivien and Bouchon 1976), 96%W also in Hawaii (Dierking et al. 2009), 95%N in the Red Sea (Shpigel and Fishelson 1989), and 79%N in French Polynesia (Meyer 2008). At a smaller spatial scale (tens of km), St John et al. (2000) do not find any differences in the diet of *P. leopardus* on different reefs on the Great Barrier Reef. Conversely, difference in diet is observed in *E. itajara* between Brazil (Gerhardinger et al. 2006), French Guyana (Artero 2014), and the Caribbean (Randall 1967, Bullock et al. 1992, Koenig and Coleman 2009). While the main preys of the jewfish are always fishes and crustaceans, this giant grouper feeds more on fishes, particularly catfishes, in Brazil and French Guyana, and presents a more diversified diet in the Caribbean with the consumption of lobsters, octopi, and even sea turtles in addition to fish and a high amount of crabs.

Role of Groupers in Ecosystem Trophic Functioning: Insights from Stable Isotope Analysis

Groupers are often considered top-predators in benthic ecosystems, with a prominent role in ecosystem functioning (Mumby et al. 2006). If the role of epinephelids as piscivores and consumers of large crustaceans is evident according to their feeding habits, may they always be viewed as top predators? Such a place in the food webs of coastal benthic ecosystems may be true for the giant groupers *E. itajara, E. lanceolatus*, and *E. malabaricus*, which have no known predators, except some sharks (Sadovy and Eklund 1999). However, many epinephelid species which consume mostly fish are considered as apex predators in temperate (Stergiou and Karpouzi 2002, Hackradt et al. 2014) and tropical environments (Page et al. 2013). But does the consumption of fish necessarily put a piscivore at the top of an ecosystem food web? Such a question is not trivial and the use of stable isotopes in addition to stomach content analysis brings new insights on the relative trophic place of predators in ecosystems (Badalamenti et al. 2002, Cresson et al. 2014). A fish species preying on carnivorous invertebrates can be positioned at a higher trophic level than a piscivorous fish species preying mainly on zooplanktivorous fishes (Cresson et al. 2014). Stomach contents give a snapshot of the food ingested before sampling, and the knowledge of fish diet variation implies the analysis of a large number of fish, but is necessary to get the specific composition of fish diets (Hyslop 1980, Baker et al. 2014). On the other hand, stable isotope analysis provides an integrated view of the dietary components assimilated by fishes over a longer period, generally months, but no accurate indication on the species ingested (e.g., Vander Zanden and Rasmussen 1999, 2001, Fry 2006).

The analysis of stable isotopes of carbon and nitrogen is now a common way to determine the place of organisms in food webs and apprehend the trophic functioning of ecosystems (e.g., Wada et al. 1991, Post 2002, Middelburg 2014). The stable isotope ratios of a consumer are dependent on those of the food items ingested, plus a trophic fractioning, mainly due to complex metabolism processes (Perga and Grey 2010). Stable isotope ratios of carbon (expressed as $\delta^{13}C$) give an information of the sources at the base of the food chain, while stable isotope ratios of nitrogen (expressed as $\delta^{15}N$) are used to determine the trophic level of the organisms (e.g., Cabana and Rasmussen 1996, Post 2002, Fry 2006, and many others). Complementary information is given by the combined use of stomach content and stable isotope analyses, allowing to better determine the real role of organisms in the functioning of trophic webs (e.g., Cocheret de la Morinière et al. 2003).

Only a few studies provide information on stable isotope signatures of groupers, both in temperate (Reñones et al. 2002, Condini et al. 2015) and tropical ecosystems (Carassou et al. 2008, Meyer 2008, Nelson et al. 2012, Nelson 2011, Wyatt et al. 2012, Letourneur et al. 2013, Page et al. 2013, Artero et al. 2014, Tzadik et al. 2015).

Ontogenetic Increase in Trophic Level

An increase in $\delta^{15}N$ value, thus in trophic level, with the size of individuals is generally evidenced in epinephelids in relation to the consumption of larger prey and of higher trophic level. This trend is observed in the Goliath grouper, *Epinephelus itajara,* in French Guyana (Artero et al. 2014) and in Florida (Tzadik et al. 2015) (Fig. 8), in *E. marginatus* in the Balearic Islands

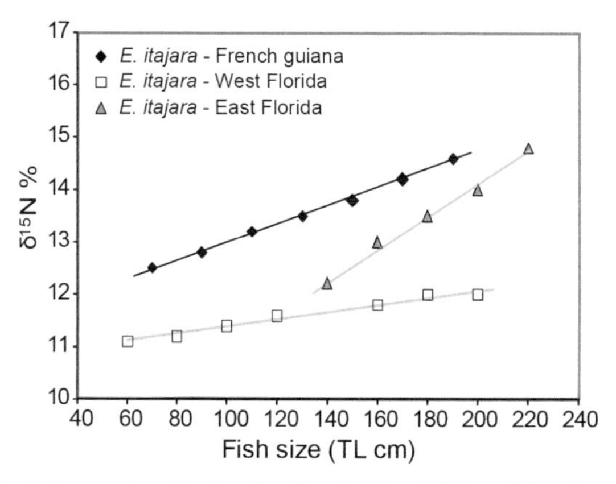

Fig. 8 Relationship between trophic level estimated by $\delta^{15}N$ and fish length in the Goliath grouper, *Epinephelus itajara*, in French Guiana (Artero et al. 2014), and West and East Florida (Tzadik et al. 2015).

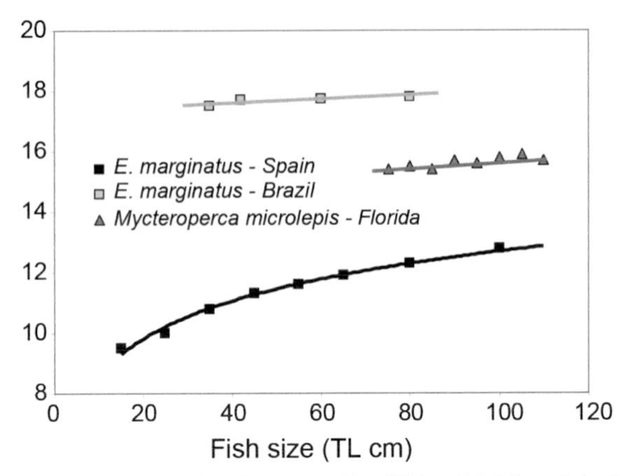

Fig. 9 Relationship between trophic level estimated by $\delta^{15}N$ and fish length in the dusky grouper, *Epinephelus marginatus*, in the Balearic Islands (Reñones et al. 2002) and in Brazil (Condini et al. 2015), and in the gag in Florida (Nelson et al. 2012).

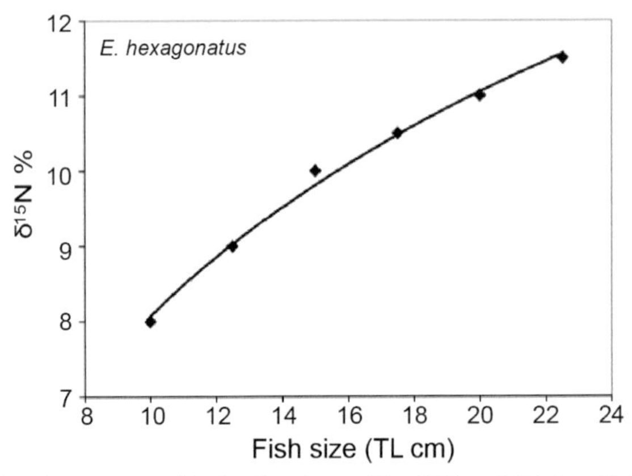

Fig. 10 Relationship between trophic level estimated by $\delta^{15}N$ and fish length in the starspotted grouper, *Epinephelus hexagonatus*, in Mururoa, French Polynesia (Page et al. 2013).

(Reñones et al. 2002) (Fig. 9), *E. hexagonatus* in Mururoa, French Polynesia (Page et al. 2013) (Fig. 10), and *Cephalopholis argus* in Moorea, French Polynesia (Meyer 2008), as well as in juveniles of *E. fuscoguttatus* reared in aquaculture in China (Chen et al. 2012). Differences in $\delta^{15}N$ values of one species between sites, as observed for *E. itajara* and *E. marginatus* (Fig. 9) is mainly due to isotopic differences of regional baselines (Reñones et al. 2002, Tzadik et al. 2015). The increase of $\delta^{15}N$ with size generally slows down in the largest size classes, which tend to feed on similar prey types as in *E. hexagonatus* (Pages et al. 2013) or in *E. marginatus* in Spain (Reñones et al. 2002). However, no change in $\delta^{15}N$ and trophic level with length is recorded in the gag grouper,

Mycteroperca microlepis, in Florida (Nelson et al. 2012, Nelson 2011), and in the brown grouper *E. marginatus* on the coasts of Brazil (Condini et al. 2015) (Fig. 9). *M. microlepis* is mainly a piscivore and preys only on fishes for the size classes analysed by Nelson (2011). The consumption of fishes located at both high and low trophic levels by all gag grouper size classes may explain such an absence of relationship between $\delta^{15}N$ and fish length. In Brazil, *E. marginatus* displays similar ontogenetic shifts in prey types and size as those observed for this species in the Mediterranean (increase in fish consumption and decrease in crustacean consumption with size) without any significant changes in $\delta^{15}N$ due to a high dispersion of values (Condini et al. 2015). Using Bayesian mixing models, these authors distinguished between ingested (stomach contents) and assimilated (isotopic signatures) dietary components. They explain the absence of trophic level increase with size by a preferential assimilation of fish material across all size classes and a higher amount of refractory material in crustaceans. Similarly, in Moorea, mixing models using stable isotope values of prey and predator indicate a higher assimilation of fish compared to crustaceans in *E. merra*, while crustaceans are more numerous and important in weight in stomach contents (Letourneur et al. 2013). These examples demonstrate the need to perform stomach content analysis to properly interpret stable isotope data.

Trophic Position of Groupers in Ecosystems

Properly evaluating the trophic position of groupers within food webs would imply the obtainment of stable isotopic data not only of their prey but also of competitors and predators, and such a data set is presently non-existent. Only a small number of studies give isotopic values of different ecosystem compartments including groupers, which are of small size in most cases. Small-sized epinephelids or the juveniles of large-size groupers presented generally $\delta^{15}N$ values intermediate between those of strict piscivorous and secondary carnivorous fish species. On western Australian coral reefs, Wyatt et al. (2012) find that *Cephalopholis sexmaculata*, a medium-sized species (max 50 cm TL), has 3% lower $\delta^{15}N$ values than *Lutjanus sebae*, *Lethrinus* spp., and *Euthynnus affinis*, and was positioned at one trophic level lower than these species. Similarly, *E. merra*, a small-sized species (max 32 cm TL), presents lower $\delta^{15}N$ values than the smaller *Synodus variegatus*, a strict piscivore, and occupies a lower trophic level than the large pelagic piscivore *Scomberomorus commerson* on the coral reefs of New Caledonia (Carassou et al. 2008). But even the goliath grouper (*E. itajara*) (> 250 cm TL) was found to be positioned at an intermediate trophic level in the food webs of Florida reefs (Koenig and Coleman 2009), similar to that of *Lagodon rhomboides* or *Lutjanus griseus*, and lower than that of sharks and jacks (Chasar et al. 2005). Most groupers may be thus regarded as 'mesopredators', i.e., predators in the mid-trophic levels

(Gathaz et al. 2013). Such a position at a mid-trophic level is in accordance with the consumption of benthic crustaceans and relatively small-sized or low-trophic level fishes. But even strict piscivores can be positioned at a mid-trophic level if they mainly ingest low level planktivorous fishes (Cresson et al. 2014).

When examining the trophic categories of the fish species ingested by groupers, it appears that the species positioned at low trophic levels (herbivores and zooplanktivores) constitute a large part of their diet (Table 2). On coral reefs, where they are abundant, herbivorous and zooplanktivorous fishes combined represent from 58% in *Cephalopholis argus* to 97% in *Cephalopholis miniata* of the prey fish ingested. On the West African coast, *Mycteroperca rubra* feeds essentially on *Sardinella maderensis*, an abundant planktivorous fish species in this area (Diatta et al. 2003). In the Mediterranean Sea, herbivorous fishes are less numerous and do not represent an important component of groupers' diet. Zooplanktivorous fish can be an important prey for *E. costae* which largely feed on *Boops boops* (Gracia López and Castelló I Orvay 2005), but are not targeted by *E. marginatus* which rely mainly on benthic carnivores and omnivores (labrids, sparids). The diet of *E. marginatus* is mainly based on the benthic food web in the Balearic Islands (Reñones et al. 2002) as well as in Brazil (Condini et al. 2015). But groupers integrate not only benthic but also pelagic trophic pathways via the consumption of zooplanktivorous prey. Thus, if groupers feed on a large variety of fish species, they rarely prey on high level predators, which implies that they can hardly be viewed as top or apex predators, but rather as mid-trophic level predators, may be with the exception of the largest individuals of the large-sized piscivorous species.

Trophic Migrations

Stable isotope analysis is useful in determining the major pathways of the organic matter in ecosystems and better apprehending their trophic functioning. They are also useful to better know the trophic migrations realized by groupers between adjacent ecosystems like mangroves, seagrass beds, and coral reefs, or between estuaries and offshore rocky habitats. The main carbon sources of these ecosystems generally present different isotopic signatures that can be traced during the ontogenetic and/or trophic migrations of fishes (Nagelkerken and van der Velde 2004). But such studies are scarce concerning groupers. Juveniles of *E. itajara* growing in mangroves exhibit lower $\delta^{13}C$ values than those settling on rocky marine habitats, as mangal tree organic matter is depleted in ^{13}C compared to marine primary producers (Artero et al. 2014, Tzadik et al. 2015). In Florida, Nelson et al. (2012) trace with stable isotopes the transfer of secondary production from coastal seagrass habitats to offshore reefs linked to the ontogenetic trophic migration of the gag grouper, *Mycteroperca microlepis*.

Table 2. Importance of low-trophic level fish (herbivores and planktivores) in the diet of groupers. Total % Fish: total percentage of fish in the diet; %N: percentage in number; %V: percentage in volume; %W: percentage in weight; % Herbivores: percentage of herbivorous fish in the total % Fish; % Planktivores: percentage of planktivorous fish in the total % Fish; % Low TL Fish: percentage of low-trophic level fish (sum of % Herbivores and % Planktivores) in the total % Fish.

Species	Location	Total % Fish	% Herbivores	% Planktivores	% Low TL Fish	Reference
Tropical zones						
Cephalopholis argus	Gulf of Aqaba, Red Sea	96 %N	50	17	67	Shpigel and Fishelson (1989)
Cephalopholis argus	Moorea, French Polynesia	79 %N	26	32	58	Meyer (2008)
Cephalopholis argus	Hawaii	96 %W	43	17	60	Dierking et al. (2009)
Cephalopholis boenak	GBR, Australia	74 %V	0	85	85	Beukers-Stewart and Jones (2004)
Cephalopholis hemistiktos	Gulf of Aqaba, Red Sea	64 %N	11	67	78	Shpigel and Fishelson (1989)
Cephalopholis miniata	Gulf of Aqaba, Red Sea	86 %N	6	91	97	Shpigel and Fishelson (1989)
Mycteroperca rubra	Senegal, West Africa	100 %N	0	90	90	Diatta et al. (2003)
Plectropomus leopardus	GBR, Australia	85 %N	38	53	91	Kingsford (1992)
Plectropomus leopardus	GBR, Australia	95 %N	8	55	63	St John et al. (2000)
Temperate zones						
Epinephelus costae	Spain, Mediterranean	97 %N	0	72	72	Gracia-López and Castelló i Orvay (2005)
Epinephelus marginatus	Spain, Mediterranean	33 %N	2	4	6	Gracia-López and Castelló i Orvay (2005)
Epinephelus marginatus	Balearic Is., Mediterranean	41 %N	0	10	10	Reñones et al. (2002)
Epinephelus marginatus	Azores, Atlantic	40 %N	7	0	7	Barreiros and Santos (1998)

Conclusion

Groupers are active predators feeding mainly on fish, crustaceans, and cephalopods. According to the resources' utilization, fish species can be grouped into generalists that utilise diverse sets of the available resources, and specialists, adapted to very specific resources within a narrow range of habitats (Fishelson 1980). Following this broad definition, groupers were often considered as generalist, opportunistic predators whose diet reflects the abundance of prey available (Harmelin-Vivien and Bouchon 1976, Parrish 1987, Shpigel and Fishelson 1989). However, even if groupers, as many fish species, adapt their diet to the resources available, they nevertheless select certain size and types of prey like crabs, stomatopods, lobsters (and other large crustaceans), and certain types of prey fish like pelagic schooling species (St John et al. 2000, Stewart and Jones 2001, Beukers-Stewart and Jones 2004).

A clear understanding of the role of carnivorous and piscivorous fish in the ecosystem functioning requires detailed descriptions of spatial, temporal, and ontogenetic changes in their diet, and thus generally the analysis of stomach contents. Such analyses often imply to catch and sacrifice the individuals, which is not acceptable for many groupers recognized as endangered species (Sadovy de Mitcheson et al. 2013). Non-destructive methods can be employed to study groupers' diet, like the use of regurgitated gut content by gently flushing their stomach with sea water and collecting prey on sieves or small-mesh nets before releasing them (Stewart and Jones 2001, Koenig and Coleman 2009, Artero et al. 2014). In the same manner, stable isotope analyses can be performed on dorsal rays that can be cut from the fish after it is captured and then released unharmed (Koenig and Coleman 2009, Tzadik et al. 2015).

Groupers are commonly considered as voracious predators positioned at a high trophic level in coastal food webs. However, the combined used of stable isotope analysis and the specific composition of their diet, indicate that most epinephelid species may be considered mid-trophic level predators (Carassou et al. 2008, Koenig and Coleman 2009, Wyatt et al. 2012, Gathaz et al. 2013), while only the largest individuals of the piscivorous species may be viewed as apex-predators and placed at the upper trophic levels of marine food webs. However, stable isotope data are still scarce and the knowledge of the ontogenetic and spatial variations of their trophic level is available only for a small number of species, and often not for the largest individuals (Reñones et al. 2002, Nelson et al. 2012, Page et al. 2013, Artero et al. 2014, Tzadik et al. 2015). Information on diet variation along a depth gradient or among habitats is also lacking for most species, as their real selection of prey in different environmental conditions affect prey composition and abundance (Stewart and Jones 2001, Beukers-Stewart and Jones 2004). Finally, while groupers feed partly or largely on fish, a positive relationship between grouper abundance and the abundance of their prey fish is observed in both temperate (Harmelin-Vivien et al. 2007) and tropical ecosystems (Koenig and

Coleman 2009). A high abundance of groupers may be then viewed as an indicator of healthy and well-functioning food webs.

References

Adams, S.M. 1976. Feeding ecology of eelgrass fish communities. Trans. Am. Fish. Soc. 105: 514–519.
Allen, G.R. 2004. Handy Pocket Guide to Tropical Coral Reef Fishes. Periplus Editions/Berkeley Books Pte Ltd, Singapore.
Anderson, C. and A. Hafiz. 1989. Common Reef Fishes of the Maldives. Part 2. Novelty Printers and Publishers, Male, Republic of Maldives.
Aronov, A. and M. Goren. 2008. Ecology of the mottled grouper (*Mycteroperca rubra*) in the Eastern Mediterranean. Electronic Journal of Ichthyology 2: 43–55.
Artero, C. 2014. Biologie et écologie du mérou géant, *Epinephelus itajara*, en Guyane française. PhD Université des Antilles et de la Guyane: 249 pp.
Artero, C., C.C. Koenig, P. Richard, R. Berzins, G. Guillou, C. Bouchon and L. Lampert. 2015. Ontogenetic dietary and habitat shifts in goliath grouper *Epinephelus itajara* from French Guiana. Endang. Species Res. 27: 155–168. https://doi.org/10.3354/esr00661.
Azevedo, J.M.N., J.B. Rodrigues, M. Mendizabal and L.M. Arruda. 1995. Study of a sample of dusky groupers, *Epinephelus marginatus* (Lowe, 1834), caught in a tide pool at Lajes do Pico, Azores. Bol. Mus. Mun. Funchal, Supp. 4: 55–64.
Azzurro, E., J. Aguzzi, F. Maynou, J.J. Chiesa and D. Savini. 2013. Diel rhythms in shallow Mediterranean rocky-reef fishes: a chronobiological approach with the help of trained volunteers. J. Mar. Biol. Assoc. U.K. 93: 461–470.
Badalamenti, F., G. D'Anna and N.V.C. Polunin. 2002. Size-related trophodynamic changes in three target fish species recovering from intensive trawling. Mar. Biol. 141: 561–570.
Baker, R., A. Buckland and M. Sheaves. 2014. Fish gut content analysis: robust measures of diet composition. Fish Fish. 15: 170–177.
Barreiros, J.P. and R.S. Santos. 1998. Notes on the food habits and predatory behaviour of the dusky grouper, *Epinephelus marginatus* (Lowe, 1834) (Pisces: Serranidae) in the Azores. Arquipélago-Life and Marine Sciences 16A: 29–35.
Beukers-Stewart, B.D. and G.P. Jones. 2004. The influence of prey abundance on the feeding ecology of two piscivorous species of coral reef fish. J. Exp. Mar. Biol. Ecol. 299: 155–184.
Blaber, S.J.M., D.A. Milton, N.J.F. Rawlinson, G. Tiroba and P.V. Nichols. 1990. Diets of lagoon fishes of the Solomon Islands: Predators of tuna baitfish and trophic effects of baitfishing on the subsistence fishery. Fish. Res. 8: 263–286.
Bowman, R.E., C.E. Stillwell, W.L. Michaels and M.D. Grosslein. 2000. Food of northwest Atlantic fishes and two common species of squid. NOAA Tech. Memo. NMFS-NE 155: 138 pp.
Bruggemann, J.H., M.W.M. Kuyper and A.M. Breeman. 1994. Comparative analysis of foraging and habitat use by the sympatric Caribbean parrotfish *Scarus vetula* and *Sparisoma viride* (Scaridae). Mar. Ecol. Prog. Ser. 112: 51–66.
Brulé, T. and L.G. Rodriguez Canche. 1993. Food habits of juvenile red groupers, *Epinephelus morio* (Valenciennes, 1928), from Campeche Bank, Yucatan, Mexico. Bull. Mar. Sci. 52: 772–779.
Brulé, T., D.O. Avila, M.S. Crespo and C. Déniel. 1999. Variación en el regimen alimenticio de juveniles de Mero, *Epinephelus morio* (V.), del banco de Campeche, México. Proc. Gulf Caribb. Fish. Inst. 45: 401–412.
Brulé, T., E. Puerto-Novelo, E. Pérez-Díaz and X. Renán-Galindo. 2005. Diet composition of juvenile black grouper (*Mycteroperca bonaci*) from coastal nursery areas of the Yucatan peninsula, Mexico. Bull. Mar. Sci. 77: 441–452.
Brulé, T., A. Mena-Loria, E. Pérez-Diaz and X. Renan. 2011. Diet of juvenile gag *Mycteroperca microlepis* from a non-estuarine seagrass bed habitat in the southern Gulf of Mexico. Bull. Mar. Sci. 87: 31–43.

Bullock, L.H. and G.B. Smith. 1991. Memoirs of the Hourglass cruises. Seabasses (Pisces: Serranidae). Marine Research Laboratory, Florida Department of Natural Resources, St. Petersburg, Florida. 8: 1–243.

Bullock, L.H., M.F. Godcharles and M.E. Mitchell. 1992. Age, growth and reproduction of jewfish *Epinephelus itajara* in the eastern Gulf of Mexico. Fish. Bull. U.S. 90: 243–249.

Bullock, L.H. and M.D. Murphy. 1994. Aspects of the life history of the yellowmouth grouper, *Mycteroperca interstitialis*, in the eastern gulf of Mexico. Bull. Mar. Sci. 55: 30–45.

Cabana, G. and J.B. Rasmussen. 1996. Comparison of aquatic food chains using nitrogen isotopes. Proc. Nation. Acad. Sc. U.S.A. 93: 10844–10847.

Carassou, L., M. Kulbicki, T.J.R. Nicola and N.V.C. Polunin. 2008. Assessment of fish trophic status and relationships by stable isotope data in the coral reef lagoon of New Caledonia, southwest Pacific. Aquat. Living. Resour. 21: 1–12.

Chasar, L.C., J.P. Chanton, C.C. Koenig and F.C. Coleman. 2005. Evaluating the effect of environmental disturbance on the trophic structure of Florida Bay, USA: multiple stable isotope analysis of contemporary and historical specimens. Limnol. Oceanogr. 50: 1059–1072.

Chen, G., H. Zhou, D. Ji and B. Gu. 2012. Stable isotope enrichment in muscle, liver, and whole fish tissues of brown-marbled groupers (*Epinephelus fuscoguttatus*). Ecological Processes 1: 7. doi:10.1186/2192-1709-1-7.

Cocheret de la Morinière, E., B.J.A. Pollux, I. Nagelkerken, M.A. Hemminga, A.H.L. Huiskes and G. van der Velde. 2003. Ontogenetic dietary changes of coral reef fishes in the mangrove-seagrass-reef continuum: stable isotopes and gut-content analysis. Mar. Ecol. Prog. Ser. 246: 279–289.

Condini, M.V., E. Seyboth, J.P. Vieira and A.M. Garcia. 2011. Diet and feeding strategy of the dusky grouper *Mycteroperca marginata* (Actinopterygii: Epinephelidae) in a man-made rocky habitat in southern Brazil. Neotrop. Ichthyol. 9: 161–168.

Condini, M.V., D.J. Hoeinghaus and A.M. Garcia. 2015. Trophic ecology of dusky grouper *Epinephelus marginatus* (Actinopterygii, Epinephelidae) in littoral and neritic habitats of Southern Brazil as elucidated by stomach contents and stable isotope analyses. Hydrobiol. 743: 109–125.

Connell, S.D. 1998. Patterns of pisciviory by resident predatory reef fish at One Tree Reef, Great Barrier Reef. Mar. Freshwat. Res. 49: 25–30.

Craig, M.T. 2007. Preliminary observations on the life history of the white-streaked grouper, *Epinephelus ongus*, from Okinawa Japan. Ichthyol. Res. 54: 81–84.

Craig, M., Y.J.S. de Mitcheson and P.C. Heemstra. 2011. Groupers of the World: A Field and Market Guide. North America: CRC Press/Taylor & Francis Group.

Cresson, P., S. Ruitton, M. Ourgaud and M. Harmelin-Vivien. 2014. Contrasting perception of fish trophic level from stomach content and stable isotope analyses: a Mediterranean artificial reef experience. J. Exp. Mar. Biol. Ecol. 452: 54–62.

Da Silva Monteiro, V.M. 1998. Peixes de Cabo Verde. Ministério do Mar, Gabinete do Secretário de Estado da Cultura. M2-Artes Gráficas, Lda., Lisbon.

Derbal, F. and M.H. Kara. 1996. Alimentation estivale du mérou, *Epinephelus marginatus* (Serranidae), des côtes est algériennes. Cybium 20: 295–301.

Derbal, F. and M.H. Kara. 2007. Régime alimentaire de la badèche *Epinephelus costae* (Steindachner, 1875) (Serranidae) des côtes de l'est algérien. pp. 67–69. *In*: Francour, P. and J. Gratiot (eds.). Second International Symposium on the Mediterranean Groupers. Nice University publ., 154 pp.

Diatta, Y., A. Bouaïn, F.L. Clotilde-Ba and C. Capapé. 2003. Diet of four serranid species from the Senegalese coast (eastern tropical Atlantic). Acta Adriat. 44: 175–182.

Dierking, J., I.D. Williams and W.J. Walsh. 2009. Diet composition and prey selection of the introduced grouper species peacock hind (*Cephalopholis argus*) in Hawaii. Fish. Bull. 107: 464–476.

Dodrill, J., C.S. Manooch, III and A.B. Manooch. 1993. Food and feeding behaviour of adult snowy grouper, *Epinephelus niveatus* (Valenciennes) (Pisces: Serranidae), collected off the central North Carolina coast with ecological notes on major food groups. Brimleyana 19: 101–135.

Eggleston, D.B., J.J. Grover and R.N. Lipcius. 1998. Ontogenetic diet shifts in Nassau grouper: Trophic linkages and predatory impact. Bull. Mar. Sci. 63: 111–126.

Fishelson, L. 1980. Partitioning and sharing of space and food resources by fishes. pp. 415–445. *In*: Bardach, J.E., J.J. Magnuson and J.M. Reinhart (eds.). Fish Behavior and its Use in the Capture and Culture of Fishes Manila, Philippines: International Center for Living Aquatic Resources Management.

Froese, R. and D. Pauly (eds.). 2015. FishBase. World Wide Web Electronic Publication. www.fishbase.org, version (02/2015).

Fry, B. 2006. Stable Isotope Ecology. Springer, New York, USA.

Gathaz, J.R., R. Goitein, M.O. Freitas, H. Bornatowski and R.L. Moura. 2013. Diet of *Cephalopholis fulva* (Perciformes: Serranidae) in the Abrolhos Bank, Northeastern Brazil. Braz. J. Aquat. Sci. Technol. 17: 61–63.

Gerhardinger, L.C., R.C. Marenzi, Á.A. Bertoncini, R.P. Medeiros and M. Hostim-Silva. 2006. Local ecological knowledge on the Goliath Grouper *Epinephelus itajara* (teleostei: serranidae) in southern Brazil. Neotrop. Ichthyol. 4: 441–450.

Gibran, F.Z. 2007. Activity, habitat use, feeding behavior, and diet of four sympatric species of Serranidae (Actinopterygii: Perciformes) in southeastern Brazil. Neotrop. Ichthyol. 5: 387–398.

Gómez-Canchong, P., L.M. Manjarrés, L.O. Duarte and J. Altamar. 2004. Atlas pesquero del area norte del Mar Caribe de Colombia. Universidad del Magadalena, Santa Marta.

Göthel, H. 1992. Fauna marina del Mediterráneo. Ediciones Omega, S.A., Barcelona.

Gracia López, V. and F. Castelló i Orvay. 2005. Food habits of groupers *Epinephelus marginatus* (Lowe, 1834) and *Epinephelus costae* (Steindachner, 1878) in the Mediterranean Coast of Spain. Hidrobiológica 15: 27–34.

Grove, J.S. and R.J. Lavenberg. 1997. The Fishes of the Galápagos Islands. Stanford University Press, Stanford.

Grover, J.J., D.B. Eggleston and J.M. Shenker. 1998. Transition from pelagic to demersal phase in early-juvenile Nassau Grouper *Ephinephelus striatus*: pigmentation, squamation and ontogeny of diet. Bull. Mar. Sci. 62: 97–113.

Hackradt, C.W., J.A. García-Charton, A. Pérez-Ruzafa, M. Harmelin-Vivien, L. Le Diréach, J. Bayle-Sempere, E. Charbonnel, D. Ody, O. Reñones, P. Sánchez-Jerez and C. Valle. 2014. Response of rocky reef top predators (Serranidae: Epinephelinae) in and around marine protected areas in the Western Mediterranean Sea. PloS ONE 9(6): e98206. doi:10.1371/journal.pone.0098206.

Harmelin, J.G. and M. Harmelin-Vivien. 1999. A review on habitat, diet and growth of the Dusky Grouper, *Epinephelus marginatus* (Lowe, 1834). Mar. Life 9: 11–20.

Harmelin-Vivien, M.L. and C. Bouchon. 1976. Feeding behaviour of some carnivorous fishes (Serranidae and Scorpaenidae) from Tulear (Madagascar). Mar. Biol. 37: 329–340.

Harmelin-Vivien, M.L. 2002. Energetics and fish diversity on coral reefs. pp. 265–274. *In*: Sale, P.F. (ed.). Coral Reef Fishes. Dynamics and Diversity in a Complex Ecosystem. Academic Press, San Diego.

Harmelin-Vivien, M., J.A. Garcia-Charton, J. Bayle-Sempere, E. Charbonnel, L. Le Direach, A. Pérez-Ruzafa, O. Reñones, P. Sanchez-Jerez and C. Valle. 2007. Importance of marine reserves for the population dynamics of groupers (Epinephelinae) in the western Mediterranean. pp. 91–93. *In*: Francour, P. and J. Gratiot (eds.). 2nd Intern. Symp. on Mediterranean Groupers, May 2007, Nice.

Heemstra, P.C. and J.E. Randall. 1993. FAO Species Catalogue. Vol. 16. Groupers of the world (family Serranidae, subfamily Epinephelinae). An annotated and illustrated catalogue of the grouper, rockcod, hind, coral grouper and lyretail species known to date. Rome: FAO. FAO Fisheries Synopsis 125(16): 1–382.

Heemstra, P.C. 1995. Serranidae. Meros, serranos, guasetas, enjambres, baquetas, indios, loros, gallinas, cabrillas, garropas. pp. 1565–1613. *In*: Fischer, W., F. Krupp, W. Schneider, C.

Sommer, K.E. Carpenter and V. Niem (eds.). Guia FAO para Identification de Especies para lo Fines de la Pesca. Pacifico Centro-Oriental. 3 Vols. FAO, Rome.

Hiatt, R.W. and D.W. Strasburg. 1960. Ecological relationships of the fish fauna on coral reefs of the Marshall Islands. Ecol. Monogr. 30: 65–127.

Hobson, E.S. 1974. Feeding relationships of teleostean fishes on coral reefs in Kona, Hawaii. Fish. Bull. 72: 915–1031.

Honebrink, R. 1990. Fishing in Hawaii: A Student Manual. Education Program, Division of Aquatic Resources, Honolulu, Hawaii. 79 pp.

Hyslop, E.J. 1980. Stomach contents analysis: a review of methods and their application. J. Fish Biol. 17: 411–429.

Jeyaseelan, M.J.P. 1998. Manual of fish eggs and larvae from Asian mangrove waters. United Nations Educational, Scientific and Cultural Organization. Paris.

Kailola, P.J., M.J. Williams, P.C. Stewart, R.E. Reichelt, A. McNee and C. Grieve. 1993. Australian fisheries resources. Bureau of Resource Sciences, Canberra, Australia.

Kingsford, M.J. 1992. Spatial and temporal variation in predation of reef fish by coral trout (*Plectropomus leopardus*, Serranidae). Coral Reefs 11: 193–198.

Koeck, B., J. Pastor, G. Saragoni, N. Dalias, J. Payrot and P. Lenfant. 2014. Diel and seasonal movement pattern of the dusky grouper *Epinephelus marginatus* inside a marine reserve. Mar. Environ. Res. 94: 38–47.

Koenig, C.C. and F.C. Coleman. 2009. Population density, demographics, and predation effects of adult goliath grouper (*Epinephelus itajara*). Final Report to NOAA MARFIN for Project NA05NMF4540045.

Kulbicki, M., Y.M. Bozec, P. Labrosse, Y. Letourneur, G. Mou-Tham and L. Wantiez. 2005. Diet composition of carnivorous fishes from coral reef lagoons of New Caledonia. Aquat. Living Resour. 18: 231–250.

Kyrtatos, N.A. 1982. Investigation on fishing and biology of the most important fishes of the region around the Aegean Sea. Island of Tinos. Thalassographica 5(specl. publ.): 1–88.

Letourneur, Y., T. Lison de Loma, P. Richard, M.L. Harmelin-Vivien, P. Cresson, D. Banaru, M.F. Fontaine, T. Gref and S. Planes. 2013. Identifying carbon sources and trophic position of coral reef fishes using diet and stable isotope ($\delta^{15}N$ and $\delta^{13}C$) analyses in two contrasted bays in Moorea, French Polynesia. Coral Reefs 32: 1091–1102.

Liao, H., C.L. Pierce and J.G. Larscheid. 2001. Empirical assessment of indices of prey importance in the diets of predacious fish. Trans. Amer. Fish. Soc. 130: 583–591.

Lieske, E. and R. Myers. 1994. Collins Pocket Guide. Coral Reef Fishes. Indo-Pacific & Caribbean including the Red Sea. Harper Collins Publishers, 400 p.

Linde, M., A.M. Grau, F. Riera and E. Massutí-Pascual. 2004. Analysis of trophic ontogeny in *Epinephelus marginatus* (Serranidae). Cybium 28: 27–35.

Liu, F.H. 1971. Studies on the biology of *Epinephelus awoara* (Temminck and Schlegel) with discovery of the metapterygoid for age-determination. New Asia Coll. Acad. Ann. 13: 49–65.

Liu, M. and Y. Sadovy. 2005. Habitat association and social structure of the chocolate hind, *Cephalopholis boenak* (Pisces: Serranidae: Epinephelinae), at Ping Chau Island, northeastern Hong Kong waters. Environ. Biol. Fish. 74: 9–18.

Lobato, F.L., D.R. Barneche, A.C. Siqueira, A.M.R. Liedke, A. Lindner, M.R. Pie, D.R. Bellwood and S.R. Floeter. 2014. Diet and diversification in the evolution of coral reef fishes. PLoS ONE 9(7): e102094. Doi:10.1371/journal.pone.0102094.

Machado, L.F., F.A.M.L. Daros, Á. Andrade Bertoncini, M. Hostim-Silva and J.P. Barreiros. 2008. Feeding strategy and trophic ontogeny in *Epinephelus marginatus* (Serranidae) from Southern Brazil. Cybium 32: 33–41.

Masuda, H. and G.R. Allen. 1993. Meeresfische der Welt - Groß-Indopazifische Region. Tetra Verlag, Herrenteich, Melle. 528 p.

Matheson III, R.H., G.R. Huntsman and C.S. Manooch, III. 1986. Age, growth, mortality, food and reproduction of the scamp, *Mycteroperca phenax*, collected off North Carolina and South Carolina. Bull. Mar. Sci. 38: 300–312.

Menzel, D.W. 1960. Utilization of food by a Bermuda reef fish, *Epinephelus guttatus*. J. Cons. Int. Exp. Mer. 24: 308–313.

Meyer, A.L. 2008. An ecological comparison of *Cephalopholis argus* between native and introduced populations. University of Hawaii, PhD, 137 pp.

Meyer, A.L. and J. Dierking. 2011. Elevated size and body condition and altered feeding ecology of the grouper *Cephalopholis argus* in non-native habitats. Mar. Ecol. Progr. Ser. 439: 203–212.

Middelburg, J.J. 2014. Stable isotopes dissect aquatic food webs from the top to the bottom. Biogeosci. 11: 2357–2371.

Morgans, J.F.C. 1982. Serranid fishes of Tanzania and Kenya. Ichthyol. Bull. J.L.B. Smith Inst. Ichthyol. 46: 1–44.

Mullaney, M.D., Jr. 1994. Ontogenetic shifts in the diet of Gag, *Mycteroperca microlepis* (Goode and Bean), (Pisces: Serranidae). Proc. Gulf Caribb. Fish. Inst. 43: 432–445.

Mullaney, M.D., Jr. and L.D. Gale. 1996. Ecomorphological relationship in ontogeny: Anatomy and diet in Gag, *Mycteroperca microlepis*. Copeia 1996: 167–180.

Mumby, P.J., C.P. Dahlgren, A.R. Harborne, C.V. Kappel, F. Micheli, D.R. Brumbaugh, K.E. Holmes, J.M. Mendes, K. Broad, J.N. Sanchirico, K. Buch, S. Box, R.W. Stoffle and A.B. Gill. 2006. Fishing, trophic cascades, and the process of grazing on coral reefs. Science 311: 98–101.

Myers, R.F. 1999. Micronesian reef fishes: a comprehensive guide to the coral reef fishes of Micronesia, 3rd revised and expanded edition. Coral Graphics, Barrigada, Guam.

Nagelkerken, I. and G. van der Velde. 2004. Relative importance of interlinked mangroves and seagrass beds as feeding habitats for juvenile reef fish on a Caribbean island. Mar. Ecol. Prog. Ser. 274: 153–159.

Nagelkerken, W.P. 1979. Biology of the graysby, *Epinephelus cruentatus*, of the coral reef of Curaçao. Studies Fauna Curaçao 60: 1–118.

Nakai, T., M. Sano and H. Kurokura. 2001. Feeding habits of the darkfin hind *Cephalopholis urodeta* (Serranidae) at Iriomote Island, southern Japan. Fish. Sci. 67: 640–643.

Nelson, J.A. 2011. On the use of stable isotopes to elucidate energy flow pathways from organisms to ecosystems. The Florida state University, Electronic Theses, Treatises and Dissertations. Paper 5067.

Nelson, J.A., R. Wilson, F.C. Coleman, C.C. Koenig, D. De Vries, C. Gardner and J.P. Chanton. 2012. Flux by fin: fish-mediated carbon and nutrient flux in the northeastern Gulf of Mexico. Mar. Biol. 159: 365–372.

Page, H.M., A.J. Brooks, M. Kulbicki, R. Galzin, R.J. Miller, D.C. Reed, R.J. Schmitt, S.J. Holbrook and C. Koenigs. 2013. Stable isotopes reveal trophic relationships and diet consumers in temperate kelp forest and coral reef ecosystems. Oceanography 26: 180–189.

Parrish, J.D. 1987. The trophic biology of snappers and groupers. pp. 405–463. *In*: Polovina, J.J. and S. Ralston (eds.). Tropical Snappers and Groupers: Biology and Fisheries Management. Westview Press, Boulder, Colorado.

Pattengill, C.V., B.X. Semmens and S.R. Gittings. 1997. Reef fish trophic structure at the Flower Gardens and Stetson Bank, NW Gulf of Mexico. Proc. 8th Int. Coral Reef Sym. 1: 1023–1028.

Perga, M.L. and Grey. 2010. Laboratory measures of isotope discrimination factors: Comments on Caut, Angulo and Courchamp (2008, 2009). J. Appl. Ecol. 47: 942–947.

Popple, I.D. and W. Hunte. 2005. Movement patterns of *Cephalopholis cruentata* in a marine reserve in St. Lucia, W.I., obtained from ultrasonic telemetry. J. Fish Biol. 67(4): 981–992.

Post, D.M. 2002. Using stable isotopes to estimate trophic position: models, methods and assumptions. Ecology 83: 703–710.

Reñones, O., N.V.C. Polunin and R. Goni. 2002. Size related dietary shifts of *Epinephelus marginatus* in a western Mediterranean littoral ecosystem: an isotope and stomach content analysis. J. Fish Biol. 61: 122–137.

Randall, J.E. and V.E. Brock. 1960. Observations on the ecology of epinepheline and lutjanid fishes of the Society Islands, with emphasis on food habits. Trans. Am. Fish. Soc. 89: 9–16.

Randall, J.E. 1965. Food habits of the nassau grouper (*Epinephelus striatus*). Assoc. Is. Mar. Lab. Caribbean, 6th meeting, Estación de Investigaciones Marinas de Margarita, Isla Margarita, Venezuela, Jan. 20–22, 1965. pp. 13–16.

Randall, J.E. 1967. Food habits of reef fishes of West Indies. Studies in Tropical Oceanography, Miami 5: 665–847.

Rodriguez, W.T. 1984. Estudio preliminar para evaluar las caracteristicas biologicas pesqueras de *Mycteroperca olfax* en las islas Galapagos (Ecuador). Bol. Cient. Téc., Inst. Nac. de Pesca, Guayaquil-Ecuador 6(3): 3–66.

Ross, S.W. and M.L. Moser. 1995. Life history of juvenile gag, *Mycteroperca microlepis*, in North Carolina estuaries. Bull. Mar. Sci. 56: 222–237.

Sadovy de Mitcheson, Y., M.T. Craig, A.A. Bertoncini, K.E. Carpenter, W.W.L. Cheung, J.H. Choat, A.S. Cornish, S.T. Fennessy, B.P. Ferreira, P.C. Heemstra, M. Liu, R.F. Myers, D.A. Pollard, K.L. Rhodes, L.A. Rocha, B.C. Russell, M.A. Samoilys and J. Sanciangco. 2013. Fishing groupers towards extinction: a global assessment of threats and extinction risks in a billion dollar fishery. Fish Fish. 14: 119–136.

Sadovy, Y. and A.M. Eklund. 1999. Synopsis of biological data on the Nassau grouper, *Epinephelus striatus* (Bloch, 1792), and the jewfish, *E. itajara* (Lichtenstein, 1822). NOAA Technical Report NMFS 146: 1–65.

Salini, J.P., S.J.M. Blaber and D.T. Brewer. 1994. Diets of trawled predatory fish of the Gulf of Carpentaria, Australia, with particular reference to predation on prawns. Aust. J. Mar. Freshwat. Res. 45: 397–411.

Sano, M., M. Shimizu and Y. Nose. 1984. Food habits of teleostean reef fishes in Okinawa Island, southern Japan. Univ. Tokyo Bull. 25: 1–128.

Sazima, I. 1986. Similarities in feeding behaviour between some marine and freshwater fishes in two tropical communities. J. Fish Biol. 29: 53–65.

Silva Lee, A.S. 1974. Habitos alimentarios de la cherna criolla *Epinephelus striatus* Bloch y algunos datos sobre su biologia. Academia de Ciencias de Cuba, Instituto de Oceanologia, Serie Oceanologica No. 25, La Habana, Cuba. 14 p.

Sluka, R. and K.M. Sullivan. 1996. Daily activity patterns of groupers in the Exuma Cays land and sea park, central Bahamas. Bahamas J. Sc. 3(2): 17–22.

Shpigel, M. and L. Fishelson. 1989. Food habits and prey selection of three species of groupers from the genus *Cephalopholis* (Serranidae: Teleostei). Environ. Biol. Fish. 24: 67–73.

Smale, M.J. 1986. The feeding biology of four predatory reef fishes off the south-eastern Cape coast, South Africa. S. Afr. J. Zool. 21: 111–130.

Stergiou, K.I. and V.S. Karpouzi. 2002. Feeding habits and trophic levels of Mediterranean fish. Rev. Fish. Biol. Fish. 11: 217–254.

Stewart, B.D. and G.P. Jones. 2001. Associations between the abundance of piscivorous fishes and their prey on coral reefs: implications for pre-fish mortality. Mar. Bio 138: 383–397.

St John, J. 1999. Ontogenetic changes in the diet of the coral reef grouper *Plectropomus leopardus* (Serranidae): patterns in taxa, size and habitat of prey. Mar. Ecol. Prog. Ser. 180: 233–246.

St John, J., G.R. Russ, I.W. Brown and L.C. Squire. 2000. The diet of the large coral reef serranid *Plectropomus leopardus* in two fishing zones on the Great Barrier Reef, Australia. Fish. Bull. 99: 180–192.

Tortonese, E. 1986. Serranidae. pp. 780–792. *In*: Whitehead, P.J.P., M.-L. Bauchot, J.-C. Hureau, J. Nielsen and E. Tortonese (eds.). Fishes of the North-eastern Atlantic and the Mediterranean. UNESCO, Paris. vol. 2.

Tzadik, O.E., E.A. Goddard, D.J. Hollander, C.C. Koenig and C.D. Stallings. 2015. Non-lethal approach identifies variability of δ^{15}N values in the fin rays of Atlantic Goliath Grouper, *Epinephelus itajara*. PeerJPrePrints 3: e1184 https://dx.doi.org/10.7287/peerj.preprints.960v1.

van der Elst, R.P. and F. Adkin (eds.). 1991. Marine linefish: priority species and research objectives in southern Africa. Oceanogr. Res. Inst., Spec. Publ. No.1. 132 p.

Vander Zanden, M.J. and J.B. Rasmussen. 1999. Primary consumer d^{13}C and d^{15}N and the trophic position of aquatic consumers. Ecology 80: 1395–1404.

Vander Zanden, M.J. and J.B. Rasmussen. 2001. Variation in $\delta^{15}N$ and $\delta^{13}C$ trophic fractionation: Implications for aquatic food web studies. Limnol. Oceanogr. 46: 2061–2066.
Wada, E., H. Mizutani and M. Minagawa. 1991. The use of stable isotopes for food web analysis. Critical Rev. Food Sci. Nutr. 30: 361–371.
Wainwright, P. and B. Richards. 1995. Predicting prey use from morphology. Env. Biol. Fishes 44: 97–113.
Wainwright, P. and D.R. Bellwood. 2002. Ecomorphology of feeding in coral reef fishes. pp. 33–55. *In*: Sale, P.F. (ed.). Coral Reef Fishes. Dynamics and Diversity in a Complex Ecosystem. Academic Press, San Diego.
Weaver, D.C. 1996. Feeding ecology and ecomorphology of three sea basses (Pisces: Serranidae) in the northeastern Gulf of Mexico. M.S. Thesis, University of Florida, Gainesville, FL.
Wootton, R.J. 1999. Ecology of Teleost Fishes, 2nd ed. Fish and Fisheries, Series 24. Kluwer Academic Press, New York.
Wyatt, A.S.J., A.M. Waite and S. Humphries. 2012. Stable isotope analysis reveals community-level variation in fish trophodynamics across a fringing coral reefs. Coral Reefs 31: 1029–1044.
Zaidi, R., F. Derbal and M.H. Kara. 2013. Régime alimentaire d'*Epinephelus costae* (Pisces, Serranidae) des côtes Est de l'Algérie. Rapp. Comm. Int. Mer Médit. 40: 902.
Zeller, D.C. 1997. Home range and activity patterns of the coral trout *Plectropomus leopardus* (Serranidae). Mar. Ecol. Prog. Ser. 154: 65–77.

Chapter 1.5

Interspecific Relationships

Patrice Francour[1,2] and *Giulia Prato*[1,*]

Introduction

The Species Interactions

The activity of any organism changes the environment in which it lives. It may alter conditions, as when the leaves curtain of a seagrass meadow traps sediment and involves an elevation of sea-bottom, or it may add or subtract resources from the environment that might have been available to other organisms, as coral growth builds habitat that numerous species inhabit. In addition, organisms interact when individuals enter the lives of others (Begon et al. 2006). Several main categories of species interactions are recognized, among which are competition, predation, and mutualism.

Competition is roughly an interaction in which one organism consumes a resource that would have been available to or consumed by another one. The competition can occur between two members of the same species or between individuals of different species. The essence of interspecific competition is that individuals of one species suffer a reduction in fecundity, growth, or survivorship as a result of resource exploitation or interference by individuals of another species. This competition is likely to affect the population dynamics of the competing species, and the dynamics, in their turn, can influence the species' distributions and their evolution (Begon et al. 2006). Predation is an interaction in which one organism eats another, thus killing it, or in which the consumer eats only part of its prey, which may then regrow to provide another bite another day (e.g., grazing). The two previous

[1] CNRS, UMR 7035 ECOSEAS, Université Côte d›Azur, Parc Valrose, 28 Avenue Valrose, 06108 Nice, France.
[2] Groupe d'Etude du Mérou (GEM). France.
Email: francour@unice.fr
* Corresponding author: g.prato@wwf.it

interactions are based upon conflict between species. However, mutualistic interactions also exist, where both organisms experience a net benefit. A mutualistic relationship is one in which organisms of different species interact to their mutual benefit. It usually involves the direct exchange of goods or services (e.g., food, defence, or transport) and typically results in the acquisition of novel capabilities by at least one partner (Begon et al. 2006).

These direct interspecific interactions (competition, predation, and mutualistic interaction) can result in ecologically significant indirect interactions between species. In indirect interactions, one species affects another through a third, intermediary species. Indirect interactions include trophic cascades, apparent competition, and indirect mutualism (Molles 2016).

We will consider in this chapter the interspecific relationships involving groupers. Predation, and more precisely the description of grouper's diet, are presented in another chapter (in this book). We will firstly focus on the ecological consequences on ecosystem functioning of the (direct or indirect) competition due to the groupers. Indeed, according to their importance in density or biomass, groupers and other high trophic level predators (HLP) can modify interspecific relationships throughout competition. In a second part, we will consider the mutualistic relationship between groupers and other species.

Ecological Consequences of Competition

Groupers are High-level Predators

High-level predators (HLP), a category including top or apex predators, are generally large-sized long-living animals such as marine mammals, sharks, and large teleosts that occupy the higher trophic levels in the food web. They are commonly characterized by late sexual maturity and their abundance, at adult stage, is usually not subject to predator control. Together these characteristics result in low resilience to demographic perturbation and high risk of extinction, conditions making them highly vulnerable to fishing (Duffy 2002). According to their biological and ecological characteristics, including longevity, late sexual maturation, sex-change (Castro and Huber 2003), and aggregation spawning (Sadovy de Mitcheson and Colin 2012), the groupers are considered as high-level predators (Sala 2004, Sadovy de Mitcheson et al. 2013).

High-level predators are predators residing at the top of a food chain and are trophically interrelated to all the species of a subweb. Paine (1966) suggested that 'the local animal diversity is related to the number of predators in the system and their efficiency in preventing single species from monopolizing some important, limiting, requisite'. In other words, he predicted that some predators may increase species diversity by reducing the probability of competitive exclusion (Molles 2016). In this context, a clear understanding of the HLP ecological role is mandatory to encompass ecosystem structure and functioning (Molles 2016).

The strength of carnivore effects generally depends on the strength of the link between the predator and its prey and often relates to the predator's body size (Paine 1980). In a system of strongly interacting links, large top predators frequently initiate the top-down control, leading to indirect effects on food webs (i.e., trophic cascades) (Ray et al. 2005). The amplitude of such phenomenon was assessed in several benthic marine ecosystems (Pinnegar et al. 2000), showing that trophic cascades range from Mediterranean rocky sublittoral, kelp forests, and rocky subtidal to coral reefs, rocky intertidal, and soft bottoms. Evidence of oceanic top-down control from large high trophic level piscivores was also found (Baum and Worm 2009).

The observation of the few examples left in the world of pristine ecosystems has provided fundamental information on the shape of an ecosystem in the presence of top predators. Recent studies revealed the structure of two pristine ecosystems, the Palmyra and Kingman atolls in the Line Islands (central Pacific) and the North Western Hawaiian Islands (Friedlander and DeMartini 2002, DeMartini et al. 2008). At both locations, large high-level predators (specifically large piscivorous snappers, groupers, carangids, and sharks) account for 55% to 85% of total fish biomass, with sharks accounting for 57% and 74% of total piscivore biomass in the Line Islands. Despite enhanced predation, high biomass of herbivores is also supported by the coral reefs, together with higher coral cover when compared to nearby fished islands of the same archipelago. The Palmyra and Kingman atolls and the North Western Hawaiian Islands ecosystems have been described as characterized by an inverted trophic pyramid, with most fish biomass at top levels, a structure that, due to historical overfishing of our oceans, had never been observed before by ecologists. Even if the existence of inverted pyramids has recently been questioned due to size-based constraints (Trebilco et al. 2013), it is undeniable that these pristine ecosystems set new baselines for evaluating present and historical human impacts, and provide new targets for MPA conservation efforts (Prato et al. 2013).

Similarly, Williams et al. (2011) compiled data on fish assemblages at 39 US flag coral reef-areas distributed across the Pacific. They showed that total reef fish biomass varied by more than an order of magnitude with the lowest values at densely-populated islands and highest biomass on reefs distant from human populations. Remote reefs (< 50 people within 100 km) averaged ca. 4 times the total fish biomass and 15 times the piscivore biomass compared to reefs near populated areas. Biomass of sharks and groupers was also considerably higher at remote reef-areas, being more than four times that observed at populated islands. The authors concluded that these results highlight the importance of the extremely remote reefs now contained within the system of Pacific Marine National Monuments as ecological reference areas.

As HLP, the large-bodied groupers can exert a control on invasive species due to the biotic resistance hypothesis, the process by which new colonists are excluded from a community by predation from and/or competition with resident species. However, an effective control would require that grouper predation exert a significant net impact on the density of their prey, which might not be the case if predation rates are low and/or invasive species recruitment rates high. A 20-year ban on fishing in the Exuma Cays Land and Sea Park has allowed predatory groupers (*Epinephelus striatus* and others) to attain some of the highest biomasses reported anywhere in the Caribbean (Mumby et al. 2011). In this area, the authors tested if groupers can prevent or limit lionfish (*Pterois volitans*/*P. miles*) invasion. They reported that the biomass of lionfish was significantly negatively correlated with the biomass of groupers (predator biomass explains 56% of the variance of prey biomass), while other smaller predatory fishes such as *Cephalopholis* spp., lutjanids, carangids, and aulostomids did not show significant correlation. On the other hand, in a broader spatial scale study including Belize, Cuba, and Bahamas, Hackerott et al. (2013) showed that groupers had no effect on lionfish abundance. In Little Cayman, Bejarano et al. (2015) suggested that in relatively complex environments, biocontrol of lionfish by large groupers may be limited and only evident as a narrowing of the range of variability of lionfish density.

The Consequences of High-Level Predator Depletion

The depletions of substantial marine mammals such as, sharks and large piscivorous fish, led to mesopredator and invertebrate predator increases, and in some cases to trophic cascades negatively impacting commercial species. When fishery data or ecologists' observations are available from a time when top predators were still present, the far-reaching impacts of high-level marine carnivore depletion on the ecosystem appear clear. One of the most well studied examples of such phenomena comes from the Aleutian Islands, where variations in sea otter abundances due to overfishing and subsequent protection have been responsible for dramatic variations of sea urchin population density. These changes have determined the alternation between the natural kelp forest systems and the impoverished condition of overgrazed rocky reefs. Moreover, diet switching of killer whales in this area and subsequent increased predation on sea otters has protected sea urchins from otter predation, ultimately causing the destruction of kelp forests (Estes and Palmisano 1974, Estes et al. 1998).

Groupers, as HLP, are a valuable fishery resource of reef ecosystems worldwide and are among those species most vulnerable to fishing pressure (Craig et al. 2011, Sadovy de Mitcheson et al. 2013). In a number of regions worldwide, the almost complete extirpation of HLP, and groupers, from marine ecosystems is a direct consequence of fishing that has disproportionately targeted them for centuries (Myers and Worm 2005). Today we face a situation where almost no pristine marine ecosystems are left, and where historical information on pre-exploitation abundance of HLP is very rare. In many places, high-level predators have been absent or rare for so long that scientists and managers have never realized how important they were in the ecosystem (Prato et al. 2013). These changes could have huge effects on interspecific relationships and ecosystem functioning, as groupers are important predators of marine ecosystems.

A long history of fishing down the food web has left Caribbean coral reefs with low species diversity and few functional players at each trophic level (low functional redundancy). Predators such as sharks, large groupers, and snappers have been extirpated from many reefs, and many herbivorous fish have been removed by selective fishing. In a six to seven year period (2002–2008/09), Mumby et al. (2012) showed that fishing has directly reduced the biomasses of large groupers, snapper and parrotfish in Belize. They also provide insights into the indirect effects of fishing. The first insight highlights the time scale needed for prey release from predators. The decline in large-bodied grouper was accompanied by at least an 8-fold increase in mesopredators over a six to seven year period, i.e., a rapid numeric response of mesopredators, requiring only a few years. This issue is consistent with mesopredator release hypothesis following removal of apex predators in a range of ecosystems (Ritchie and Johnson 2009). The mechanisms behind such shifts in mesopredators might be complex. Direct predation on small serranids by large groupers has been observed and inferred from inventories of stomach contents (Mumby et al. 2012).

In the Mediterranean, the megafauna has been virtually eliminated by overfishing (Sala 2004). This impoverished state is supposedly very far from the pristine conditions, and probably since long before the onset of industrial fishing. Archaeozoological reconstructions based on the study of fish bone remains (i.e., osteometry) allowed investigation of the history of fishing in times preceding the advent of writing. The data has revealed how Mediterranean shallow waters were once dominated by large sized piscivores, which attained much bigger sizes than nowadays (Desse and Desse-Berset 1993). Desse and Desse-Berset proposed that these observations allow us to chronologically set the beginning of overexploitation (Desse and Desse-Berset 1993). The analysis of a large amount of fish bones recovered from a Neolithic coastal site of Cap Andreas Kastros, Cyprus, revealed evidences of early exploitation of pristine populations. Here, selective fishing conducted from the coast was directed to large specimens of tunas and groupers (Desse and Desse-Berset 1994). Similarly, other Neolithic coastal Mediterranean sites revealed large specimens, attaining sizes that are not comparable with the mediocre dimensions of fish captured by fishermen today (Desse and Desse-Berset 1993). Until the end of the Mesolithic and during Neolithic eras, the groupers were very abundant in the coastal systems between the 35th and 40th parallel, accounting for 30–80% of the examined bony remains. Sites in Spain, Tunisia, Corse, Cyprus, Sicily, and other Italian sites revealed the presence of healthy populations of *Epinephelus* spp., with all size ranges represented (Desse and Desse-Berset 1999).

Anecdotal research has also led to very interesting discoveries on this topic. A survey of ancient Greek, Etruscan, and Roman mosaics and paintings depicted large groupers often reaching the size of a man, being caught at the water surface by fishermen using poles or harpoons from boats, a technique that would yield no grouper catch today (Guidetti and Micheli 2011). An especially striking implication of these studies is that not only in ancient times were much larger individuals commonly fished, but that their abundance in coastal waters was high, allowing humans to fish them directly from land or from little boats (Guidetti and Micheli 2011). Current groupers' bathymetric distribution shows well how populations actively respond to human exploitations. Largest individuals of this species indeed find refuge at depths that exceed the diving limit of most of the recreational spearfishermen (La Mesa and Vacchi 1999, Di Franco et al. 2009, Hackradt et al. 2014).

A recent study coupling historical reconstruction and modelling delivered a detailed account of successive waves of fish depletions in the Adriatic sea, and shows well the trajectory of degradation undergone by the Mediterranean ecosystem (Lotze et al. 2011). Lotze's results show that large predators and consumers > 1 m in length were reduced to 11% of former abundance, which is a far more drastic reduction than smaller macrofauna (47%), especially in the last century. Among depletion of top predators, the dusky grouper (*Epinephelus marginatus*) may be locally extinct (Lotze et al. 2011). As a result, a process of trophic downgrading was observed in the Adriatic (Lotze et al. 2011), with diversity shifting towards smaller, lower trophic level species. Increased exploitation and functional extinctions have altered and largely simplified food webs by changing the proportions of top predators, intermediate consumers, and basal species.

A concluding remark could be that wherever high-level predators (groupers and others) have been extirpated, ecosystems have consequently become degraded and simplified. In this context, a clear understanding of their ecological role is limited by the fact that our observations are restricted to already altered ecosystems, affected by the decline and, in some cases, disappearance of top predators (Coll and Libralato 2012). Historical data from coastal ecosystems are more abundant and suggest that losses of large predatory fish and mammals were especially pronounced and led to marked changes in coastal ecosystems structure and function (Sala 2004, Myers and Worm 2005). In fact, the fauna of predators we have today in many coastal ecosystems is a 'ghost' (Dayton et al. 1998) of what it was before human impacts. Such ecosystems nowadays are often controlled by medium-sized predators (i.e., mesopredators), although larger carnivores originally preying upon them likely controlled the trophic web in the past. It is then plausible to assume that a return of high-level carnivores to a system will allow degraded systems to recover (Ray et al. 2005). A tempting question is whether conservation of these predators could restore biodiversity and ecological functioning (Ray et al. 2005), and consequently modify the interspecific relationships. The science of marine reserves can give insights on this potential.

The Recovery of High-Level Predators Population in Marine Protected Areas

General Consideration

Several studies have demonstrated that marine reserves are an effective tool for the recovery of large piscivorous fish and upper trophic levels (direct effects), but have also shown a large variability of effects in terms of triggered trophic cascades (indirect effects). In some temperate ecosystems it was possible to demonstrate that recovery of high-level predators (sea otters, groupers, snappers) can lead to the re-establishment of lost trophic interactions (e.g., sea urchins and macroalgae) (Estes 1990, Shears and Babcock 2002). In more diverse ecosystems like coral reefs, a more variable response is observed, depending on conditions such as duration of protection and possible compensation effects due to functional redundancy (e.g., Russ and Alcala 1996, Shears and Babcock 2003, McClanahan et al. 2007).

A global meta-analysis based on data from 124 reserves demonstrated that protection yielded significant average increases of density, biomass, average organism size, and species' richness of the communities within reserves (Lester et al. 2009). Differential responses were observed among taxonomic groups, with large fish and invertebrates targeted by fishing showing significant increases in density and biomass overall, while algal cover increased in temperate reserves and decreased in tropical reefs, due to the recovery of exploited large herbivores. Other meta-analytical studies revealed that commercial species, including many top predatory fish, were observed to increase in density in many southern Europe MPAs (Claudet et al. 2008, 2010).

Thus, reserves reveal initial trajectories towards recovery, but if compared with the few studied pristine ecosystems, it appears clear that the levels of piscivore biomass observed in recent marine reserves across the world are well below what the ecosystems could sustain. Moreover, for indirect changes to occur in marine reserves, an absolute increase in abundance, mean size, or biomass of target species, i.e., a restoration or build up to some (unknown) former level, is necessary (Micheli et al. 2004, Babcock et al. 2010).

Modification of the Interspecific Relationships

As worldwide groupers populations are under some level of threat (Craig et al. 2011) owing to their high economic value to fisheries (Sadovy de Michelson et al. 2013), stable and high-density populations could only be found nowadays inside marine reserves (Craig et al. 2011, Valls et al. 2012, Hackradt et al. 2014), even if they are still generally far away from their carrying capacity in terms of high-level predators.

Anderson et al. (2014) assessed the assemblage of groupers and sea basses (Epinephelidae and Serranidae) inside and outside the Arvoredo Marine Reserve, Santa Catarina State, southern Brazil. Density and biomass of Epinephelidae (HLP) and Serranidae (mesopredators) were recorded.

The most abundant groupers were *Epinephelus marginatus* and *Mycteroperca acutirostris*, while *Serranus flaviventris* and *S. balwini* were the most abundant sea combers. Grouper biomass was significantly higher inside the reserve, indicating the effectiveness of this MPA for target and threatened species, such as *E. marginatus*. In contrast, biomass of sea combers was higher outside the MPA, as a possible result of prey release effect. Historical data of spearfishing indicate that there is still a long way to a full recovery of the biomass of top predators, especially groupers and sharks.

The introduction of the shallow-water grouper *Cephalopholis argus* from Moorea (French Polynesia) to the Main Hawaiian Islands allowed Meyer and Dierking (2011) to test the effect of interspecific competition on groupers. In Moorea, the grouper diversity is high with 14 species, but lower in Hawaii, where only two rare native deep-water groupers occur. In this non-native environment, *C. argus* has flourished and has become the dominant apex predator on many reefs. A comparison with the native populations in Moorea showed that mean total length, mass, growth, and body condition were each significantly elevated in Hawaii. In addition, while small *C. argus* of comparable size in the two locations consumed similar-sized prey, large *C. argus* in Moorea consumed significantly longer and deeper-bodied prey than their counterparts in Hawaii. This pattern was unrelated to the size distributions of available prey, and may thus reflect stronger intra- and interspecific competition for small prey in Moorea.

In Mediterranean marine protected areas, several studies highlighted a recovery of groupers and other HLP underwater using visual censuses (e.g., Harmelin-Vivien et al. 2008, Di Franco et al. 2009, Sala et al. 2012, Garcia-Rubies et al. 2013, Guidetti et al. 2014) or food web modelling with Ecopath with Ecosim (Albouy et al. 2010, Coll and Libralato 2012, Valls et al. 2012). However, relatively few papers deal with ecological consequences of recovery of HLP, but several other ones consider mesopredator (sea breams) recovery (e.g., Sala et al. 1998, Guidetti 2007, Guidetti and Sala 2007). Hackradt et al. 2020 observed a negative correlation in abundance between groupers (*Epinephelus marginatus*, *E. costae*, and *Mycteroperca rubra*) and a sea comber species (*Serranus scriba*), suggesting some degree of interspecific relationship. However, the correlations between grouper species were positive. As the abundance of mesopredators may increase in the absence of top predators, the existence of a controlling mechanism on *S. scriba* assemblage through cascading top-down effects can be suggested in marine protected areas, even if there is no direct predation from groupers upon combers. A study of the coastal fish assemblages in the marine park of Ustica Island, Italy, showed more abundant groupers (*Epinephelus marginatus*) in the no-take area than outside, but a reverse pattern was highlighted for *Serranus cabrilla* (La Mesa and Vacchi 1999). Along the Mediterranean coast of Spain, species diversity and trophic diversity have been quantified in five MPAs and adjacent non-protected areas (Villamor and Becerro 2012). Top predators were significantly more abundant inside than outside MPAs, but the higher abundance of herbivores outside MPAs was not statistically different. A recent large-scale study covering several MPAs and fishing sites across the Mediterranean revealed a trajectory of degradation and recovery, with high-level predator biomass being significantly larger at protected than at non protected sites (Sala et al. 2012). A gradient of 31-fold range increase in fish biomass was observed, reaching a maximum of 115-fold. This is the largest fish biomass gradient ever reported for reef fish assemblages, and is probably indicative of the large impact of historical and current fishing pressure in the Mediterranean (Sala et al. 2012). Continuous increase of high-level predators (particularly groupers) at the Medes islands, where they reached 49% of fish biomass after 27 years of protection, show that the potential for recovery in Mediterranean MPAs is comparable to other parts of the world, and that aiming at achieving fish biomass values similar to those observed in the pristine tropical systems is possible. However, no clear pattern in the structure of benthic community was associated with the gradient on fish biomass, but three alternative community states were revealed in the trajectory of recovery: large fish biomass and reef dominated by non-canopy algae, lower fish biomass but abundant algal canopies and suspension feeders, and low fish biomass and extensive barrens.

Mutualistic Relationships

Interspecific Behavioural Association Between Grouper and Other Species

Interspecific foraging associations have been thought to bring benefits to individuals, as they might increase feeding success by making normally inaccessible food resources available and/or decrease in susceptibility to predation (see references in Gerhardinger et al. 2006). As groupers lay amongst the most inquisitive fishes, this behaviour is well documented on several studies where their opportunistic and learning behaviour turns into feeding tactics, such as following octopuses, moray and snake eels in Red Sea: *Epinephelus fasciatus* with *Gymnothorax griseus* (Karplus 1978); *E. fasciatus* with *Octopus macropus*, *O. cyaneus* or *Siderea grisea*, *Grammistes sexlineatus* or *Variola louti* with *Siderea grisea*, *Cephalopis miniatus*, *C. argus* or *C. hemistiktos* with *Octopus cyaneus* (Diamant and Shpigel 1985). Similarly, Gerhardinger et al. (2006) video-recorded an association between a juvenile of *Epinephelus marginatus* and an adult of *Myrichthys ocellatus* in a shallow rocky reef shore of Santa Catarina (Brazil). The opportunistic association reportedly allowed the groupers to have access to small cryptic organisms made available when disturbed by the nuclear predators (eels or octopuses). Diamant and Shpigel (1985) believe that this behaviour is learnt, because in a given area some groupers may not visibly react to a nuclear predator passing nearby, while others readily follow it.

Cooperative hunting, i.e., the increase in successful prey capture observed when two or more individuals engage in a hunt, has been demonstrated in a wide variety of species (see references in Bshary et al. 2006). Recent evidence has found that such interspecific cooperative hunting (based on true coordination), that had previously only been reported for a small number of species (most of which are mammals or birds), extends to cooperation between the groupers and others species. Bshary et al. (2006) reported cooperation between reef grouper *Plectropomus pessuliferus* (a diurnal predator) and the giant moray eel, *Gymnothorax javanicus* (a nocturnal hunter that usually rests in crevices during the day), in the coral reefs of the Red Sea. Unsworth and Cullen-Unsworth (2012) reported cooperation between the highfin grouper (*Epinephelus maculatus*) and the reef octopus (*Octopus cyanea*) in the Great Barrier Reef, which are two economically and ecologically important predators on Indo-Pacific coral reefs, previously known as solitary hunters.

Successful hunting and feeding were not always observed, but most of the authors hypothesize that the observed non-random interactions are based upon the mutual benefits of increased hunting success. The simplest interspecific associations involve groupers following some nuclear predators. However, cooperative hunting implying interspecific communicative, coordinated, and cooperative hunting between the grouper and the other species have also been reported (Vail et al. 2013). Vail et al. (2014) showed that *Plectopomus leopardus* determine appropriately when a situation requires a collaborator and quickly learns to choose the more effective collaborator. Bshary et al. (2006) reported that groupers signal to moray eels in order to initiate joint searching and recruit moray eels to prey hiding places. The signalling is dependent on grouper hunger level, and both partners benefit from the association: in order to avoid predatory groupers, reef fish hide in corals. Moray eels and octopuses sneak through crevices in the reef and attempt to corner their prey into holes. Consequently, the best strategy for prey to adopt in order to avoid moray eel or octopus predation is to swim into open water. The hunting strategies of the two predators are therefore complementary, and a coordinated hunt between individuals of the two species confronts prey with a multi-predator attack that is difficult to avoid (Bshary et al. 2006).

Given the geographical and species variations between the different reported associations (simple interspecific foraging association or cooperative hunting), probably many more such interactions exist. However, these interspecific cooperations are frequently reported within highly protected and well managed reef areas, where marine life potentially exists in an environment closer to its 'natural' state (Unsworth and Cullen-Unsworth 2012).

Conclusion

The grouper species that occur worldwide are extremely important ecologically, and many of them are important economically wherever they occur. They include most of the important flagship species and many of the top-level predators in warm-temperate and tropical ecosystems associated with rocky and coral reefs. They appear to be acting as keystone species, and by their presence or behaviour enhance the diversity of the communities within which they live and global ecosystem functioning. Their biological and ecological characteristics make them highly vulnerable to exploitation and habitat loss. Their depletion has been reported in most of the rocky and coral reef world-wide, with clear and striking ecological consequences due to the numerous interspecific relationships between groupers and other species inhabiting the same habitat.

We did not consider in this chapter the relationship with human beings. However, the economic value of groupers as flag species for the Scuba divers and snorkelers, as potential valuable aquaculture species, and as target species for artisanal fishermen, deserved a review in full.

References

Albouy, C., D. Mouillot, J. Rocklin, J.M. Culioli and F. Le Loc'h. 2010. Simulation of the combined effects of artisanal and recreational fisheries on a Mediterranean MPA ecosystem using a trophic model. Marine Ecology Progress Series 412: 207–221.

Anderson, A.B., R.M. Bonaldo, D.R. Barneche, C.W. Hackradt, F.C. Félix-Hackradt, J.A. Garcia-Charton and S.R. Floeter. 2014. Recovery of grouper assemblages indicates effectiveness of a marine protected area in Southern Brazil. Mar. Ecol. Prog. Ser. 514: 207–215.

Babcock, R.C., N.T. Shears, A.C. Alcala, N.S. Barrett, G.J. Edgar, K.D. Lafferty, T.R. McClanahan and G.R. Russ. 2010. Decadal trends in marine reserves reveal differential rates of change in direct and indirect effects. Proceedings of the National Academy of Sciences 107: 18256–18261.

Baum, J.K. and B. Worm. 2009. Cascading top-down effects of changing oceanic predator abundances. Journal of Animal Ecology 78: 699–714.

Begon, M., C.R. Townsend and J.L. Harper. 2006. Ecology. From Individuals to Ecosystems. Fourth edition. Blackwell Publishing Ltd.

Bejarano, S., K. Lohr, S. Hamilton and C. Manfrino. 2015. Relationships of invasive lionfish with topographic complexity, groupers, and native prey fishes in Little Cayman. Mar. Biol. 162: 253–266.

Bshary, R., A. Hohner, K. Ait-el-Djoudi and H. Fricke. 2006. Interspecific communicative and coordinated hunting between groupers and giant moray eels in the Red Sea. Plos Biol. 4(12): e431.

Castro, P. and M.E. Huber. 2003. Marine Biology, Fourth Edition. The McGraw-Hill Companies.

Claudet, J., C.W. Osenberg, L. Benedetti-Cecchi, P. Domenici, J.A. Garcia-Charton, A. Pérez-Ruzafa, F. Badalamenti, J. Bayle-Sempere, A. Brito and F. Bulleri. 2008. Marine reserves: Size and age do matter. Ecology Letters 11: 481–489.

Claudet, J., C.W. Osenberg, P. Domenici, F. Badalamenti, M. Milazzo, J.M. Falcon, I. Bertocci, L. Benedetti-Cecchi, J.A. Garcia-Charton and R. Goñi. 2010. Marine reserves: Fish life history and ecological traits matter. Ecological Applications 20: 830–839.

Coll, M. and S. Libralato. 2012. Contributions of food web modelling to the ecosystem approach to marine resource management in the Mediterranean Sea. Fish and Fisheries 13: 60–88.

Craig, M.T., Y.J. Sadovy de Mitcheson and P.C. Heemstra. 2011. Groupers of the World: A Field and Market Guide. CRC Press, Boca Raton, Florida.

Dayton, P.K., M.J. Tegner, P.B. Edwards and K.L. Riser. 1998. Sliding baselines, ghosts, and reduced expectations in kelp forest communities. Ecological Applications 8: 309–322.

DeMartini, E.E., A.M. Friedlander, S.A. Sandin and E. Sala. 2008. Differences in fish-assemblage structure between fished and unfished atolls in the northern Line Islands, central Pacific. Marine Ecology Progress Series 365: 199–215.

Desse, J. and N. Desse-Berset. 1993. Pêche et surpêche en Mé'diterrané'e: le té'moignage des os. Exploitation des Animaux Sauvages à Travers le Temps 13: 332–333.

Desse, J. and N. Desse-Berset. 1994. Osteometry and fishing strategies at Cape Andreas Kastros (Cyprus, 8th millennium BP). Annalen Koninklijk Museum voor Midden Afrika Zoologische Wetenschappen 274.

Desse, J. and N. Desse-Berset. 1999. Préhistoire du mérou. Mar. Life 9: 19–30.

Di Franco, A., S. Bussotti, A. Navone, P. Panzalis and P. Guidetti. 2009. Evaluating effects of total and partial restrictions to fishing on Mediterranean rocky-reef fish assemblages. Marine Ecology Progress Series 387: 275–285.

Diamant, A. and M. Shpigel. 1985. Interspecific feeding associations of groupers (Teleostei: Serranidae) with octopuses and moray eels in the Gulf of Eilat (Aqaba). Environ. Biol. Fish. 13: 153–159.

Duffy, J.E. 2002. Biodiversity and ecosystem function: The consumer connection. Oikos 99: 201–219.

Estes, J.A. and J.F. Palmisano. 1974. Sea otters: Their role in structuring nearshore communities. Science 185: 1058–1060.

Estes, J.A. 1990. Growth and equilibrium in sea otter populations. The Journal of Animal Ecology 385–401.

Estes, J.A., M.T. Tinker, T.M. Williams and D.F. Doak. 1998. Killer whale predation on sea otters linking oceanic and nearshore ecosystems. Science 282: 473–476.

Friedlander, A.M. and E.E. DeMartini. 2002. Contrasts in density, size, and biomass of reef fishes between the northwestern and the main Hawaiian islands: The effects of fishing down apex predators. Marine Ecology Progress Series 230: 253–264.

Garcia-Rubies, A., B. Hereu and M. Zabala. 2013. Long-Term recovery patterns and limited spillover of large predatory fish in a Mediterranean MPA. PLoS One 8(9): e73922.

Gerhardinger, L.C., M. Hostim-Silva, R. Samagaia and J.P. Barreiros. 2006. A following association between juvenile *Epinephelus marginatus* (Serranidae) and *Myrichthys ocellatus* (Ophichthidae). Cybium 30: 82–84.

Guidetti, P. 2007. Predator diversity and density affect levels of predation upon strongly interactive species in temperate rocky reefs. Oecologia 154: 513–520.

Guidetti, P. and E. Sala. 2007. Community-wide effects of marine reserves in the Mediterranean Sea. Mar. Ecol. Prog. Ser. 335: 43–56.

Guidetti, P. and F. Micheli. 2011. Ancient art serving marine conservation. Frontiers in Ecology and the Environment 9: 374–375.

Guidetti, P., P. Baiata, E. Ballesteros, A. Di Franco, B. Hereu, E. Macpherson, F. Micheli, A. Pais, P. Panzalis, A.A. Rosenberg, M. Zabala and E. Sala. 2014. Large-scale assessment of Mediterranean marine protected areas effects on fish assemblages. PLoS ONE 9(4): e91841.

Hackerott, S., A. Valdivia, S.J. Green, I.M. Côté, C.E. Cox, L. Akins, C.A. Layman, W.F. Precht and J.F. Bruno. 2013. Native predators do not influence invasion success of Pacific Lionfish on Caribbean reefs. PLoS One 8(7): e68259.

Hackradt, C.W., J.A. Garcia-Charton, M. Harmelin-Vivien, A. Perez-Ruzafa, L. Le Direac'h, J. Bayle-Sempere, E. Charbonnel, D. Ody, O. Reñones, P. Sanchez-Jerez and C. Valle. 2014. Response of rocky reef top predators (Serranidae: Epinephelinae) in and Around marine protected areas in the western Mediterranean sea. PLoS One 9(6): e98206.

Hackradt, C.W., F.C. Félix-Hackradt, J. Treviño-Otón, A. Pérez-Ruzafa and J.A. García-Charton. 2020. Density-driven habitat use differences across fishing zones by predator fishes (Serranidae) in south-western Mediterranean rocky reefs. Hydrobiologia 847: 757–770.

Harmelin-Vivien, M., L. Le Direac'h, J. Bayle-Sempere, E. Charbonnel, J.A. Garcia-Charton, D. Ody, A. Pérez-Ruzafa, O. Reñones, P. Sanchez-Jerez and C. Valle. 2008. Gradients of abundance and biomass across reserve boundaries in six Mediterranean marine protected areas: Evidence of fish spillover? Biol. Conserv. 141: 1829–1839.

Karplus, I. 1978. A feeding association between the grouper *Epinephelus fasciatus* and the moray eel *Gymnothorax griseus*. Copeia 1978: 164.

La Mesa, G. and M. Vacchi. 1999. An analysis of the coastal fish assemblage of the Ustica Island marine reserve (Mediterranean Sea). Marine Ecology 20: 147–165.

Lester, S.E., B.S. Halpern, K. Grorud-Colvert, J. Lubchenco, B.I. Ruttenberg, S.D. Gaines, S. Airamé and R.R. Warner. 2009. Biological effects within no-take marine reserves: A global synthesis. Marine Ecology Progress Series 384: 33–46.

Lotze, H.K., M. Coll and J.A. Dunne. 2011. Historical changes in marine resources, food-web structure and ecosystem functioning in the Adriatic Sea, Mediterranean. Ecosystems 14: 198–222.

McClanahan, T.R., N.A.J. Graham, J.M. Calnan and M.A. MacNeil. 2007. Toward pristine biomass: Reef fish recovery in coral reef marine protected areas in Kenya. Ecological Applications 17: 1055–1067.

Meyer, A.L. and J. Dierking. 2011. Elevated size and body condition and altered feeding ecology of the grouper *Cephalopholis argus* in non-native habitats. Mar. Ecol. Prog. Ser. 439: 203–212.

Micheli, F., B.S. Halpern, L.W. Botsford and R.R. Warner. 2004. Trajectories and correlates of community change in no-take marine reserves. Ecological Applications 14: 1709–1723.

Molles, M.C. Jr. 2016. Ecology: Concepts and Applications. Seventh edition. McGraw-Hill Education, New York.

Mumby, P.J., A.R. Harborne and D.R. Brumbaugh. 2011. Grouper as a natural biocontrol of invasive Lionfish. PLoS One 6(6): e21510.

Mumby, P.J., R.S. Steneck, A.J. Edwards, R. Ferrari, R. Coleman, A.R. Harborne and J.P. Gibson. 2012. Fishing down a Caribbean food web relaxes trophic cascades. Mar. Ecol. Prog. Ser. 445: 13–24.

Myers, R.A. and B. Worm. 2005. Extinction, survival or recovery of large predatory fishes. Philosophical Transactions of the Royal Society B: Biological Sciences 360: 13–20.

Paine, R.T. 1966. Food web complexity and species diversity. American Naturalist 100: 65–75.

Paine, R.T. 1980. Food webs: Linkage, interaction strength and community infrastructure. Journal of Animal Ecology 49: 667–685.

Pinnegar, J.K., N.V.C. Polunin, P. Francour, F. Badalamenti, R. Chemello, M.L. Harmelin-Vivien, B. Hereu, M. Milazzo, M. Zabala and G. d'Anna. 2000. Trophic cascades in benthic marine ecosystems: Lessons for fisheries and protected-area management. Environmental Conservation 27: 179–200.

Prato, G., P. Guidetti, F. Bartolini, L. Mangialajo and P. Francour. 2013. The importance of high-level predators in Marine Protected Areas management: consequences of their decline and their potential recovery in the Mediterranean context. Advances in Oceanography and Limnology 4: 176–193.

Ray, J., K.H. Redford, R. Steneck and J. Berger. 2005. Large Carnivores and the Conservation of Biodiversity, Island Press.

Ritchie, E.G. and C.N. Johnson. 2009. Predator interactions, mesopredator release and biodiversity conservation. Ecol. Lett. 12: 982–998.

Russ, G.R. and A.C. Alcala. 1996. Marine reserves: Rates and patterns of recovery and decline of large predatory fish. Ecological Applications 947–961.

Sadovy de Mitcheson, Y. and P.L. Colin. 2012. Reef Fish Spawning Aggregations: Biology, Research and Management. Springer, Dordrecht Heidelberg London New-York.

Sadovy de Mitcheson, Y., M.T. Craig, A.A. Bertoncini, K.E. Carpenter, W.W.L. Cheung, J.H. Choat, A.S. Cornish, S.T. Fennessy, B.P. Ferreira, P.C. Heemstra, M. Liu, R.F. Myers, D.A. Pollard, K.L. Rhodes, L.A. Rocha, B.C. Russell, M.A. Samoilys and J. Sanciangco. 2013. Fishing groupers towards extinction: a global assessment of threats and extinction risks in a billion dollar fishery. Fish and Fisheries 14: 119–136.

Sala, E., C.F. Boudouresque and M. Harmelin-Vivien. 1998. Fishing, trophic cascades, and the structure of algal assemblages: evaluation of an old but untested paradigm. Oikos 82: 425–439.

Sala, E. 2004. The past and present topology and structure of Mediterranean subtidal rocky-shore food webs. Ecosystems 7: 333–340.

Sala, E., E. Ballesteros, P. Dendrinos, A. Di Franco, F. Ferretti, D. Foley, S. Fraschetti, A. Friedlander, J. Garrabou, H. Güçlüsoy, P. Guidetti, B.S. Halpern, B. Hereu, A.A. Karamanlidis, Z. Kizilkaya, E. Macpherson, L. Mangialajo, S. Mariani, F. Micheli, A. Pais, K. Riser, A. Rosenberg, M. Sales, K.A. Selkoe, R. Starr, F. Tomas and M. Zabala. 2012. The structure of Mediterranean rocky reef ecosystems across environmental and human gradients, and conservation implications. PLoS ONE 7(2): e32742.

Shears, N.T. and R.C. Babcock. 2002. Marine reserves demonstrate top-down control of community structure on temperate reefs. Oecologia 132: 131–142.

Shears, N.T. and R.C. Babcock. 2003. Continuing trophic cascade effects after 25 years of no-take marine reserve protection. Marine Ecology Progress Series 246: 1–16.

Trebilco, R., J.K. Baum, A.K. Salomon and N.K. Dulvy. 2013. Ecosystem ecology: Size-based constraints on the pyramids of life. Trends in Ecology & Evolution 28: 423–431.

Unsworth, R.K.F. and L.C. Cullen-Unsworth. 2012. An inter-specific behavioural association between a highfin grouper (*Epinephelus maculatus*) and a reef octopus (*Octopus cyanea*). Marine Biodiversity Records 5(e97): 1–3.

Vail, A.L., A. Manica and R. Bshary. 2013. Referential gestures in fish collaborative hunting. Nature Communications 4: 1765.

Vail, A.L., A. Manica and R. Bshary. 2014. Fish choose appropriately when and with whom to collaborate. Curr. Biol. 24: R791–R793.

Valls, A., D. Gascuel, S. Guénette and P. Francour. 2012. Modeling trophic interactions to assess the potential effects of a marine protected area: case study in the NW Mediterranean Sea. Marine Ecology Progress Series 456: 201–214.

Villamor, A. and M.A. Becerro. 2012. Species, trophic, and functional diversity in marine protected and non-protected areas. J. Sea Res. 73: 109–116.

Williams, I.D., B.L. Richards, S.A. Sandin, J.K. Baum, R.E. Schroeder, M.O. Nadon, B. Zgliczynski, P. Craig, J.L. McIlwain and R.E. Brainard. 2011. Differences in reef fish assemblages between populated and remote reefs spanning multiple archipelagos across the Central and Western Pacific. J. Mar. Biol. 826234: 1–14.

Section 2
Conservation and Management

CHAPTER 2.1

Fisheries Regulation

Groupers' Management and Conservation

D. Rocklin,[1] *I. Rojo,*[1] *M. Muntoni,*[2] *D. Mateos-Molina,*[1,3] *I. Bejarano,*[4] *A. Caló,*[5] *M. Russell,*[6] *J. Garcia,*[7] *F.C. Félix-Hackradt,*[8] *C.W. Hackradt*[8] and *J.A. García-Charton*[1,*]

Introduction

Marine populations have been fished since time immemorial (Roberts 2007). However, the widespread development of fishing technologies in recent decades has led to an unprecedented level of fishing effort in coastal marine ecosystems (McGoodwin 1990, Kennelly and Broadhurst 2002, Ormerod 2003). Overfishing often focuses on high trophic level, large predatory fishes (Pauly et al. 1998, Myers and Worm 2003) including groupers, which are

[1] Departamento de Ecología e Hidrología, Universidad de Murcia, Campus de Espinardo, 30100 Murcia, Spain.
[2] GIP Seine-Aval, Espace des marégraphes – Hangar C, Quai de Boisguilbert – CS 41174, 76176 Rouen Cedex 1, France.
[3] Emirates Nature in Association with World Wide Fund for Nature (Emirates Nature-WWF), The Sustainable City (main entrance), P.O. Box 454891, Dubai, United Arab Emirates.
[4] Department of Biology, Chemistry and Environmental Sciences, American University of Sharjah, PO Box 26666, Sharjah, United Arab Emirates.
[5] Department of Earth and Marine Sciences (DiSTeM), University of Palermo, Via Archirafi 20-22, Palermo, Italy.
[6] Science and Conservation of Fish Aggregations, c/o 212/88 Macquarie Street, Brisbane, Queensland 4005, Australia.
[7] UMS CNRS 3514 STELLA MARE, Università di Corsica Pasquale Paoli, Biguglia, France.
[8] Centre for Environmental Sciences, Institute of Humanities, Arts and Science, Federal University of Southern Bahia, Campus Sosígenes Costa, Porto Seguro, Brazil.
* Corresponding author: jcharton@um.es

highly valued for their organoleptic qualities (Sadovy de Mitcheson et al. 2012, Sujatha Kandula et al. 2015). They are found to be a major target for artisanal, recreational, and commercial fisheries worldwide, for both local consumption and export (Craig et al. 2011, Giglio et al. 2014, Schiller et al. 2015), and are among the most desired fish in the live reef food fish trade (Sadovy et al. 2003, Frisch et al. 2016a). Moreover, these territorial high trophic-level predators can be easily approached by spearfishers or caught in nets, traps, or lines (Heemstra and Randall 1993, Morris et al. 2000, Porch et al. 2006). Their life-history traits, characterized by slow growth, a long life span that can include sexual transition (often protogynous hermaphrodites), late sexual maturity, and spawning aggregation behaviour (Shapiro 1987, Domeier and Colin 1997, Arreguin-Sánchez et al. 1996), make them particularly sensitive to overfishing. Groupers provide income and food security for many coastal communities and local economies (Rhodes and Tupper 2007, Sadovy de Mitcheson et al. 2020a, IUCN 2020). Globally, while their annual market value has been estimated between US $350 million (Pauly and Zeller 2015) and US $1 billion (Sadovy de Mitcheson et al. 2012, 2020a), their stocks are still lacking effective management.

Therefore, being one of the most targeted fish species in the world, groupers are also among the most endangered (Morris et al. 2000). According to Sadovy de Mitcheson et al. (2020a), about 13% of the 167 grouper species are considered as 'Threatened' comprising the 'Vulnerable', 'Endangered', and 'Critically endangered' subcategories. Besides, the function played by groupers in the ecosystem is crucial (Friedlander and DeMartini 2002, Heithaus et al. 2008, Rizzari et al. 2014). They are apex predators who have been found to fulfil a similar functional role to sharks and snappers (Frisch et al. 2016b): they can regulate species abundance, distribution, and diversity and they provide essential food sources for scavengers and remove the sick and weak individuals from populations of prey species. Groupers can also provide natural direct and indirect biocontrol against invasive species such as the lionfish (Mumby et al. 2011, Ellis and Faletti 2017), as well as acting as ecosystem engineers by shaping their environment through their burrowing behaviour (Coleman and Williams 2002, Coleman et al. 2010).

For these reasons, protecting and restoring depleted grouper stocks to a healthy status is critical. However, evaluating their effective vulnerability and finding the management strategies that could effectively protect the viability and sustainability of their populations is challenging. To be successful, management strategies must be based on highly reliable knowledge about the status of particular stocks (Hilborn et al. 2020). However, such a task is particularly difficult for groupers, since catch monitoring and fisheries statistics data are often lacking, and the degree of uncertainty around stock estimations is usually high. Most grouper fisheries are considered data-poor, and about fifty grouper species are classified as 'Data Deficient', which precludes accurate evaluation (Sadovy de Mitcheson et al. 2012, 2020a, Luiz et al. 2016). According to IUCN criteria (Polidoro et al. 2008, IUCN 2020), the available biological and/or ecological information on these species is insufficient for proper analysis, representing a major limitation for consistently assessing grouper stocks status around the world.

In this chapter, we aim to give a global overview of how grouper fisheries are regulated, based on specific case studies. The first section of this chapter describes various methods used to estimate grouper stocks status by reviewing traditional stock assessment models, exploring alternative solutions specifically developed for data-poor cases, and considering the increased application of ecosystem modelling as a decision-support tool for fisheries management. The second section of this chapter focuses on the diverse management methods employed worldwide for protecting, maintaining, and restoring grouper populations to a healthy status. We thus describe conventional management measures, mainly focusing on fishing effort limitation, and the use of Marine Protected Areas (MPAs) as tools for fisheries management. We also address the particular case of the protection of grouper spawning aggregations. We intend to bring to light the extensive efforts taking place worldwide to protect grouper populations and support the development of sustainable local fisheries. On the other side, we intend to point out the difficulties and lack of information inherent to grouper fisheries studies and to provide useful insights for future research and management applications.

Estimating Stocks Health as a Basis for Implementing Appropriate Management

Before implementing management decisions for regulation, it is essential to estimate the state of the considered population through stock assessments. Stock assessments integrate several typologies of information often obtained during fisheries monitoring and population surveys, aiming to estimate the stock size and harvest rates (Hilborn and Walters 1992). By this approach, it is possible to identify reference points for sustainable fisheries and to forecast the response of the resource to alternative management scenarios (Hoggarth et al. 2006, Cadrin and Dickey-Collas 2015). The main objective of fish stock assessment is to predict the temporal dynamics of the stock in terms of future yields and biomass levels as a function of different natural (e.g., climate change) and anthropic (e.g., fishing effort) variables, lastly giving insights about their sustainability (Maunder and Piner 2015). The growing demand for information about stock dynamics to support fishery conservation and management poses important challenges (Pita et al. 2016), and has led to continuous improvements of fish stock assessments methods (Dichmont et al. 2021). Notwithstanding, one of the major limits for stock assessment is related to the availability and the quality of the data (Sparr and Venema 1998). For example, methods used for determining limit reference points and consequently appraising the stock status (Froese et al. 2014), are usually based on total catch or absolute abundance data; however, these so-called fishery dependent-data are not always easily available (Maunder and Piner 2015). In the specific case of groupers, and considering these limitations, several stock assessments methods have been proposed according to the data availability.

Estimating Groupers Stocks Status using Stock Assessment Modelling

The simpler way to treat fishing catch data is to model catch per unit effort (CPUE) and landing records time series by regression-type analyses. For example, Mavruk (2020) combined multiple sources of both fishery-independent (trawl surveys) and dependent (landing statistics) data for the white grouper (*Epinephelus aeneus*) in Turkey, to set the basis for a monitoring program of this species. More elaborated analyses were done by Burgos and Defeo (2004), by performing an exponential surplus production model on CPUE data of the red grouper (*Epinephelus morio*) of the Campeche Bank (Gulf of Mexico), concluding that the Red grouper fishery was overexploited, and that immediate management actions were needed to prevent stock collapse.

For well-known groupers fisheries, stock assessments are generally carried out using Virtual Population Analysis (VPA). This powerful method used for estimating the stock abundances per year-class requires a deep knowledge of age-structured catch data and natural mortality rates (Anderson 1978, Lassen and Medley 2001). Early works using this approach with grouper populations were the studies performed on Red grouper in the Gulf of Mexico by Goodyear and Schirripa (1993) and Contreras et al. (1994). VPA has been also largely used to estimate the population status of the gag grouper *Mycteroperca microlepis* in the Gulf of Mexico (Schirripa and Goodyear 1994, Turner et al. 2001, Nowlis 2006), for which information on landings and discards, size composition, size at age and catch rate, from both recreational and commercial fisheries, is available. Although such models are largely used, the uncertainty related to the lack of information about groupers life history characteristics represents a critical concern. For example, Turner et al. (2001) determined that the gag grouper stock status was challenging to assess because of the complexity in estimating the stock-recruitment relationship. Usseglio et al. (2015) pointed out the difficulties in applying VPA for the Galapagos sailfin grouper *Mycteroperca olfax*, for which the sexual and longevity patterns were not completely resolved. Important constraints for groupers stock assessment are related to the complexity of the biological and ecological traits of these species. Particularly, the groupers' reproductive aspects related to sex changes and the disproportional fishing effort targeting the biggest individuals (often males) notably influence sex ratios (Armsworth 2001, Grüss et al. 2014), which can be hardly accountable in stock assessment models. Thus, developing stock assessment methods that incorporate the reproductive plasticity of these species should reduce the uncertainty, and consequently, improve the reliability of groupers stock assessments (Alonzo et al. 2008).

In this context, progress has been done for developing integrated analyses able to link life-history processes, such as natural mortality and recruitment, to external stressors within a single-species stock assessment framework (Young et al. 2006, Methot Jr. and Wetzel 2013). For example, Abdel Barr et al. (2010) applied a recruit model to assess the greasy grouper *Epinephelus tauvina* stock health in the gulf waters off Qatar also studied by El-Sayed et al. (2007), evidencing that the depletion of adult groupers was related to the overfishing of the recruits. Other several examples of the application of recruit model can be found in Richu et al. (2018) on duskytail grouper *Epinephelus bleekeri* in the Arabian Sea coast of southern India, Agustina et al. (2019) on leopard coral grouper *Plectropomus leopardus* in Indonesia, or Manojkumar et al. (2019) on malabar grouper *Epinephelus malabaricus* in the southeast coast of India.

Accounting for environmental influences on population processes can improve the stock assessment models parameters estimations and enhance their predictions. The "SouthEast Data, Assessment, and Review" group (SEDAR 2015) conducted an assessment for the red grouper *Epinephelus morio* using the Stock Synthesis model (Methot Jr. and Wetzel 2013). This catch-at-age model integrates multiple data sources and enables the inclusion of biological and environmental processes. The results indicated that red grouper stocks were not overfished, neither undergoing overfishing. The study also considered environmental factor effects for estimating the mortality and explaining historic trends in abundance of this species in the Gulf of Mexico, finding that the drastic decrease of their abundance between 2005 and 2006 was due to a severe red tide caused by blooms of the dinoflagellate *Karenia brevis*, underlying the importance to consider environmental events for enhancing the assessment model's plausibility. The recent study by Sagarese et al. (2021) on the same species further showed the importance of considering all sources of natural mortality in stock assessment models to provide realistic fisheries advice.

When catch data or other measures of absolute abundance are unavailable, or when the available age data is too sparse to develop an explicit age-based assessment to estimate cohort strength, other solutions must be considered. Porch et al. (2006) developed a catch-free assessment model for estimating the goliath grouper *Epinephelus itajara* stock status off southern Florida, through estimation of its relative abundance. However, due to the lack of information on biological parameters, the researchers used a Bayesian estimation scheme integrating auxiliary information extracted from similar case studies (named "sister species/systems" in Honey et al. (2010)), as well as expert judgements by interviewing local fishers. The model provided results consistent with the trends estimated and described by the fishermen. The goliath grouper population had less than a 40% chance to recover to levels where the spawning biomass would be above 50% of the virgin level within the next 15 years.

All these results only represent a small fraction of the work carried out on grouper species, and point out the increasing sophistication of traditional stock assessment methodology, which is certainly improving our ability in describing and forecasting the stock status. On the other side, these studies highlight the limits and the need to focus the efforts toward the achievement of more accurate, precise, and realistic assessments, by improving the resolution and implementing ecological, ecosystem, and environmental data. Stock assessment embodies a powerful decision support tool for policymaking, by suggesting acceptable and sustainable harvest strategies within an ecological framework. This is particularly relevant for groupers, considering their ecological and socio-economic importance and their global decline. Therefore, developing reliable grouper stock assessments is essential for implementing proper management strategies.

Estimating Groupers Stock Status in a Data-Poor Context

For many fish stocks, there is not enough data for developing data-rich conventional stock assessment methods such as those previously described (Carruthers et al. 2014). The large amount of necessary information, which is usually available for regularly monitored fisheries, does not exist for data-poor (also called data-less or data-limited) species and fisheries, which is generally the case in tropical areas (Pauly et al. 2002, Ehrhardt and Deleveaux 2007, Silvano and Valbo-Jørgensen 2008). This is particularly true for data-poor grouper fisheries, with little empirical information available (Newman et al. 2015b).

However, assessing the stock health of grouper populations affected by data-poor fisheries is still necessary for implementing effective and sustainable management measures (Honey et al. 2010). In this context, several studies promoting a data-less approach to fisheries management were undertaken to overcome this issue. Various methods used for estimating stock health in a data-limited context, their functioning, and the required input parameters, have been already reviewed and compared, considering the quantity of indispensable data (Honey et al. 2010, Carruthers et al. 2014, Newman et al. 2015a, Chrysafi and Kuparinen 2015, Dowling et al. 2015). These methods are based on models or empirical indicators, but landings or catches trends, species life history parameters, and experts' estimates of depletion are always required (Chrysafi and Kuparinen 2015). Thus, these methodologies can only be applied to moderately known groupers fisheries, where historical catches and some of the species' life-history parameters are acknowledged (Meissa et al. 2013, Ndiaye et al. 2013).

In the case of data-poor fisheries, Johannes (1998) considered that "managing fisheries sub-optimally is preferable to not managing them at all". Consequently, this author proposed alternative solutions by considering all type of available data and using them as proxies for helping in management decisions. For example, it has been shown that collecting the Fishers

Ecological Knowledge (FEK, and its variants, the so-called Local Ecological Knowledge—LEK, Traditional Ecological Knowledge—TEK, or Indigenous Ecological Knowledge—IEK (see Davis and Ruddle 2010)), using interviews is a valuable way to obtain information on barely known stocks, thanks to the fishers' abilities to identify slight changes in the fisheries (Silvano and Valbo-Jørgensen 2008, Carr and Heyman 2012). The "Robin-Hood" approach can be also applied: in this method, data-rich species and fisheries assessments are used to inform the data-poor ones. Thus, information on better-known stocks of the same species or information on other related species may be considered (Smith et al. 2009).

Ehrhardt and Deleveaux (2007) evaluated the Nassau grouper *Epinephelus striatus* stock abundance using scarce and known-to-be largely underestimated information on landings statistics for three fishing seasons. They used a calibrated length-cohort analysis algorithm applicable to tropical species, considering groupers aggregating behaviour for reproduction. The study combined this analysis with hydroacoustic studies completed during the spawning aggregations events. The results showed that the two techniques gave approximately similar estimates of the relatively high abundant Bahamas stock, which did not suffer from overexploitation at the time of the study, but was probably exploited at a level close to the maximum sustainable yield.

In another example, Weilgus et al. (2007) developed a matrix population model to conduct a population viability analysis (PVA) on the leopard grouper *Mycteroperca rosacea* in Loreto Bay National Park (Gulf of California, Mexico). The analysis used density data obtained from underwater visual censuses (UVC) from multiple sites—a type of information usually available for reef fishes worldwide. However, biological data were obtained from other related species (gag grouper *M. microlepis* and scamp grouper *Mycteroperca phenax*). These authors also related the biological data to environmental factors by incorporating the effects of El Niño/La Niña Southern Oscillation on population dynamics. These methods permitted the authors to conclude that the leopard grouper population off the main island was not viable under the considered conditions.

In Belize, where no quantitative data on the goliath grouper *E. itajara* fishery was available, Graham et al. (2009) combined various methods to assess the stock status of this species. Their methodology included a 2-years market survey, aiming to obtain the groupers' population structure information, as well as community meetings and face-to-face interviews with fishers. These interviews were displayed both for obtaining historical information on the grouper population abundances and for designating valuable sites for scientific fisheries' sampling. Whereas this study did not permit to identify spawning aggregation sites, the paucity of adult individuals found during the surveys was proof of an overfishing situation, with a clear need for implementation of strong management policies.

The work of Usseglio et al. (2016) gathered data on population change over 30 years of Galapagos grouper (*Mycteroperca olfax*) within the Galapagos Marine Reserve from different sources (historical photographs, reconstructing data from graphs of previous studies, landing data taken by observers in different ports and for different periods, and sampling onboard in some islands); their study concluded that new fishing regulations should be implemented.

For their part, Silvano et al. (2017) used data issued from fish UVC combined with landing data to assess the effects of fishing pressure on two grouper species (*E. marginatus* and *Mycteroperca acutirostris*) (together with a non-target grunt, *Haemulon aurolineatum*) in 21 islands outside and inside an MPA at the Paraty Bay, Brazilian southeastern coast. This work allowed the authors to propose new measures (including a redesign of the MPA limits) to ensure the conservation of the endangered grouper species.

As an additional example, Rhodes et al. (2020) made a fish market survey in Chuuk (Federated States of Micronesia) to examine the age-based reproductive life history of camouflage grouper (*Epinephelus polyphekadion*) and squaretail coral grouper (*Plectropomus areolatus*), to assess whether current management acts conserve them.

When very little field data is available, ethno-ichthyological studies can complement fishery data information gathered by conventional approaches (Silvano and Valbo-Jørgensen 2008). In 2005, Sáenz-Arroyo et al. (2005) gathered a bibliographic review of the 20th-century grey literature and interviews of three generations of fishers for obtaining better estimates of the early abundances of the Gulf grouper *Mycteroperca jordani* in the Gulf of California. While such a method did not provide actual estimates of the population stock, it helped to shift the baseline to more realistic estimations than the optimistic ones based on the contemporary official statistics. Such an approach was also tested by Bunce et al. (2008) in a small African island fishery, where data from fisheries was rare and unreliable. Interviews conducted with three generations of fishers, permitted to highlight the decline of large predators, including the white-blotched grouper *Epinephelus multinotatus*, and the former presence of the eightbar grouper *Epinephelus octofasciatus*, indicative of the ecosystem deterioration. In Brazil, fishers known as "grouper experts" were interviewed and provided insightful information on *E. itajara* ecological attributes (Gerhardinger et al. 2009). Fishers were asked to represent on a satellite image the groupers' abundance in the sites they know. These data permitted to locate previously unknown aggregation sites, which is a crucial information for the groupers' conservation plan, and therefore to identify priority areas for future research and conservation. Historical photos and newspaper articles analysis can also contribute to clarifying the importance of the resource decline, as demonstrated by McClenachan (2009). The study used "trophy" photos of goliath groupers in South Florida to evaluate the groupers' relative biomass, length frequencies, and the number of individuals caught per fishing trip. A long time series of newspaper articles also highlighted the temporal trend of the large caught fish lengths. The study found a clear decline in the absolute and relative abundance of the *E. itajara* in the early 1960s and during the 1970s, but underlined that before the 1950s, the stock, while considered pristine at this time, already suffered a significant decline. Analogously, Anderson et al. (2014) reported the abrupt decay in grouper abundance around Florianópolis (Santa Catarina State, southern Brazil) since the early 1940s based on the historical photographs taken in the area—documented by Souza (2000), combined with contemporary UVC data.

Silvano and Valbo-Jørgensen (2008), interested in estimating the reliability of ethno-ichthyological surveys such as described before, argued to formulate hypotheses based on LEK studies. The study confronts LEK studies, when possible, to available information in the biological literature; for instance, they cited the case of fishers from Southern Brazil that affirmed that *Epinephelus* spp. and *Mycteroperca* spp. spawn in their home reef instead of migrating to the open ocean, classifying this LEK data as of "low likelihood", because such behaviour is contrary to what is generally observed for groupers. Gaspare et al. (2015) applied this comparative approach to a grouper fishery in Tanzania, by interviewing fishers about the ecology, the biology, and the fishing practices related to numerous groupers species, and LEK results were compared with conventional scientific knowledge displayed locally. Such an approach permits the authors to agree or to disagree with LEK or with other literature results, and in the latter case, highlights the need for more research on the considered topic. Mavruk et al. (2018) elaborated a fisheries' assessment study by interviewing fishers (both professional and recreational) operating in the Gulf of Iskenderun and Mersin (Turkey, Mediterranean Sea) about the fishing gears used, the fishing periods, the depth and bottom type of the fishing area, the best day's catch, and the minimum, maximum, and usual sizes of the catch regarding six native (*Epinephelus marginatus, E. aeneus, E. costae, E. caninus, Hyporthodus haifensis*, and *Mycteroperca rubra*) and seven non-indigenous species (*Cephalopholis taeniops, E. aerollatus, E. coioides, E. fasciatus, E. geoffroyi, E. malabaricus*, and *M. fusca*). The population of goliath grouper (*Epinephelus itajara*) was the subject of a FEK study combined with local catch data in Abrolhos Bank (southeastern Brazil) by Zapelini et al. (2017), in the same area assessed by Giglio et al. (2018) based on interviews to consumers in a local market to understand consumption preferences with groupers and sharks. Also, Begossi et al. (2019) combined literature review and fieldwork (including visits to fish markets, systematic collection of fish landing data over short periods in several fishing ports, fish plankton larvae surveys, and fishers' interviews) to establish the conservation status of *E. marginatus* and *E. morio* along the coast of Brazil. Analogously, Ribeiro et al. (2021) interviewed professional fishers about five grouper species (*Mycteroperca bonaci, E. adcensionis, E. itajara, Cephalopholis fulva*, and *E. morio*) in three fishing communities in the State of Rio Grande do Norte (northern coast of Brazil). These studies made it possible to characterise in very efficient ways fisheries for which very little data had previously been available.

Citizen science is another source of useful data to be used in population studies (Forrester et al. 2015), as exemplified by the study made by Thorson et al. (2014). This study analysed citizen science data issued from the REEF database (Schmitt and Sullivan 1996) reported in the Florida Keys between 1993–2012 to model the populations of goliath grouper (*E. itajara*). This species was also assessed by Koenig et al. (2011) using the same REEF dataset based on volunteer divers, combined with data obtained from surveys performed by the authors, and a third dataset issued from a tagging program, to characterize recovery patterns and informing the management of this important grouper species. As another example, Begossi et al. (2016)—see also Begossi and Salivonchyk (2019)—trained fishers of Copacabana (Rio de Janeiro, Brazil) to collect data on dusky grouper (*E. marginatus*) catches and morphometry (length and weight), monitor grouper gonads and supply information on fishing spots and prices.

As a novel approach to estimate grouper abundance and size using non-conventional methodologies, recently Sbragaglia et al. (2021) applied data mining of videos published on YouTube related to recreational fishing of four grouper species (*Epinephelus marginatus, E. aeneus, E. costae*, and *E. caninus*) in Italy between 2011 and 2017. This innovative method helps to mitigate data deficiencies and inform about harvesting patterns shown by recreational anglers and spearfishers.

Models for Ecosystem Management

In the early 2000s, there has been growing attention among fisheries scientists to consider the marine ecosystem as a whole and to include all the interactions involved in the ecosystem (Glazier 2011), highlighting that fish stocks management cannot disregard the ecosystem where the fish live and in which the fisheries operate. This led to the development of the Ecosystem-Based Fishery Management (EBFM) approach (Pikitch et al. 2004) and the Ecosystem Approach to Fisheries (EAF) (Garcia et al. 2003). Those approaches encompass the fundamental principles to consider holistic management of the marine systems, including both the human and the natural dimensions (Botsford et al. 1997, Murawski 2000, Fanning et al. 2011). The shift from a single-species to an EAF approach has fostered the development of a set of different tools (i.e., models and indicators), providing a framework for integrating ecosystem issues into the management programs (Coll and Libralato 2012). Even if ecosystem modelling approaches present some limitations (Christensen and Walters 2004) and would gain from additional improvements (Grüss et al. 2015), it is now commonly used to explore the ecological consequences of the decline of high trophic level species (Heithaus et al. 2008) or to compare the benefits of diverse management scenarios on the whole ecosystem (Albouy et al. 2010). The EAF approach nowadays thus represents one of the most valuable tools for providing wide spectra of scientific information needed to help managers find optimal management solutions (FAO 2008).

Various ecosystem-based models were developed for supporting EAF, and vary in terms of complexity (Reum et al. 2021). Among these models, ATLANTIS (Fulton and Smith 2004), ECOPATH with ECOSIM (EwE) (Pauly et al. 2000), Multispecies Virtual Population Analysis–Extended (MSVPA-X) (Garrison et al. 2010), and OSMOSE (Shin and Cury 2001, 2004) are considered 'whole' ecosystem models and attempt to account for all the considered ecosystem trophic levels, from the primary producers to the top predators (Plagányi 2007). ATLANTIS is a modelling framework used for policy and management evaluations, and is based on a series of sub-models (biophysical, fisheries, monitoring, assessment, management, and socioeconomic) (Grafton et al. 2010). Although ATLANTIS is thought to be one of the most useful operating models within a simulation testing framework, the huge amount of data needed for its optimal implementation limits its use in most marine systems (Plagányi 2007). The widely used EwE is a useful tool for the development of scenarios for exploring future trends of marine biodiversity and changes in ecosystem services (Coll and Libralato 2012). EwE allows to address several ecological questions, such as marine ecosystem structure and functioning or human impacts evaluation, and to test management strategy options, analyzing different scenarios for resource conservation (Christensen and Walters 2004, Coll and Libralato 2012). OSMOSE (Object-oriented Simulator of Marine ecosystem Exploitation) is spatially structured and individual-based and considers the trophic interactions based on the species sizes. It can be used to predict ecosystem changes or to test management scenarios of fisheries (Marzloff et al. 2009, Travers-Trolet et al. 2014). As a further example, Sagarese et al. (2015) introduced the deleterious influence of red tides on red grouper (*Epinephelus morio*) population mortality in the Gulf of Mexico on modelled projection scenarios, by using the OSMOSE modelling approach, and showed the importance of considering explicitly the likely effects of other sources of fish mortality in addition to natural and fishing mortality estimates.

Since these models aim to include the whole ecosystem components, they consider that a wide range of species and groupers are one of these multiple components, generally included in the 'top predator' category (Chen et al. 2008, Tsehaye and Nagelkerke 2008). For this reason, ecosystem modelling studies specifically focusing on groupers are still scarce. Arreguín-Sánchez et al. (2004) used EwE to model optimized harvesting strategies in a Mexican fishery, mainly targeting snappers and groupers. The study considered both socio-economic and ecological criteria for evaluating the optimal management scenarios. They found that slightly increasing the fishing effort on snappers and groupers would enhance the socio-economic performance of the fisheries while presenting a limited ecological risk for the grouper populations.

Later, Ainsworth et al. (2008) analyzed several EwE ecosystem simulations for supporting ecosystem-based fisheries management in an Indonesian archipelago. This study evaluated the most probable ecological and economic impacts of limiting the commercial exploitation of groupers, mainly through its impact on the artisanal fisheries; they concluded that artisanal fisheries would slightly benefit from the commercial fishing restriction and that groupers biomass would increase in the system, enhancing its diversity.

More recently, Prato et al. (2014) proposed a simplified and standardized EwE model by aggregating functional groups and comparing the corresponding outputs in a French Marine National Park. The study observed that the most influential species on the model were all characterized by high biomass, a high trophic level, and a diversified diet. Among them, the large dusky grouper *Epinephelus marginatus* was found to play a very significant role in the ecosystem and accurately estimating its population biomass is considered crucial for developing effective management strategies. These results have been then used to design optimized monitoring surveys in data-poor marine areas to obtain a coherent picture of the ecosystem functioning through a cost-efficient process (Prato 2016, Prato et al. 2016).

Finally, research was done in the Gulf of Mexico (West Florida Shelf), bringing attention to the groupers populations' management and conservation using ecosystem modelling. The gag grouper *M. microlepis* was one of the most important fished species in the West Florida Shelf and was undergoing overfishing. Consequently, strong management policies were implemented to end the overfishing and to allow the recovery of its population. To help understand the role of this species in the ecosystem, two models were developed. An OSMOSE model helped to enhance the knowledge about the whole trophic structure of this ecosystem and to learn more about the gag grouper diet patterns and natural mortality rates (Grüss et al. 2015), important characteristics to consider when implementing management policies. Simultaneously, an EwE model was developed to evaluate the impact of the expected biomass increase of this species on the ecosystem (Chagaris et al. 2015). The results indicated contrasted impacts: the protection of the gag grouper would lead to biomass declines for some other groupers species, such as the Black grouper *Mycteroperca bonaci*—but without collapse risks, while species targeting benthic-associated prey would see their biomass increasing. Related research focused on the red grouper (Grüss et al. 2016). In this study, an OSMOSE model was developed to test the results of changes in the fishing mortality of the red grouper *E. morio* on the ecosystem. It was found that increasing or decreasing its biomass in the ecosystem through limited or increased fishing effort, would inversely impact the biomass of its major competitors, the gag grouper and the red snapper (*Lutjanus campechanus*). However, the importance of such results for a management point-of-view was moderated by the fact that some predicted model parameters showed discrepancies with the observed ones.

Protecting and Managing the Groupers Populations from Overfishing

All the methodologies described previously have their strengths and weaknesses, but have to be considered as a first step for addressing resource-management issues and are valuable decision-support tools aiming to develop optimal ecosystem-based conservation strategies for groupers. When stocks are found to be in decline or overexploited, it is crucial to implement management measures for protecting them. Many possibilities exist to protect endangered species (Selig et al. 2017), and the choice of the most effective measures highly relies on the context. Then, the type of management held will vary depending on the severity of exploitation, the local social and economic contexts, and the fisheries' characteristics.

Traditional fisheries management measures, such as fishing quotas, gear restrictions, seasonal closures, or minimum/maximum size limits can be used to restrict the fishing pressure on groupers' stocks. MPAs can also be considered to protect not only groupers, but the whole communities and habitats found in the considered areas. Finally, based on the groupers' life history characteristics, protecting them during their spawning aggregations, when they are found to be particularly vulnerable, appears necessary and is being increasingly implemented.

Using Conventional Fisheries Management Measures for Protecting Groupers Stocks

Monitoring fishing effort is a conventional and traditional method for managing marine resources and has been commonly applied to grouper fisheries. In the Great Barrier Reef (Australia), fishing gears and practices were regulated in early 1900s for coral reef fishes in general. As described by Leigh et al. (2014), the first grouper-specific management measure there consisted of the implementation in 1976—i.e., more than 10 years after the beginning of its commercial catches monitoring—of a minimum legal size for the Coral trout *Plectropomus leopardus*, and bag limits in the 1990s for recreational fishing. Maximum legal sizes, aiming at protecting the largest individuals having a greater fecundity, were added in 2003. Additionally, in 2004, an eight-days "spawning-closure" fishing ban around the spawning season was also applied. All this suite of measures allowed the coral trout population to stay in a good status (Leigh et al. 2014).

Similarly, in the Gulf of Mexico, various management measures were implemented since 1990 to help maintain the groupers populations in good status. These measures, including among others minimum size and daily bag limits, quotas, gears type's restrictions, and seasonal closures were regularly adjusted following subsequent stock evaluations (Gulf of Mexico Fishery Management Council 2016a). These measures were applied to both commercial and recreational fishing activities, the latter accounting for a significant part of the total catches. As a result, in 2016, it was considered that the red grouper *E. morio* populations were not overfished and were not experiencing overfishing (Gulf of Mexico Fishery Management Council 2016a), a situation that has remained unchanged since then (SEDAR 2019). Strong management measures were also implemented for restoring the gag grouper *M. microlepis* populations: while they suffered high mortalities in 2005 due to a harmful algal bloom, their populations recovered and were not anymore considered overfished in 2014 (Gulf of Mexico Fishery Management Council 2016b).

Partial or total prohibition of specific fishing gears or practices targeting groupers can be applied to manage their populations. In France, recreational spearfishing of the dusky grouper *E. marginatus*, classified among the endangered grouper species in the IUCN Red List of Threatened Species (Cornish and Harmelin-Vivien 2004), is banned since 1980 in Corsica and since 1993 along the whole French Mediterranean coasts. This moratorium, together with other factors, helped their populations to slightly recover, but it was still considered insufficient in 2013, and the moratorium was extended for 10 more years. Moreover, another four grouper species were added to this moratorium (the goldblotch grouper *Epinephelus costae*, the dogtooth grouper *Epinephelus caninus*, the mottled grouper *Mycteroperca rubra*, and the wreckfish *Polyprion americanus*), and recreational and professional fishing using hooks is prohibited (Préfecture PACA 2013). However, grouper recreational fishing was not banned in Italy and Spain or the north of Africa, all bordering countries of the Western Mediterranean basin, which could counteract their population's recovery. The protection of groupers in these locations is only possible in the existent MPAs, but the estimated low connectivity level between them highlighted the limitations of their network effectiveness as a tool for grouper populations management (Andrello et al. 2013).

In the western Atlantic, a total harvest ban has been applied to the highly fished goliath grouper *E. itajara* (Craig 2011). This ban was implemented in 1990 in the federal waters of southeastern United States (including Gulf of Mexico), since 1993 in the US Caribbean, and since 2002 in Brazil (CassCalay and Schmidt 2009, Craig 2011). In the former area, anglers were pushing for the end of the moratorium, but it has been demonstrated that goliath groupers may have greater economic value when they are alive: recreational divers were willing to pay much more for diving in their spawning aggregations than anglers for fishing them (Shideler et al. 2015, Shideler and Pierce 2016). In Brazil, the moratorium was renewed four times (the last one until 2023) (Giglio et al. 2014, Bueno et al. 2016), but important goliath grouper catches by poachers are still recorded, due to a lack of moratorium awareness and ban enforcement (Giglio et al. 2014, 2017, Bentes et al. 2019), decreasing the effectiveness of such measure to protect them.

Conventional management measures are commonly applied to commercial fishing, but are not always suitable for recreational fishing. Indeed, recreational fisheries are characterized by a large number of participants, the variability of gears, time, and space, and the non-centralized landing sites; they are badly known, rarely monitored, and their management remains a challenge (Rocklin et al. 2014). However, as estimated by Morris et al. (2000), 36% of the fully tropical Epinephelidae species undergo recreational fishing pressure. Thus, specific management measures applicable to recreational activities must be considered. In recreational fishing, catch-and-release practices are common and can represent a large part of the catches (Ferter et al. 2013), but these non-targeted released individuals (under the legal size or over quota) can suffer post-release mortalities (Bartholomew and Bohnsack 2005). Decreasing the injuries and the number of non-targeted catches by using specific hook size and shape has been explored as a possible management measure for groupers targeted by recreational fishing. In the Campeche Bank (Mexico) red grouper fishery, Brulé et al. (2015) tested the impact of using circle hooks, which sizes were below and above the official size required by the federal government. They found that the legal hook size implementation was inefficient to improve the red grouper populations, since the number of undersized caught individuals was similar as when using smaller hooks. In North Carolina (U.S.A.), there are various grouper species, among which the red grouper, the gag grouper, and the scamp grouper *M. phenax*, are commonly fished by both recreational and commercial fishers using hook-and-line (Bacheler and Buckel 2004). To reduce involuntary mortalities, they also tested different types and sizes of hooks and found that although the number of sub-legal individuals did not differ, injuries made to the fish were lower when using circle hooks, contributing to increasing their survival rates.

An innovative solution adopted recently to try to reduce mortality of released groupers is the use of descender devices to recompress barotraumatized fishes (Eberts and Somers 2017, Runde and Buckel 2018). A recent study by Runde et al. (2020) off North Carolina (USA) using acoustic tracking was able to measure improved survival of groupers recompressed using descender devices compared to groupers released from the surface.

Another method increasingly informing the catch-and-release approach is the use of fish harvest tags in recreational fishing, like that proposed in the South Atlantic Region of USA for grouper species presenting low recreational annual catch limits (SAFMC 2013, 2015). Commonly used in hunting, they are found to offer various advantages, such as improving the harvesting season, promoting equitable tags allocation, offering a way to obtain reliable information on catch data, and can also be a source of revenues (Johnston et al. 2007). For example, Shideler et al. (2015) found that anglers in Florida would be willing to pay nearly $80 for a goliath grouper harvest tag. Also, Shertzer et al. (2018) modelled release mortality of Warsaw grouper *Hyporthodus nigritus* in catch-and-release recreational fishing in South Carolina (USA), through a state-space capture-recapture model, using data collected in the context of a game fish tagging program.

Very recently, Bonney et al. (2021) reported a pilot study to collect discard information in the South Atlantic region of the USA, by using a mobile app to provide the length of released scamp groupers and other supplemental data that can be collected during regular fishing (both commercial and recreational).

Marine Protected Areas as a Tool for Protecting Groupers

The previously described conventional management measures of fisheries resources often focus on a target species by developing *ad hoc* conservation actions aiming at preserving its populations. Although single-species management approaches are particularly adapted in the case of highly mobile species (e.g., tunas or swordfish), its limitation in protecting coastal species characterized by a low mobility was pointed out in the last decade. Natural refuges, such as remote areas or deep waters, hardly accessible for fisheries, play a key role in the protection and conservation of many species (Bohnsack 1994, Friedlander and DeMartini 2002, Goetze et al. 2011, D'Agata et al. 2016, Valdivia et al. 2017, McClanahan 2019), including groupers (Bejarano et al. 2014). But these areas are now also threatened by the continuous improvement of fishing technologies (Morato et al. 2006, Valdemarsen 2001). MPAs, by offering protected places profitable to marine species, have been strongly recommended as an alternative to traditional fisheries regulation methods (Pauly et al. 2002, Edgar et al. 2019), and thus they have been developed worldwide for both biodiversity conservation and fisheries management purposes (Boersma and Parrish 1999, Halpern 2003, Lubchenco et al. 2003, FAO 2011, Coll et al. 2012, Lubchenco and Grorud-Colvert 2015, Pérez-Ruzafa et al. 2017). MPAs aim to reduce or eliminate fishing mortality within their boundaries through reductions in fishing effort and restrictions in the use of fishing gears. Consequently, habitat damage is reduced, which will in turn increase habitat carrying capacity, juvenile survival, growth, and recruitment (Coleman et al. 2000, Rodwell et al. 2003, Cheng et al. 2019). Therefore, effective MPAs are expected to increase the density and biomass of protected fish species (Russ 2002, Halpern 2003, Edgar et al. 2014, Giakoumi et al. 2017, Rojo et al. 2019). MPAs are generally recognized to provide greater biomass and sizes of protected species (Halpern 2003, Hackradt et al. 2014), enhancement of larval production (Almany et al. 2013), and to participate in the export of biomass profiting to larger fisheries yields (Russ and Alcala 1996, Hilborn et al. 2004, Harmelin-Vivien et al. 2008, Gaines et al. 2010, Di Lorenzo et al. 2016, 2020). Thus, MPAs can represent an effective fisheries management tool benefiting both grouper populations and habitats critical to support all their life stages, e.g., nursery, migration, and spawning aggregation habitats (Craig et al. 2011, Sadovy de Mitcheson and Colin 2012, Hackradt et al. 2014). Moreover, MPAs can represent an interesting option in places where conventional fisheries management measures can be harder to implement due to few or no fishing effort monitoring, even if poor compliance, limited enforcement, and poaching may counteract their effectiveness (Frisch et al. 2016a).

MPAs as Groupers Sanctuaries

The design of MPAs depends on their objectives. Their zoning can be adjusted to favour specific fishing activities and/or to restrict other ones, and can include different sub-areas with distinct protection levels (Day 2002, Pérez-Ruzafa et al. 2017). The most common design of an MPA integrates: (i) fully protected area(s) (also called "no-take zones" or "marine reserves"), where all extractive activities are forbidden, and/or (ii) partially protected area(s) (or buffer zones if surrounding the no-take areas), with restricted or strongly regulated commercial and/or recreational fisheries (Agardy et al. 2003, Claudet et al. 2008, Lester and Halpern 2008, Di Franco et al. 2009, Rocklin et al. 2011, García-Rubies et al. 2013, Pérez-Ruzafa et al. 2017, Rojo et al. 2019).

Whereas not all species positively respond to protection (Polunin and Roberts 1993, Floeter et al. 2006, Claudet et al. 2010, Anderson et al. 2014, Edgar et al. 2014, Rojo et al. 2019), heavily exploited ones such as groupers are more likely to benefit from the implementation of an MPA. There are many examples of direct benefits of the prohibition of fishing activities on the abundance, biomass, and/or size of groupers inside MPAs worldwide, both in temperate and tropical areas (Nemeth 2005, García-Charton et al. 2008, Sadovy de Mitcheson and Colin 2012, Mateos-Molina et al. 2014, Hackradt et al. 2014, Anderson et al. 2014).

An early study on the benefits of MPAs for groupers showed that in the Ras Mohammed Marine Park (Egypt), groupers were three times heavier in comparison with the ones in non-protected areas (Roberts and Polunin 1993). Later, in the Philippines, the biomass of groupers was estimated to be 6 to 31 times higher in the Sumilon Island Reserve than in fished areas, with an average weight of twice the one of non-protected groupers (Alcala 1998). In the Bahamas, Sluka et al. (1997) found that the biomass and length of the Nassau grouper were higher in the protected areas than outside, with more than 35% of the protected individuals reaching more than 50 cm length. In that same area, Chiappone et al. (2000) found that zones displaying low fishing pressure presented higher density, biomass, and diversity of grouper species than in less regulated areas, being particularly noticeable for large grouper species in no-take zones. In Southern Brazil, Anderson et al. (2014) found that the dusky grouper *E. marginatus* and the comb grouper *Mycteroperca acutirostris* biomass was higher inside the Arvoredo Marine Biological Reserve than in the fished areas. They also observed large individuals of the targeted grouper species (the gag grouper, the black grouper, the red grouper, and the yellowmouth grouper *Mycteroperca interstitialis*) exclusively in the marine reserve. The "Jardines de la Reina" archipelago in Cuba showed higher densities and frequencies of some grouper species inside the marine reserve (particularly for the yellowfin grouper *Mycteroperca venenosa* and the Nassau grouper *E. striatus*) while others, such as the tiger grouper *Mycteroperca tigris* or the black grouper *M. bonaci* did not respond positively to protection; this has been explained by the fact that, as toxic species for human consumption, they were not previously targeted by fishers (Pina-Amargós et al. 2014).

However, all these positive trends in grouper abundance, biomass, and size within MPAs can be slightly variable depending on the species considered (Micheli et al. 2004, Claudet et al. 2010), the previously exerted fishing pressure (Hackradt et al. 2014), the size of the MPA (Claudet et al. 2008, Edgar et al. 2014, Rojo et al. 2019), the time since the MPA was established (Coll et al. 2012, García-Rubies et al. 2013, Edgar et al. 2014, Friedlander et al. 2017, Rojo et al. 2021) and the diversity of habitats included in the protected areas (García-Charton and Pérez-Ruzafa 2001, García-Charton et al. 2004), being greater for large species that live aggregated, and in MPAs strongly enforced and protected for long times (Rojo et al. 2019, 2021).

MPAs Benefits to Adjacent Fisheries Areas

The increase in density and biomass of groupers populations build up within MPAs can benefit both adjacent and remote non-protected areas through two main processes: (i) spillover, the active movement of young and adult individuals from MPAs to areas beyond their limits (Harmelin-Vivien et al. 2008, Di Lorenzo et al. 2016), and (ii) eggs and larvae seeding, produced inside MPAs and exported to outside areas (Christie et al. 2010, Almany et al. 2013). When MPAs are effective for groupers inside their boundaries, increased abundance of larger and older individuals producing a larger number of eggs (Francis et al. 2007) results in greater reproductive outputs (Sluka et al. 1997, Nemeth 2005, Sadovy de Mitcheson and Colin 2012), potentially drifting to surrounding areas through dispersal.

The magnitude of spillover is greatly dependent on the grouper abundance and mobility (Claudet et al. 2008, Abecasis et al. 2014). Some grouper species are highly sedentary and present low mobility ranges, varying from hundreds of meters to few kilometres (e.g., dusky grouper *E. marginatus;* Afonso et al. 2016, Di Franco et al. 2018), having, therefore, a relatively limited spillover scale. Other species, however, have extensive

Table 1. Summary of the potential benefits of MPAs for groupers (↓: Reduce, ↑: Increase).

Potential benefits of MPAs for groupers	
Inside MPAs	**Outside MPAs**
↓ of fishing mortality (or even elimination)	↑ Spillover
↓ gear impacts on critical habitats	↑ Larval export
↑ grouper density	
↑ grouper mean size/age	
↑ grouper biomass	
↑ reproductive output (eggs/larvae)	

mobility ranges (from dozens to hundreds of kilometres). In such cases, they can travel long distances to reproduce in spawning aggregations sites (spawning migrations) (Nemeth et al. 2007), or migrate during their life cycle across several habitats, from inshore to offshore areas (ontogenetic migrations, as observed for the Nassau grouper *E. striatus*) (Sluka et al. 1997, Eklund and Schull 2001, Kaunda-Arara and Rose 2004, Mumby et al. 2004). The main trigger for spillover also varies among grouper species: for sedentary low mobile ones, this process is density-dependent and only occurs when the groupers are sufficiently abundant within MPAs (Sánchez Lizaso et al. 2000, Hackradt et al. 2014, 2020). The great spillover events of highly mobile species, instead, are seasonal and mainly triggered by natural clues, such as lunar phases (Fukunaga et al. 2020). Habitat composition, heterogeneity, and continuity in areas surrounding MPAs may also influence the magnitude of spillover (Forcada et al. 2009). MPAs surrounded by low structural habitats (e.g., sandy bottom) display smaller spillover scales compared to MPAs surrounded by complex structural areas (e.g., rocky bottom) (Edgar et al. 2014). Moreover, highly structurally complex habitats help to connect protected to non-protected areas by providing shelter and refuge to groupers during their displacement (García-Charton et al. 2004). This has been, for example, observed for the dusky grouper *E. marginatus*, the goldblotch grouper *E. costae*, and the mottled grouper *M. rubra* in the Mediterranean Sea (Hackradt et al. 2014).

Likewise, egg and larvae seeding occurs when their greater production within MPAs benefit non-protected fished areas through larval dispersal (Pelc et al. 2009, 2010, Planes et al. 2009). Non-adjacent fisheries can also benefit from increased outputs of eggs when adult groupers migrate to spawn in aggregations located far from MPAs (Roberts et al. 2000). For example, in Manus Island (Papua New Guinea), at least 50% of the planktonic eggs and larvae of the squaretail coral grouper Plectropomus areolatus were found to settle within 14 km from the spawning aggregation sites (Almany et al. 2013), which may increase the groupers biomass in nearby areas. For their part, Crec'hriou et al. (2010) documented the existence of a gradient of Epinephelus spp. eggs abundance across the limits of Cabrera National Park (Western Mediterranean), indicating that this MPA may act as a source of grouper propagules to neighbouring unprotected areas.

Through the spillover and seeding effects, adults, juveniles, eggs, and larvae of groupers are provided to areas located outside MPAs, available for fisheries' exploitation. In that way, MPAs may benefit local fisheries and participate to increase groupers catches (Roberts et al. 2001). In the Western Mediterranean Sea, different studies demonstrated higher abundance, size, and mean weight of groupers (*E. marginatus*, *E. costae*, and *M. rubra*) not only inside the reserve borders, but also in nearby unprotected areas (García-Charton et al. 2004, Di Franco et al. 2009, Hackradt et al. 2014). Such benefits have been demonstrated in the artisanal longline fishery targeting groupers in the Cabrera National Park (Balearic Islands, Mediterranean) (Goñi et al. 2008).

In this area, fishing effort, catch-per-unit-area, and revenues significantly decreased with distance from the MPA limits, indirectly demonstrating evidence of spillover. It must be underlined that the benefits of MPAs to adjacent and non-adjacent grouper fisheries depend on their effectiveness to protect groupers stocks. Grouper species displaying long-distance migratory behaviour will benefit from large MPAs, or MPAs strategically located within an effective network (Rolim et al. 2019). In such conditions, the home range of the considered stock is covered, including essential groupers habitats and spawning aggregation areas (Sweeting and Polunin 2005, Sadovy de Mitcheson et al. 2012, Schärer-Umpierre et al. 2014).

It follows from the above that MPAs are an essential tool for halting the decline of grouper stocks, as corroborated by recent studies. For example, a metapopulation model combined with habitat (bathymetry) distribution and connectivity estimates was developed by Belharet et al. (2020) to assess the likely effect on dusky grouper stocks in the Mediterranean Sea of either modifying (increasing or decreasing) the current fishing mortality rate or increasing the existing MPA coverage, and the economic implications of these changes; these authors concluded that achieving fisheries sustainability requires either a significant reduction of fishing mortality in unprotected areas, or/and an increase in the size of fully protected areas while keeping the overall fishing effort constant.

The Special Case of Spawning Aggregations

Most grouper species, like many other reef fishes, temporarily aggregate to spawn in large numbers at specific places and times in reproductive events known as spawning aggregations (Domeier and Colin 1997, Domeier 2012). Grouper spawning aggregations occur all over the world and support some of the most valuable and productive fisheries (Sadovy and Colin 2012, Hughes et al. 2020, IUCN 2020, Bezerra et al. 2021). Large adult fish from different reefs, and sometimes different countries, travel to spawn together at these aggregations, where they invest their total annual or semi-annual reproductive effort (Domeier and Colin 1997). Fisheries sustainability requires reproductive success (Mumby et al. 2006). Therefore, spawning aggregations are critical for replenishing and maintaining the productivity of many grouper populations, as well as the multiple ecosystem services they provide (Sadovy de Mitcheson et al. 2020c).

The predictability of mass spawning events makes aggregating grouper populations particularly vulnerable to intensive fishing (Sadovy 1997, Sala et al. 2001, Sadovy and Domeier 2005, Sadovy de Mitcheson 2016). Some sites and dates of these fish reproductive gatherings are historically known by fishers and represent opportunities for high catch rates of large-sized fish with low fishing effort (Sadovy de Mitcheson and Erisman 2012, Choat 2012, Erisman et al. 2017, Sadovy and Eklund 1999, Chérubin et al. 2020). Fisheries in spawning aggregations extract the larger and older fish during their reproductive time. Therefore, if not properly managed, this fishery can significantly affect the fertility and structure of the population (Shapiro et al. 1998, Armsworth 2001, Sala et al. 2001, Sadovy and Domeier 2005, Sadovy and Erisman 2012, Rhodes et al. 2013). Moreover, fishing on spawning aggregations is the principal threat to the recovery of the grouper species currently classified as threatened or endangered by the International Union for Conservation of Nature (IUCN Red List). A major example is the critical endangered Nassau grouper *Epinephelus striatus*, whose spawning aggregations have reduced in number and size after being subjected to high-fishing pressure for decades (50 CFR 2016). In addition to fishing, climate change has been identified as a threat for grouper spawning aggregations provoking changes in the spawning phenology, contracting the spawning season, and reducing the spawning probability (Asch and Erisman 2018, Pitt et al. 2017).

Many groupers spawning aggregations are in decline or have even disappeared (Domeier and Colin 1997, Sala et al. 2001, Claydon 2004, Aguilar-Perera 2006, Ojeda-Serrano et al. 2007, Sadovy de Mitcheson et al. 2008, 2012, Rhodes et al. 2013, FAO 2019), and there is a crucial need of appropriate management globally to help sustain important fisheries, livelihoods, food security, and to help ensure healthy and resilient coral reefs (Sadovy de Mitcheson and Colin 2012, Russell et al. 2012). For example, direct management measures focused on spawning aggregation sites (Erisman et al. 2017, Sadovy de Mitcheson et al. 2020b). However, the scarce accurate and up to date data of these populations limit our knowledge of the current status of the stocks and their spawning areas (Chérubin et al. 2020, Sadovy de Mitcheson et al. 2020c), which is critical for the implementation of meaningful management.

Spawning aggregations are challenging to study because most occur at remote locations (Claro and Lindeman 2003, Kobara and Heyman 2010, Kobara et al. 2013), in relatively deep waters (between 30 and 80 m), and are often in full activity at dusk (Chérubin et al. 2020). Some approaches used to fill in grouper spawning aggregations data gaps include fishing and diving techniques to estimate densities and delineate the area (Nemeth et al. 2007), the use of local ecological knowledge to identify the location of spawning aggregation sites unknown to science but known by locals (Gerhardinger et al. 2009, Bezerra et al. 2021), the application of genetic approaches to understand genetic diversity and external recruitment (Bernard et al. 2016), and the use of innovative robotic technology able to detect the sound produced by some groupers during reproductive events (Chérubin et al. 2020). It is known that the combination of academic research, local ecological knowledge, and citizen science participation is one of the most successful ways to achieve conservation outcomes in terms of identifying and monitoring spawning sites. Such a pathway also helps to pave the road to have the stakeholders' support in the implementation of management measures (Fulton et al. 2018).

But the merging of all those actors is not always possible, and other ways need to be considered.

Little management has been directed to fish spawning aggregation sites (Sadovy de Mitcheson et al. 2008). Deciding the most effective tool to protect and manage a spawning aggregation can be challenging, especially when there is little or no knowledge of the complex patterns of behaviour and movement associated with the aggregations. Precautionary management and the ecosystem approach to fisheries can be used where little data exists, as well as where scientific data is available and meaningful management is implemented (Sadovy de Mitcheson et al. 2020a).

Numerous successful management initiatives showcase the benefits of the effective protection of grouper spawning aggregations to their fisheries and conservation. On the Australian Great Barrier Reef, groupers, particularly the coral trout *P. leopardus*, are targeted by both commercial and recreational fisheries, using mostly hook and line fishing gear (Russell 2006). The coral trout is known to aggregate to spawn at least on the reefs they inhabit and has been afforded some proactive management to help ensure it is not overfished. In addition to MPAs covering 30% of the reefs throughout the Great Barrier Reef Marine Park, seasonal fishing closures were implemented to prevent fishing during the key spawning times. Groupers are also protected using minimum size, maximum size, and catch limits. The combination of these management measures gives the grouper population wide protection and the fisheries a reasonable chance of being sustainable (Russell 2006).

In Bermuda, the 50-years seasonal fishing closures of the red hind *Epinephelus guttatus* spawning aggregation sites has allowed a significant increase in the mean fish size over time, and therefore in fish biomass (Luckhurst and Trott 2008). This management regime appears to have allowed this Red hind population to recover after suffering overfishing for decades (Pitt et al. 2017). Likewise, the establishment of the Red Hind Bank Marine Conservation District in St. Thomas (US Virgin Islands), an MPA closed to fishing since 1999, had led to increases in the density, mean size, and biomass of the Red hinds in one of the most important spawning aggregation sites for this species (Beets and Friedlander 1999, Nemeth 2005, Luckhurst and Nemeth 2014). In Little Cayman, the Nassau grouper *Epinephelus striatus* has undergone a remarkable recovery after the implementation of science-based conservation strategies (consisting of a combination of a seasonal closure during the spawning period, bag limits during the open season, and gear restrictions), to reconstruct their abundance in spawning aggregations. Little Cayman is now home to the largest remaining identified aggregation of this species anywhere in the world (Waterhouse et al. 2020). In western Palau (Micronesia), the combination of seasonal grouper fishing closures during spawning months and permanent spatial protection from MPAs that cover their major spawning sites have led to increases in the number of spawners of the brown-marbled grouper *Epinephelus fuscoguttatus* (Gouezo et al. 2015) and the squaretailed coral grouper *Plectropomus areolatus* (Sadovy de

Mitcheson and Nemeth 2014, Sadovy et al. 2020), as well as to the stability of the camouflage grouper *Epinephelus polyphekadion* (Sadovy et al. 2020).

As a singular case of success of conservation measures, three artificial reefs built in southern Brazil—RAM and Balsa Norte (Brandini 2003) in the coast of Paraná state, and Monobóia in Santa Catarina state, provided refuge to large populations of goliath groupers which formed spawning aggregations, as characterized by Bueno et al. (2016) by conducting SCUBA diving surveys; this constitutes one of the few examples of achievement of actual ecological outcomes of artificial reefs (Félix-Hackradt and Hackradt 2008, Hackradt et al. 2011).

While these case studies showed a recovery of the groupers' populations when spawning, aggregations sites were specifically protected and effectively managed, highly depleted populations may not recover despite protection measures enforcement. In Belize, different management approaches aimed at restoring the populations of Nassau grouper have failed in maintaining or rebuilding stocks (Gibson 2007, Benedetti 2013). These approaches include marine reserves, where all types of fishing activities are prohibited, fishing seasonal closure, and minimum and maximum size limits (NOAA 2014). Lack of control or poaching behaviour can also counteract the recovery of depleted groupers populations. In Melanesia, the square-tailed coral grouper is heavily fished by small-scale commercial spearfishers targeting aggregation sites at night using underwater flashlights (Hamilton 2014). To reverse the declining aggregation numbers, MPAs have been implemented at several aggregation sites, but marked improvements have not been seen at most managed and monitored sites, because of poaching. However, it is hoped that this species life history characteristics (fast-growing, limited larval dispersal, prolonged spawning season) can offer a chance to see its aggregations recover quickly (Hamilton 2014).

The small size and site and time fidelity of the aggregations can play against but also in favour of their conservation. If well managed, this nature expedites spawning aggregations monitoring, assessment, and enforcement, and the benefits reach the entire population (Erisman et al. 2017). Many threatened populations of spawning aggregating groupers in the world may still have a chance to recover.

Conclusions and Perspectives

Groupers are highly valuable commercial species that play a crucial ecological role in coastal ecosystems (Sadovy de Mitcheson et al. 2012, 2020a). Recent studies have demonstrated that these species represent the majority of fish biomass in remote pristine reefs, together with sharks and other marine top predators (Sandin et al. 2008, Williams et al. 2011, Friedlander et al. 2014). Gaps in historical fishing and ecological data can start to be filled by gathering LEK in coastal areas from marine resources users (mainly professional and recreational fishers and recreational divers), which can help reconstruct past abundances and catches (Gerhardinger et al. 2006, Coll et al. 2014, Bender et al. 2014, Gaspare et al. 2015, Lima et al. 2016). Other valuable data can also be collected from paleontological, archaeological, historical, and even artistic records (Morales and Rosello 2004, McClenachan 2009, Fortibuoni et al. 2010, Guidetti and Micheli 2011, Condini et al. 2018). These non-conventional approaches have allowed researchers to obtain a more precise idea about the importance of groupers for both marine ecosystems and fisheries, and have highlighted that their present abundances are far below their past levels.

From the above review, it results that various technical procedures can be applied to more sustainably managed grouper fisheries, involving either traditional approaches based on stock assessment models, including those developed in data-poor contexts, or more integrated ones (i.e., ecosystem-based fisheries management, tropho-dynamic modelling, etc.). These approaches aim either to establish conventional fisheries measures, such as fishing quotas, gear restrictions, seasonal closures, or minimum/maximum size limits, or to adopt holistic measures such as MPAs. Deciding the appropriate management regime for ensuring that groupers can aggregate, spawn, replenish fish stocks, and supply fish for consumption depends on several factors (Russell et al. 2012), among which the population structure (in relation to reproductive output) has been recently highlighted (Easter et al. 2020).

The protection of grouper spawning aggregations is crucial for the conservation of most species (Sadovy de Mitcheson and Colin 2012), including those that have been extirpated by overfishing (Chollett et al. 2020). Besides, the transboundary and shared nature of grouper spawning aggregations, which include fish and eggs/larvae movements across a wide geographical range (Cowen et al. 2006), requires national but also international cooperation for the implementation and enforcement of effective management regulations (FAO 2019, Sadovy de Mitcheson et al. 2020b). A good example is the management plan recently proposed to protect the aggregations of the Nassau grouper in the Wider Caribbean Region. This plan coordinates intersectoral efforts within and among several countries, and actively involves key stakeholders that include fishers, governments, academics, and Non-Governmental Organizations (NGOs). More international, cooperative, and adaptive management plans should lead the design, implementation, and enforcement of effective management regulations in grouper spawning aggregations.

Groupers benefit more from the implementation of protection measures such as MPAs than most species worldwide (Hackradt et al. 2014). These measures permit them to recover more 'natural' structures of coastal fish communities to counteract past declines (McClanahan et al. 2007). Despite being the object of numerous management strategies, however, groupers are among the most endangered fish groups worldwide. The reasons for this are multiple and mainly rely on their biological traits, their susceptibility to being captured by a wide number of fishing gears, the destruction of their habitats, and pollution, which make them highly vulnerable (see Sadovy de Mitcheson and Liu, this volume). In this context, single management measures alone are unlikely to be sufficient at effectively managing groupers. Ideally, spatial protection of the whole ecosystem should be coupled with fishery-oriented management approaches. In addition to being highly targeted by commercial fishing, groupers are also subjected to strong recreational fishing pressure, especially spearfishing, both in the Mediterranean (Coll et al. 2004, Morales-Nin et al. 2005, Lloret et al. 2008, Gordoa 2009, Rocklin et al. 2011, Lloret and Font 2013, Font and Lloret 2014) and worldwide (Sluka and Sullivan 1998, Rhodes and Tupper 2007, Frisch et al. 2008, 2012, 2016b, Barcellini et al. 2013, Diogo and Pereira 2013, Lindfield et al. 2014, Young et al. 2016). This implies that, in parallel to commercial fisheries management, urgent measures should also be implemented to stop the depletion of groupers by recreational fisheries. For the latter, alternative strategies could consist of banning highly endangered grouper species from being fished, such as the goliath grouper *E. itajara* in Florida (McClenachan 2009) or the dusky grouper *E. marginatus* in French Mediterranean (Cottalorda et al. 2012). Other necessary measures would be stopping trophy records (Shiffman et al. 2014), while promoting other fishing modalities, such as sustainable practices of catch-and-release (Wilson and Burns 1996, Bartholomew and Bohnsack 2005, Brownscombe et al. 2016) using descender devices (Runde et al. 2020), implementing fish harvest tags (Johnston et al. 2007), or developing underwater reef fish photo-contests.

Studies, such as that of Belharet et al. (2020) cited above, constitute additional examples that increasing the number and surface of the MPAs can contribute to maintaining the stock biomass of groupers. Although the need to increase MPA coverage has been strongly claimed—e.g., the "Thirty by Thirty" target, i.e., to make 30% of the global ocean MPAs by 2030 (IUCN 2016, O'Leary et al. 2016, Baillie and Zhang 2020, Roberts et al. 2020), or even more (Wilson 2016), often appealing to the economic benefits that this would bring (Brander et al. 2020), it is increasingly clear that not only quantitative, but qualitative targets should be addressed (Barnes et al. 2018). The current trend towards protecting very large areas (Boonzaier and Pauly 2016), which is effectively achieving very rapid progress towards compliance with international agreements (Lubchenco and Grorud-Colvert 2015, but see Sala et al. 2018), has been the subject of debate (Singleton and Roberts 2014, Wilhem et al. 2014, Devillers et al. 2015, 2020, Jones and De Santo 2016, O'Leary et al. 2018, Magris and Pressey 2018, Rocha 2018, Artis et al. 2020). Regarding the conservation status of groupers, we argue that the effective protection of coastal areas, where most groupers live, and where protection measures are more complex to incept due to the multiplicity of economic sectors operating in a necessarily limited space and therefore often subject to conflict, is key to ensure the sustainability of groupers' populations.

An important challenge to the use of spatial management measures is the adoption of co-management, participative approaches to marine governance

(Hogg et al. 2013, Kockel et al. 2019, Collier 2020, Voorberg and Van der Veer 2020). More specifically, taking perceptions and opinions from the fishing sector into account when planning management measures is key to achieve true protection of groupers by ensuring local support to both traditional management measures (e.g., Retnonigntyas et al. 2021) and MPA rules (Silva and Lopes 2015, Mancha-Cisneros et al. 2018, Muntoni et al. 2019, Semitiel-García and Noguera-Méndez 2019, Di Franco et al. 2020).

Climate change also poses a serious challenge to the conservation of fish populations (Koenigstein et al. 2016, Pratchett et al. 2017). Seawater warming, through thermal tolerance and physiological effects (Pörtner and Farrell 2008, Cheung and Pauly 2016), can affect groupers' populations by shifting their geographic range (Ben Abdallah et al. 2007, Guidetti et al. 2010, Brito et al. 2011), reducing their body size (Messmer et al. 2017), influencing their reproduction, fecundity, and spawning behaviour (Zabala et al. 1997, Hereu et al. 2006, Carter et al. 2014), affecting eggs and larvae development and dispersal (Chérubin et al. 2011, Peck et al. 2012, Marancik et al. 2012, Almany et al. 2013, Weisberg et al. 2014), altering recruitment (Félix-Hackradt et al. 2013, 2014), and consequently impacting their connectivity (Schunter et al. 2011, Andrello et al. 2014). Moreover, groupers are being indirectly threatened by climate-driven habitat alterations (Williamson et al. 2014). Thus, these individual and population changes are likely to cause shifts in community structure and trophic relationships (Francour and Prato, this volume). These projected changes in seawater temperature and other physicochemical and biological climate-driven changes are parameters to take into account when designing management measures aimed at maintaining healthy grouper populations worldwide.

Scientific investigation of groupers' biology and ecology applied to their populations and broader ecosystem management is facing an urgent need to define and delve into new research goals. It is crucial to improve the understanding of the changes taking place in marine communities (Heithaus et al. 2008, Young et al. 2015), as well as in the organisms' trajectories, such as mesopredator release and increased stress in prey species (Stallings 2008, Madin et al. 2012, Palacios et al. 2016), to discern the role of groupers in mediating and buffering biological invasions (Mumby et al. 2011, Valdivia et al. 2014), and to characterize the behaviour syndromes and personalities in both highly harvested and restored areas (Sih et al. 2012, Bergseth et al. 2016). New research must focus on the features of individual movements and spillover (Koeck et al. 2014, Matley et al. 2015), as well as habitat selection at several spatial scales and levels of protection (Lindberg et al. 2006). Additionally, forecasting connectivity schemes (Calò et al. 2013, Andrello et al. 2014), and developing reliable grouper populations modelling (Heppell et al. 2006, Wielgus et al. 2007), would provide insightful information to inform the design of MPAs networks in a manner that can address identified groupers' management concerns in a scenario of global change (McLeod et al. 2009), as well as contribute to the development of effective fisheries management strategies. An essential link of scientific research is the use of TEK-LEK (Medeiros et al. 2018, Ribeiro et al. 2021) and citizen science (Freiwald et al. 2018, Fulton et al. 2018, Bonney et al. 2021, Grol et al. 2021) as complementary sources of data, which have to be urgently and massively incorporated into research funding programmes, also as a way to ensure the support and compliance of local communities to the incepted conservation measures (Bennett et al. 2019).

Finally, another important aspect that should be considered to prevent the extinction of this important fish group (Rudd and Tupper 2002, Gill et al. 2015, Shideler and Pierce 2016) is the economic value of living vs. dead groupers, especially in light of the economic importance of underwater tourism in MPAs (Sala et al. 2013, 2016). Ultimately, efforts should be undertaken by scientists to readily translate their results into feasible, operational management options in the present urgency scenarios. Not less prominently, environmental dissemination (Grorud-Colvert et al. 2010) and education (Zorrilla-Pujana and Rossi 2014) of scientific results should be an essential part of the endeavours of ecologists (Pace et al. 2010) to ensure the agreement and compliance of stakeholders and the general public with the management actions undertaken.

Acknowledgements

We would like to thank Dr. Brennan Chapman Lowery (Memorial University of Newfoundland) for reviewing this chapter. This work is part of the research projects REDEMED (MINECO CGL2013-49039-R) and ABHACO²DE (Fundación Séneca 19516/PI/14).

References

50 C.F.R. 2016. Endangered and threatened wildlife and plants: Final listing determination on the proposal to list the Nassau grouper as threatened under the Endangered Species Act 50 § 125, p. 223.

Abdel Barr, M.A., A.F.M. El-Sayed and A.M. Osman. 2010. The use of per recruit models for stock assessment and management of greasy grouper *Epinephelus tauvina* in the Arabian Gulf waters off Qatar. Trop. Life Sci. Res. 21: 83–90.

Abecasis, D., P. Afonso and K. Erzini. 2014. Combining multispecies home range and distribution models aids assessment of MPA effectiveness. Mar. Ecol. Prog. Ser. 513: 155–169.

Afonso, P., D. Abecasis, R.S. Santos and J. Fontes. 2016. Contrasting movements and residency of two serranids in a small Macaronesian MPA. Fish. Res. 177: 59–70.

Agardy, T., P. Bridgewater, M.P. Crosby, J. Day, P.K. Dayton, R. Kenchington, D. Laffoley, P. McConney, P.A. Murray, J.E. Parks and L. Peau. 2003. Dangerous targets? Unresolved issues and ideological clashes around marine protected areas. Aquat. Conserv. Mar. Freshw. Ecosyst. 13: 353–367.

Aguilar-Perera, A. 2006. Disappearance of a Nassau grouper spawning aggregation off the southern Mexican Caribbean coast. Mar. Ecol.-Prog. Ser. 327: 289.

Agustina, S., A.S. Panggabean, M. Natsir, H. Retroningtyas and I. Yulianto. 2019. Yield-per-recruit modeling as biological reference points to provide fisheries management of Leopard Coral Grouper (*Plectropomus leopardus*) in Saleh Bay, West Nusa Tenggara. IOP Conf. Ser.: Earth Environ. Sci. 278: 012005.

Ainsworth, C.H., D.A. Varkey and T.J. Pitcher. 2008. Ecosystem simulations supporting ecosystem-based fisheries management in the Coral Triangle, Indonesia. Ecol. Model. 214: 361–374.

Albouy, C., D. Mouillot, D. Rocklin, J.M. Culioli and F.L. Loch. 2010. Simulation of the combined effects of artisanal and recreational fisheries on a Mediterranean MPA ecosystem using a trophic model. Mar. Ecol. Prog. Ser. 412: 207–221.

Alcala, A.C. 1998. Community-based coastal resource management in the Philippines: A case study. Ocean Coast. Manag. 38: 179–186.

Almany, G.R., R.J. Hamilton, M. Bode, M. Matawai, T. Potuku, P. Saenz-Agudelo, S. Planes, M.L. Berumen, K.L. Rhodes, S.R. Thorrold, G.R. Russ and G.P. Jones. 2013. Dispersal of grouper larvae drives local resource sharing in a coral reef fishery. Curr. Biol. 23: 626–630.

Alonzo, S.H., T. Ish, M. Key, A.D. MacCall and M. Mangel. 2008. The importance of incorporating protogynous sex change into stock assessments. Bull. Mar. Sci. -Miami 83: 163–179.

Anderson, A.B., R.M. Bonaldo, D.R. Barneche, C.W. Hackradt, F.C. Felix Hackradt, J.A. Garcia Charton and S.R. Floeter. 2014. Recovery of grouper assemblages indicates effectiveness of a marine protected area in Southern Brazil. Mar. Ecol. Prog. Ser. 514: 207–215.

Anderson, E.D. 1978. An Explanation of Virtual Population Analysis. National Marine Fisheries Service, Woods Hole, Massachusetts: 7pp.

Andrello, M., D. Mouillot, J. Beuvier, C. Albouy, W. Thuiller and S. Manel. 2013. Low connectivity between Mediterranean marine protected areas: A biophysical modeling approach for the dusky grouper *Epinephelus marginatus*. PLoS ONE 8: e68564.

Andrello, M., D. Mouillot, S. Somot, W. Thuiller and S. Manel. 2014. Additive effects of climate change on connectivity between marine protected areas and larval supply to fished areas. Divers. Distrib. 21: 139–150.

Armsworth, P.R. 2001. Effects of fishing on a protogynous hermaphrodite. Can. J. Fish. Aquat. Sci. 58(3): 568–578. https://doi.org/10.1139/cjfas-58-3-568.

Arreguin-Sánchez, F., J.L. Munro, M.C. Balgos and D. Pauly. 1996. Biology, fisheries and culture of tropical groupers and snappers. ICLARM Conference Proceedings.

Arreguin-Sánchez, F., A. Hernández-Herrera, M. Ramirez-Rodriguez and H. Pérez-España. 2004. Optimal management scenarios for the artisanal fisheries in the ecosystem of La Paz Bay, Baja California Sur, Mexico. Ecol. Model. 172: 373–382.

Artis, E., N.J. Gray, L.M. Campbell, R.L. Gruby, L. Acton, S.B. Zigler and L. Mitchell. 2020. Stakeholder perspectives on large-scale marine protected areas. PLoS ONE 15: e0238574.

Asch, R.G. and B. Erisman. 2018. Spawning aggregations act as a bottleneck influencing climate change impacts on a critically endangered reef fish. Diversity and Distributions 24(12): 1712–1728.

Bacheler, N.M. and J.A. Buckel. 2004. Does hook type influence the catch rate, size, and injury of grouper in a North Carolina commercial fishery? Fish. Res. 69: 303–311.

Baillie, J. and Y.P. Zhang. 2020. Space for nature. Science 361: 1051.

Barcellini, V.C., F.S. Motta, A.M. Martins and P.S. Moro. 2013. Recreational anglers and fishing guides from an estuarine protected area in southeastern Brazil: Socioeconomic characteristics and views on fisheries management. Ocean Coast. Manag. 76: 23–29.

Barnes, M.D., L. Glew, C. Wyborn and D. Craigie. 2018. Prevent perverse outcomes from global protected area policy. Nat. Ecol. Evol. 2: 759–762.

Bartholomew, A. and J.A. Bohnsack. 2005. A review of catch-and-release angling mortality with implications for no-take reserves. Rev. Fish Biol. Fish. 15: 129–154.

Beets, J. and A. Friedlander. 1999. Evaluation of a conservation strategy: A spawning aggregation closure for red hind, *Epinephelus guttatus*, in the U.S. Virgin Islands. Environ. Biol. Fishes 55: 91–98.

Begossi, A., S. Salivonchyk and R.A.M. Silvano. 2016. Collaborative research on Dusky grouper (*Epinephelus marginatus*): Catches from the small-scale fishery of Copacabana Beach, Rio de Janeiro, Brazil. J. Coast. Manag. 19: 1000428.

Begossi, A., S. Salivonchyk, B. Glamuzina, S. Pacheco de Souza, P.F.M. Lopes, R.H.G. Priolli, D.O. do Prado, M. Ramires, M. Clauzet, C. Zapelini, D.T. Schneider, L.T. Silva and R.A.M.

Silvano. 2019. Fishers and groupers (*Epinephelus marginatus* and *E. morio*) in the coast of Brazil: integrating information for conservation. J. Ethnobiol. Ethnomed. 15: 53.

Begossi, A.S. and S. Salivonchyk. 2019. Integrating science and citizen science: the dusky grouper (*Epinephelus marginatus*) sustainable fishery of Copacabana, Rio de Janeiro, Brazil. bioRxiv 759357.

Bejarano, I., R.S. Appeldoorn and M. Nemeth. 2014. Fishes associated with mesophotic coral ecosystems in La Parguera, Puerto Rico. Coral Reefs 33: 313–328.

Belharet, M., A. Di Franco, A. Calò, L. Mari, J. Claudet, R. Casagrandi, M. Gatto, J. Lloret, C. Sève, P. Guidetti and P. Melià. 2020. Extending full protection inside existing marine protected areas, or reducing fishing effort outside, can reconcile conservation and fisheries goals. J. Appl. Ecol. 57: 1948–1957.

Ben Abdallah, A., J. Ben Souissi, H. Méjri, C. Capapé and D. Golani. 2007. First record of *Cephalopholis taeniops* (Valenciennes) in the Mediterranean Sea. J. Fish Biol. 71: 610–614.

Bender, M.G., G.R. Machado, P.J. de A. Silva, S.R. Floeter, C. Monteiro-Netto, O.J. Luiz and C.E.L. Ferreira. 2014. Local ecological knowledge and scientific data reveal overexploitation by multigear artisanal fisheries in the Southwestern Atlantic. PLoS ONE 9: e110332.

Benedetti, L.S. 2013. Marine Protected Areas (MPAs) as a Fisheries Management Tool for the Nassau Grouper (*Epinephelus striatus*) in Belize. M.S. Thesis, United Nations University Institute of Water, Environment and Health (UNU-INWEH), Toronto, Canada.

Bennett, N.J., A. Di Franco, A. Calò, E. Nethery, F. Niccollini, M. Milazzo and P. Guidetti. 2019. Local support for conservation is associated with perceptions of good governance, social impacts, and ecological effectiveness. Conserv. Lett. 2019: e12640.

Bentes, B., N.C.B. Mendes, A.G.C.M. Klautau, C.S. Viana, J.G. Romao Jr., K.C.A. Silva, C.E.R. Andrade, L.J.P. Gomes and I.H.A. Cintra. 2019. Incidental catch of goliath grouper *Epinephelus itajara* (Lichtenstein, 1822) and *Epinephelus* sp. (Bloch, 1793) in industrial fisheries of Brazilian Northern coast: a critical endangerous species. Biota Amazonia – Macapá 9: 58–59.

Bergseth, B.J., D.H. Williamson, A.J. Frisch and G.R. Russ. 2016. Protected areas preserve natural behaviour of a targeted fish species on coral reefs. Biol. Conserv. 198: 202–209.

Bernard, A.M., K.A. Feldheim, R. Nemeth, E. Kadison, J. Blondeau, B.X. Semmens and M.S. Shivji. 2016. The ups and downs of coral reef fishes: the genetic characteristics of a formerly severely overfished but currently recovering Nassau grouper fish spawning aggregation. Coral Reefs 35(1): 273–284.

Bezerra, I.M., M. Hostim-Silva, J.L.S. Teixeira, C.W. Hackradt, F.C. Félix-Hackradt and A. Schiavetti. 2021. Spatial and temporal patterns of spawning aggregations of fish from the Epinephelidae and Lutjanidae families: An analysis by the local ecological knowledge of fishermen in the Tropical Southwestern Atlantic. Fisheries Research 239: 105937.

Boersma, P.D. and J.K. Parrish. 1999. Limiting abuse: Marine protected areas, a limited solution. Ecol. Econ. 31: 287–304.

Bohnsack, J.A. 1994. Marine reserves: They enhance fisheries, reduce conflicts, and protect resources. Naga 17: 4–7.

Bonney, R., J. Byrd, J.T. Carmichael, L. Cunningham, L. Oremland, J. Shirk and A. von Harten. 2021. Sea change: Using citizen science to inform fisheries management. BioScience 2021: 1–12.

Boonzaier, L. and D. Pauly. 2016. Marine protection targets: an updated assessment of global progress. Oryx 50: 27–35.

Botsford, L.W., J.C. Castilla and C.H. Peterson. 1997. The management of fisheries and marine ecosystems. Science 277: 509–515.

Brander, L.M., P. van Beukering, L. Nijsten, A. McVittie, C. Baulcomb, F.V. Eppink and J.A.C. van der Lelij. 2020. The global costs and benefits of expanding Marine Protected Areas. Mar. Pol. 116: 103953.

Brandini, F. 2003. Recifes Artificiais Marinhos: Uma proposta de conservação da biodiversidade e desenvolvimento da pesca artesanal através da criação de um parque marinho na costa do estado do Paraná. Final report. Ministério da Ciência e Tecnologia, Brasilia, 400 pp.

Brito, A., S. Clemente and R. Herrera. 2011. On the occurrence of the African hind, *Cephalopholis taeniops*, in the Canary Islands (eastern subtropical Atlantic): Introduction of large-sized demersal littoral fishes in ballast water of oil platforms? Biol. Invasions 13: 2185.

Brownscombe, J.W., A.J. Danylchuk, J.M. Chapman, L.F.G. Gutowsky and S.J. Cooke. 2016. Best practices for catch-and-release recreational fisheries—angling tools and tactics. Fish. Res. 186: 693–705.

Brulé, T., J. Montero-Muñoz, N. Morales-López and A. Mena-Loria. 2015. Influence of circle hook size on catch rate and size of red grouper in shallow waters of the Southern Gulf of Mexico. North Am. J. Fish. Manag. 35: 1196–1208.

Bueno, L.S., A.A. Bertoncini, C.C. Koenig, F.C. Coleman, M.O. Freitas, J.R. Leite, T.F. de Souza and M. Hostim-Silva. 2016. Evidence for spawning aggregations of the endangered Atlantic goliath grouper *Epinephelus itajara* in southern Brazil. J. Fish Biol. 89: 876–889.

Bunce, M., L.D. Rodwell, R. Gibb and L. Mee. 2008. Shifting baselines in fishers' perceptions of island reef fishery degradation. Ocean Coast. Manag. 51: 285–302.

Burgos, R. and O. Defeo. 2004. Long-term population structure, mortality and modeling of a tropical multi-fleet fishery: the red grouper *Epinephelus morio* of the Campeche Bank, Gulf of Mexico. Fish. Res. 66: 325–335.

Cadrin, S.X. and M. Dickey-Collas. 2015. Stock assessment methods for sustainable fisheries. ICES J. Mar. Sci. J. Cons. 72: 1–6.

Calò, A., F.C. Félix-Hackradt, J. Garcia, C.W. Hackradt, D. Rocklin, J.T. Otón and J.A.G. Charton. 2013. A review of methods to assess connectivity and dispersal between fish populations in the Mediterranean Sea. Adv. Oceanogr. Limnol. 4: 150–175.

Carr, L.M. and W.D. Heyman. 2012. "It's about seeing what's actually out there": Quantifying fishers' ecological knowledge and biases in a small-scale commercial fishery as a path toward co-management. Ocean Coast. Manag. 69: 118–132.

Carruthers, T.R., A.E. Punt, C.J. Walters, A. MacCall, M.K. McAllister, E.J. Dick and J. Cope. 2014. Evaluating methods for setting catch limits in data-limited fisheries. Fish. Res. 153: 48–68.

Carter, A.B., C.R. Davies, B.D. Mapstone, G.R. Russ, A.J. Tobin and A.J. Williams. 2014. Effects of region, demography, and protection from fishing on batch fecundity of common coral trout (*Plectropomus leopardus*). Coral Reefs 33: 751–763.

CassCalay, S.L. and T.W. Schmidt. 2009. Monitoring changes in the catch rates and abundance of juvenile goliath grouper using the ENP creel survey, 1973–2006. Endanger. Species Res. 7: 183–193.

Chagaris, D.D., B. Mahmoudi, C.J. Walters and M.S. Allen. 2015. Simulating the trophic impacts of fishery policy options on the West Florida shelf using Ecopath with Ecosim. Mar. Coast. Fish. 7: 44–58.

Chen, Z., Y. Qiu, X. Jia and S. Xu. 2008. Simulating fisheries management options for the Beibu Gulf by means of an ecological modelling optimization routine. Fish. Res. 89: 257–265.

Cheng, B.S., A.H. Altieri, M.E. Torchin and G.M. Ruiz. 2019. Can marine reserves restore lost ecosystem functioning? A global synthesis. Ecology 100(4): e02617.

Chérubin, L.M., R.S. Nemeth and N. Idrisi. 2011. Flow and transport characteristics at an *Epinephelus guttatus* (red hind grouper) spawning aggregation site in St. Thomas (US Virgin Islands). Ecol. Model. 222: 3132–3148.

Chérubin, L.M., F. Dalgleish, A.K. Ibrahim, M. Schärer-Umpierre, R.S. Nemeth, A. Matthews and R. Appeldoorn. 2020. Fish spawning aggregations dynamics as inferred from a novel, persistent presence robotic approach. Frontiers in Marine Science 6: 779.

Cheung, W.W.L. and D. Pauly. 2016. Impacts and effects of ocean warming on marine fishes. pp. 239–253. *In:* Laffoley, D. and J.M. Baxter (eds.). Explaining Ocean Warming: Causes, Scale, Effects and Consequences. IUCN, Gland, Switzerland.

Chiappone, M., R. Sluka and K.S. Sealey. 2000. Groupers (Pisces: Serranidae) in fished and protected areas of the Florida Keys, Bahamas and northern Caribbean. Mar. Ecol. Prog. Ser. 198: 261–272.

Choat, J.H. 2012. Spawning aggregations in reef fishes: Ecological and evolutionary processes. pp. 85–116. *In*: Sadovy de Mitcheson, Y. and P.L. Colin (eds.). Reef Fish Spawning Aggregations: Biology, Research and Management. Springer, Dortrecht.

Chollett, I., M. Priest, S. Fulton and W.D. Heyman. 2020. Should we protect extirpated fish spawning aggregation sites? Biol. Conserv. 241: 108395.

Christensen, V. and C.J. Walters. 2004. Ecopath with Ecosim: Methods, capabilities and limitations. Ecol. Model. 172: 109–139.

Christie, M.R., B.N. Tissot, M.A. Albins, J.P. Beets, Y. Jia, D.M. Ortiz, S.E. Thompson and M.A. Hixon. 2010. Larval connectivity in an effective network of marine protected areas. PLoS ONE 5: e15715.

Chrysafi, A. and A. Kuparinen. 2015. Assessing abundance of populations with limited data: Lessons learned from data-poor fisheries stock assessment. Environ. Rev. 24: 25–38.

Claro, R. and K.C. Lindeman. 2003. Spawning aggregation sites of snapper and grouper species (Lutjanidae and Serranidae) on the insular shelf of Cuba. Gulf Caribb. Res. 14: 91–106.

Claudet, J., C.W. Osenberg, L. Benedetti-Cecchi, P. Domenici, J.-A. García-Charton, Á. Pérez-Ruzafa, F. Badalamenti, J. Bayle-Sempere, A. Brito, F. Bulleri, J.-M. Culioli, M. Dimech, J.M. Falcón, I. Guala, M. Milazzo, J. Sánchez-Meca, P.J. Somerfield, B. Stobart, F. Vandeperre, C. Valle and S. Planes. 2008. Marine reserves: Size and age do matter. Ecol. Lett. 11: 481–489.

Claudet, J., C.W. Osenberg, P. Domenici, F. Badalamenti, M. Milazzo, J.M. Falcón, I. Bertocci, L. Benedetti-Cecchi, J.-A. García-Charton, R. Goñi, J.A. Borg, A. Forcada, G.A. de Lucia, Á. Pérez-Ruzafa, P. Afonso, A. Brito, I. Guala, L.L. Diréach, P. Sanchez-Jerez, P.J. Somerfield and S. Planes. 2010. Marine reserves: Fish life history and ecological traits matter. Ecol. Appl. 20: 830–839.

Claydon, J. 2004. Spawning aggregations of coral reef fishes: Characteristics, hypotheses, threats and management. pp. 265–302. *In:* Gibson, R.N., R.J.A. Atkinson and J.D.M. Gordon (eds.). Oceanography and Marine Biology: An Annual Review. CRC Press, Boca Raton, Florida.

Coleman, F.C., C.C. Koenig, G.R. Huntsman, J.A. Musick, A.M. Eklund, J.C. McGovern, G.R. Sedberry, R.W. Chapman and C.B. Grimes. 2000. Long-lived reef fishes: The grouper-snapper complex. Fisheries 25: 14–21.

Coleman, F.C. and S.L. Williams. 2002. Overexploiting marine ecosystem engineers: Potential consequences for biodiversity. Trends Ecol. Evol. 17: 40–44.

Coleman, F.C., C.C. Koenig, K.M. Scanlon, S. Heppell, S. Heppell and M.W. Miller. 2010. Benthic habitat modification through excavation by red grouper, *Epinephelus morio*, in the northeastern Gulf of Mexico. Open Fish Sci. J. 3: 1–15.

Coll, J., M. Linde, A. García-Rubies, F. Riera and A.M. Grau. 2004. Spear fishing in the Balearic Islands (west central Mediterranean): Species affected and catch evolution during the period 1975–2001. Fish. Res. 70: 97–111.

Coll, M. and S. Libralato. 2012. Contributions of food web modelling to the ecosystem approach to marine resource management in the Mediterranean Sea. Fish Fish. 13: 60–88.

Coll, M., C. Piroddi, C. Albouy, F. Ben Rais Lasram, W.W.L. Cheung, V. Christensen, V.S. Karpouzi, F. Guilhaumon, D. Mouillot, M. Paleczny, M.L. Palomares, J. Steenbeek, P. Trujillo, R. Watson and D. Pauly. 2012. The Mediterranean Sea under siege: Spatial overlap between marine biodiversity, cumulative threats and marine reserves. Glob. Ecol. Biogeogr. 21: 465–480.

Coll, M., M. Carreras, C. Ciércoles, M.-J. Cornax, G. Gorelli, E. Morote and R. Sáez. 2014. Assessing fishing and marine biodiversity changes using fishers' perceptions: The Spanish Mediterranean and Gulf of Cadiz case study. PLoS ONE 9: e85670.

Collier, C.E. 2020. Enabling conditions for community-based comanagement of marine protected areas in the United States. Mar. Pol. 122: 104244.

Condini, M.V., J.A. García-Charton and A.M. Garcia. 2018. A review of the biology, ecology, behavior and conservation status of the dusky grouper, *Epinephelus marginatus* (Lowe 1834). Rev. Fish Biol. Fish. 28: 301–330.

Contreras, M., F. Arreguín-Sánchez, J.A. Sánchez, V. Moreno and M.A. Cabrera. 1994. Mortality and population size of the Red grouper (*Epinephelus morio*) fishery from the Campeche Bank. Proc. 43rd Ann. Gulf Caribb. Fish. Inst., GCFI, Charleston, SC, USA 392–401.

Cornish, A.S. and M.L. Harmelin-Vivien. 2004. *Epinephelus marginatus* (Dusky Grouper). Available at: http://www.iucnredlist.org/details/7859/0 [Accessed June 15, 2016].

Cottalorda, J.-M., J.-M. Dominici, J.-G. Harmelin, M. Harmelin-Vivien, P. Louisy and P. Francour. 2012. Etude et synthèse des principales données disponibles sur les espèces de « mérous » de la Réserve naturelle de Scandola et de ses environs immédiats. Contrat Parc Naturel Régional de Corse/GIS Posidonie. Univ. Nice Sophia Antipolis, ECOMERS publ.: 48 pp.

Cowen, R.K., C.B. Paris and A. Srinivasan. 2006. Scaling connectivity in marine populations. Science 311: 522–527. doi: 10.1126/science.1122039.

Craig, M.T. 2011. *Epinephelus itajara*. The IUCN Red List of Threatened Species 2011: e.T195409A8961414.

Craig, M.T., Y.J.S. Mitcheson and P.C. Heemstra (eds.). 2011. Groupers of the World: A Field and Market Guide. NISC (Pty) Ltd, Grahamstown, South Africa.

Crec'hriou, R., F. Alemany, E. Roussel, A. Chassanite, J.Y. Marinaro, J. Mader, E. Rochel and S. Planes. 2010. Fisheries replenishment of early life taxa: Potential export of fish eggs and larvae from a temperate marine protected area. Fish. Oceanogr. 19: 135–150.

D'Agata, S., D. Mouillot, L. Wantiez, A.M. Friedlanmder, M. Kulbicki and L. Vigliola. 2016. Marine reserves lag behind wilderness in the conservation of key functional roles. Nat. Commun. 7: 12000.

Davis, A. and K. Ruddle. 2010. Constructing confidence: Rational skepticism and systematic enquiry in local ecological knowledge research. Ecol. Appl. 20: 880–894.

Day, J.C. 2002. Zoning: Lessons from the great barrier reef marine park. Ocean Coast. Manag. 45: 139–156.

Devillers, R., R.L. Pressey, A. Grech, J.N. Kittinger, G.J. Edgar, T. Ward and R. Watson. 2015. Reinventing residual reserves in the sea: are we favouring ease of establishment over need for protection? Aquat. Conserv. Freshw. Ecosyst. 25: 480–504.

Devillers, R., R.L. Pressey, T.J. Ward, A. Grech, J.N. Kittinger, G.J. Edgar and R.A. Watson. 2020. Residual marine protected areas five years on: Are we still favouring ease of establishment over need for protection? Aquat. Conserv. Freshw. Ecosyst. 30: 1758–1764.

Di Franco, A., S. Bussotti, A. Navone, P. Panzalis and P. Guidetti. 2009. Evaluating effects of total and partial restrictions to fishing on Mediterranean rocky-reef fish assemblages. Mar. Ecol. Prog. Ser. 387: 275–285.

Di Franco, A., J.G. Plass-Johnson, M. Di Lorenzo, B. Meola, J. Claudet, S.D. Gaines, J.A. García-Charton, S. Giakoumi, K. Grorud-Colvert, C.W. Hackradt, F. Micheli and P. Guidetti. 2018. Linking home ranges to protected area size: the case study of the Mediterranean Sea. Biol. Conserv. 221: 175–181.

Di Franco, A., K. Hogg, A. Calò, N.J. Bennett, M.A. Sévin-Allouet, O. Esparza-Alaminos, M. Lang, D. Koutsoubas, M. Prvan, L. Santarossa, F. Niccolini, M. Milazzo and P. Guidetti. 2020. Improving marine protected area governance through collaboration and co-production. J. Environm. Manage. 269: 110757.

Di Lorenzo, M., J. Claudet and P. Guidetti. 2016. Spillover from marine protected areas to adjacent fisheries has an ecological and a fishery component. J. Nat. Conserv. 32: 62–66.

Di Lorenzo, M., P. Guidetti, A. Di Franco, A. Calò and J. Claudet. 2020. Assessing spillover from marine protected areas and its drivers: A meta-analytical approach. Fish Fish. 21: 906–915.

Dichmont, C.M., R.A. Deng, N. Dowling and A.E. Punt. 2021. Collating stock assessment packages to improve stock assessments. Fish. Res. 236: 105844.

Diogo, H.M.C. and J.G. Pereira. 2013. Impact evaluation of spear fishing on fish communities in an urban area of São Miguel Island (Azores Archipelago). Fish. Manag. Ecol. 20: 473–483.

Domeier, M.L. and P.L. Colin. 1997. Tropical reef fish spawning aggregations: defined and reviewed. Bul. Mar. Sci. 60: 698–726.

Domeier, M.L. 2012. Revisiting spawning aggregations: Definitions and challenges. pp. 1–20. *In*: Sadovy de Mitcheson, Y. and P.L. Colin (eds.). Reef Fish Spawning Aggregations: Biology, Research and Management. Springer.

Dowling, N.A., C.M. Dichmont, M. Haddon, D.C. Smith, A.D.M. Smith and K. Sainsbury. 2015. Empirical harvest strategies for data-poor fisheries: A review of the literature. Fish. Res. 171: 141–153.

Eberts, R.L. and C.M. Somers. 2017. Venting and descending provide equivocal benefits for catch-and-release survival: Study design influences effectiveness more than barotrauma relief method. N. Am. J. Fish. Manag. 37: 612–623.

Edgar, G.J., R.D. Stuart-Smith, T.J. Willis, E. Kininmonth, S. Baker, S. Banks, N.S. Barret, M.A. Becerro, A.T.F. Bernard, J. Berkhout, C.D. Buxton, S.J. Campbell, A.T. Cooper, M. Davey, S.C. Edgar, G. Försterra, D. Galván, A.J. Irigoyen, D.J. Kushner, R. Moura, P.E. Parnell, S.C. Shears, G. Soler, E.M.A. Strain and R.J. Thomson. 2014. Global conservation outcomes depend on marine protected areas with five key features. Nature 506: 216–220.

Edgar, G.J., T.J. Ward and R.D. Stuart-Smith. 2019. Weaknesses in stock assessment modelling and management practices affect fisheries sustainability. Aquatic Conserv.: Mar. Freshw. Ecosyst. 2019: 1–7.

Ehrhardt, N.M. and V.K.W. Deleveaux. 2007. The Bahamas' Nassau grouper (*Epinephelus striatus*) fishery: Two assessment methods applied to a data-deficient coastal population. Fish. Res. 87: 17–27.

Eklund, A.-M. and J. Schull. 2001. A stepwise approach to investigating the movement patterns and habitat utilization of goliath grouper, *Epinephelus itajara*, using conventional tagging, acoustic telemetry and satellite tracking. pp. 189–216. *In*: Sibert, J.R. and J.L. Nielsen (eds.). Electronic Tagging and Tracking in Marine Fisheries. Springer, Netherlands.

El-Sayed, A.F.M., A.M.A. Osman and M.A. Barr. 2007. Estimation of total mortality rates and virtual population analysis of *Epinephelus tauvina* from Arabian Gulf Qatar. Egypt. J. Aquat. Res. 33: 362–370.

Ellis, R.D. and M.E. Falletti. 2017. Native grouper indirectly ameliorates the negative effects of invasive lionfish. Mar. Ecol. Prog. Ser. 558: 267–279.

Erisman, B.E., W. Heyman, S. Kobara, T. Ezer, S. Pittman, O. Aburto-Oropeza and R.S. Nemeth. 2017. Fish spawning aggregations: where well-placed management actions can yield big benefits for fisheries and conservation. Fish Fish. 18: 128–144.

Easter, E.E., M.S. Adreani, S.L. Hamilton, M.A. Steele, S. Pang and J.W. White. 2020. Influence of protogynous sex change on recovery of fish populations within marine protected areas. Ecol. Appl. 30: e02070.

Fanning, L., R. Mahon and P. McConney (eds.). 2011. Towards Marine Ecosystem-Based Management in the Wider Caribbean. Amsterdam University Press, Amsterdam.

FAO. 2008. Fisheries management. 2, The ecosystem approach to fisheries. 2.1, Best practices in ecosystem modelling for informing an ecosystem approach to fisheries. FAO Technical Guidelines for Responsible Fisheries, Food and Agriculture Organization of the United Nations, Rome, Italy: 93 pp

FAO. 2011. Fisheries management, v. 4: Marine protected areas and fisheries. FAO Technical Guidelines for Responsible Fisheries, Food and Agriculture Organization of the United Nations, Rome, Italy: 198 pp.

FAO. 2019. Report of the second meeting of the CFMC/WECAFC/OSPESCA/CRFM Spawning Aggregations Working Group (SAWG), Miami, Florida, 27–29 March 2018. FAO Fisheries and Aquaculture Report. No. 1261. Western Central Atlantic Fishery Commission. Bridgetown.

Félix-Hackradt, F.C. and C.W. Hackradt. 2008. Populational study and monitoring of the goliath grouper, *Epinephelus itajara* (Lichtenstein, 1822), in the coast of Paraná, Brazil. Brazil. J. Conserv. 6: 141–156.

Félix-Hackradt, F.C., C.W. Hackradt, J. Treviño-Otón, A. Pérez-Ruzafa and J.A. García-Charton. 2013. Temporal patterns of settlement, recruitment and post-settlement losses in a rocky reef fish assemblage in the South-Western Mediterranean Sea. Mar. Biol. 160: 2337–2352.

Félix-Hackradt, F.C., C.W. Hackradt, J. Treviño-Otón, A. Pérez-Ruzafa and J.A. García-Charton. 2014. Habitat use and ontogenetic shifts of fish life stages at rocky reefs in South-western Mediterranean Sea. J. Sea Res. 88: 67–77.

Ferter, K., M.S. Weltersbach, H.V. Strehlow, J.H. Vølstad, J. Alós, R. Arlinghaus, M. Armstrong, M. Dorow, M. de Graaf, T. van der Hammen, K. Hyder, H. Levrel, A. Paulrud, K. Radtke, D. Rocklin, C.R. Sparrevohn and P. Veiga. 2013. Unexpectedly high catch-and-release rates in European marine recreational fisheries: Implications for science and management. ICES J. Mar. Sci. 70: 1319–1329.

Floeter, S.R., B.S. Halpern and C.E.L. Ferreira. 2006. Effects of fishing and protection on Brazilian reef fishes. Biol. Conserv. 128: 391–402.

Font, T. and J. Lloret. 2014. Biological and ecological impacts derived from recreational fishing in Mediterranean coastal areas. Rev. Fish. Sci. Aquac. 22: 73–85.

Forcada, A., C. Valle, P. Bonhomme, G. Criquet, G. Cadiou, P. Lenfant and J.L. Sanchez Lizaso. 2009. Effects of habitat on spillover from marine protected areas to artisanal fisheries. Mar. Ecol. Prog. Ser. 379: 197–211.

Forrester, G., P. Baily, D. Conetta, L. Forrester, E. Kintzing and L. Jarecki. 2015. Comparing monitoring data collected by volunteers and professionals shows that citizen scientists can detect long-term change on coral reefs. J. Nat. Conserv. 24: 1–9.

Fortibuoni, T., S. Libralato, S. Raicevich, O. Giovanardi and C. Solidoro. 2010. Coding early naturalists' accounts into long-term fish community changes in the Adriatic Sea (1800–2000). PLoS ONE 5: e15502.

Francis, R.C., M.A. Hixon, M.E. Clarke, S.A. Murawski and S. Ralston. 2007. Ten commandments for ecosystem-based fisheries scientists. Fisheries 32: 217–233.

Freiwald, J., R. Meyer, J.E. Caselle, C.A. Blanchette, K. Hovel, D. Neilson, J. Dugan, J. Altstatt, K. Nielsen and J. Bursek. 2018. Citizen science monitoring of marine protected areas: Case studies and recommendations for integration into monitoring programs. Mar. Ecol. 39: e12470.

Friedlander, A.M. and E.E. DeMartini. 2002. Contrasts in density, size, and biomass of reef fishes between the northwestern and the main Hawaiian islands: The effects of fishing down apex predators. Mar. Ecol. Prog. Ser. 230: 253–264.

Friedlander, A.M., D. Obura, R. Aumeeruddy, E. Ballesteros, J. Church, E. Cebrian and E. Sala. 2014. Coexistence of low coral cover and high fish biomass at Farquhar Atoll, Seychelles. PLoS ONE 9: e87359.

Friedlander, A.M., Y. Golbuu, E. Ballesteros, J.E. Caselle, M. Gouezo, M. Olsudong and E. Sala. 2017. Size, age, and habitat determine effectiveness of Palau's Marine Protected Areas. PLoS One 12: e0174787.

Frisch, A.J., R. Baker, J.-P.A. Hobbs and L. Nankervis. 2008. A quantitative comparison of recreational spearfishing and linefishing on the Great Barrier Reef: Implications for management of multi-sector coral reef fisheries. Coral Reefs 27: 85–95.

Frisch, A.J., A.J. Cole, J.-P.A. Hobbs, J.R. Rizzari and K.P. Munkres. 2012. Effects of spearfishing on reef fish populations in a multi-use conservation area. PLoS ONE 7: e51938.

Frisch, A.J., D.S. Cameron, M.S. Pratchett, D.H. Williamson, A.J. Williams, A.D. Reynolds, A.S. Hoey, J.R. Rizzari, L. Evans, B. Kerrigan, G. Muldoon, D.J. Welch and J.-P.A. Hobbs. 2016a. Key aspects of the biology, fisheries and management of Coral grouper. Rev. Fish Biol. Fish. 1–23.

Frisch, A.J., M. Ireland, J.R. Rizzari, O.M. Lönnstedt, K.A. Magnenat, C.E. Mirbach and J.-P.A. Hobbs. 2016b. Reassessing the trophic role of reef sharks as apex predators on coral reefs. Coral Reefs 35: 459–472.

Froese, R., G. Coro, K. Kleisner and N. Demirel. 2014. Revisiting safe biological limits in fisheries. Fish Fish. 17: 193–209.

Fukunaga, K., F. Yamashina, Y. Takeuchi, C. Yamauchi and A. Takemura. 2020. Moonlight is a key entrainer of lunar clock in the brain of the tropical grouper with full moon preference. BMC Zool. 5: 11.

Fulton, E.A. and A.D.M. Smith. 2004. Lessons learnt from a comparison of three ecosystem models for Port Phillip Bay, Australia. Afr. J. Mar. Sci. 26: 219–243.
Fulton, S., J. Caamal-Madrigal, A. Aguilar-Pereira, L. Bourillón and W.D. Heyman. 2018. Marine conservation outcomes are more likely when fishers participate as citizen scientists: Case studies from the Mexican Mesoamerican Reef. Citizen Science: Theory and Practice 3(1): 1–12.
Gaines, S.D., C. White, M.H. Carr and S.R. Palumbi. 2010. Designing marine reserve networks for both conservation and fisheries management. Proc. Natl. Acad. Sci. 107: 18286–18293.
Garcia, S.M., A. Zerbi, C. Aliaume, T. Do Chi and G. Laserre. 2003. The Ecosystem Approach to Fisheries, Issues, Terminology, Principles, Institutional Foundations, Implementation and Outlook. FAO Fisheries Technical Paper. 443, Food & Agriculture Organization of the United Nations, Rome, Italy: 71 pp.
García-Charton, J.A. and A. Pérez-Ruzafa. 2001. Spatial pattern and the habitat structure of a Mediterranean rocky reef fish local assemblage. Mar. Biol. 138: 917–934.
García-Charton, J.A., A. Pérez-Ruzafa, P. Sánchez-Jerez, J.T. Bayle-Sempere, O. Reñones and D. Moreno. 2004. Multi-scale spatial heterogeneity, habitat structure, and the effect of marine reserves on Western Mediterranean rocky reef fish assemblages. Mar. Biol. 144: 161–182.
García-Charton, J.A., A. Pérez-Ruzafa, C. Marcos, J. Claudet, F. Badalamenti, L. Benedetti-Cecchi, J.M. Falcón, M. Milazzo, P.J. Schembri, B. Stobart, F. Vandeperre, A. Brito, R. Chemello, M. Dimech, P. Domenici, I. Guala, L. Le Diréach, E. Maggi and S. Planes. 2008. Effectiveness of European Atlanto-Mediterranean MPAs: Do they accomplish the expected effects on populations, communities and ecosystems? J. Nat. Conserv. 16: 193–221.
García-Rubies, A., B. Hereu and M. Zabala. 2013. Long-term recovery patterns and limited spillover of large predatory fish in a Mediterranean MPA. PLoS ONE 8: e73922.
Garrison, L.P., J.S. Link, P. Kilduff, M.D. Cieri, B. Muffey, D.S. Vaughan, A. Sharov, B. Mahmoudi and R.J. Latour. 2010. An expansion of the MSVPA approach for quantifying predator-prey interactions in exploited fish communities. ICES J. Mar. Sci. 67: 856–870.
Gaspare, L., I. Bryceson and K. Kulindwa. 2015. Complementarity of fishers' traditional ecological knowledge and conventional science: Contributions to the management of groupers (Epinephelinae) fisheries around Mafia Island, Tanzania. Ocean Coast. Manag. 114: 88–101.
Gerhardinger, L.C., A.A. Bertoncini and M. Hostim-Silva. 2006. Local ecological knowledge and Goliath grouper spawning aggregations in the South Atlantic Ocean: Goliath grouper spawning aggregations in Brazil. SPC Tradit. Mar. Resour. Manag. Knowl. Inf. Bull. 20: 33–34.
Gerhardinger, L.C., M. Hostim-Silva, R.P. Medeiros, J. Matarezi, Á.A. Bertoncini, M.O. Freitas and B.P. Ferreira. 2009. Fishers' resource mapping and goliath grouper *Epinephelus itajara* (Serranidae) conservation in Brazil. Neotropical Ichthyol. 7: 93–102.
Giakoumi, S., C. Scianna, J. Plass-Johnson, F. Micheli, K. Grorud-Colvert, P. Thiriet, J. Claudet, G. Di Carlo, A. Di Franco, S.D. Gaines, J.A. García-Charton, J. Lubchenco, J. Reimer, E. Sala and P. Guidetti. 2017. Ecological effects of full and partial protection in the crowded Mediterranean Sea: a regional meta-analysis. Sci. Rep. 7: 1–12.
Gibson, J. 2007. Managing a Nassau grouper fishery: A case study from Belize. Proceedings 60th Gulf Caribb. Fish. Inst. 603–604.
Giglio, V.J., Á.A. Bertoncini, B.P. Ferreira, M. Hostim-Silva and M.O. Freitas. 2014. Landings of goliath grouper, *Epinephelus itajara*, in Brazil: Despite prohibited over ten years, fishing continues. Nat. Conserv. 12: 118–123.
Giglio, V.J., M.G. Bender, C. Zapelini and C.E.L. Ferreira. 2017. The end of the line? Rapid depletion of a large-sized grouper through spearfishing in a subtropical marginal reef. Perspect. Ecol. Conserv. 15: 115–118.
Giglio, V.J., M.L.G. Ternes, O.J. Luiz, C. Zapelini and M.O. Freitas. 2018. Human consumption and popular knowledge on the conservation status of groupers and sharks caught by small-scale fisheries on Abrolhos Bank, SW Atlantic. Mar. Pol. 89: 142–146.

Gill, D.A., P.W. Schuhmann and H.A. Oxenford. 2015. Recreational diver preferences for reef fish attributes: Economic implications of future change. Ecol. Econ. 111: 48–57.
Glazier, E. 2011. Ecosystem Based Fisheries Management in the Western Pacific. John Wiley & Sons.
Goetze, J.S., T.J. Langlois, D.P. Egli and E.S. Harvey. 2011. Evidence of artisanal fishing impacts and depth refuge in assemblages of Fijian reef fish. Coral Reefs 30: 507–517.
Goñi, R., S. Adlerstein, D. Alvarez-Berastegui, A. Forcada, O. Reñones, G. Criquet, S. Polti, G. Cadiou, C. Valle, P. Lenfant and others. 2008. Spillover from six Western Mediterranean marine protected areas: Evidence from artisanal fisheries. Mar. Ecol. Prog. Ser. 366: 159–174.
Goodyear, C.P. and M.J. Schirripa. 1993. The red grouper fishery of the Gulf of Mexico. Southeast Fisheries Center, Miami Laboratory contribution No. MIA-92/93-75. 122 pp.
Gordoa, A. 2009. Characterization of the infralittoral system along the north-east Spanish coast based on sport shore-based fishing tournament catches. Estuar. Coast. Shelf Sci. 82: 41–49.
Gouezo, M., Asap Bukurrou, Mark Priest, Lincoln Rehm, Geory Mereb, Dawnette Olsudong, Arius Merep and Kevin Polloi. 2015. Grouper Spawning Aggregations: the effectiveness of protection and fishing regulations. PICRC Technical Report No. 15–13.
Grafton, R.Q., R. Hilborn, D. Squires, M. Tait and M. Williams (eds.). 2010. Handbook of Marine Fisheries Conservation and Management. Oxford University Press, USA.
Graham, R.T., K.L. Rhodes and D. Castellanos. 2009. Characterization of the goliath grouper Epinephelus itajara fishery of southern Belize for conservation planning. Endanger. Species Res. 7: 195–204.
Grol, M.G.G., J. Vercelloni, T.M. Kenyon, E. Bayraktarov, C.P. van den Berg, D. Harris, J.A. Loder, M. Mihaljević, P.I. Rowland and C.M. Roelfsema. 2021. Conservation value of a subtropical reef in south-eastern Queensland, Australia, highlighted by citizen-science efforts. Mar. Freshw. Res. 72: 1–13.
Grorud-Colvert, K., S.E. Lester, S. Airamé, E. Neeley and S.D. Gaines. 2010. Communicating marine reserve science to diverse audiences. Proc. Natl. Acad. Sci. 107: 18306–18311.
Grüss, A., J. Robinson, S.S. Heppell, S.A. Heppell and B.X. Semmens. 2014. Conservation and fisheries effects of spawning aggregation marine protected areas: What we know, where we should go, and what we need to get there. ICES J. Mar. Sci. 71: 1515–1534.
Grüss, A., M.J. Schirripa, D. Chagaris, M. Drexler, J. Simons, P. Verley, Y.-J. Shin, M. Karnauskas, R. Oliveros-Ramos and C.H. Ainsworth. 2015. Evaluation of the trophic structure of the West Florida Shelf in the 2000s using the ecosystem model OSMOSE. J. Mar. Syst. 144: 30–47.
Grüss, A., M.J. Schirripa, D. Chagaris, L. Velez, Y.-J. Shin, P. Verley, R. Oliveros-Ramos and C.H. Ainsworth. 2016. Estimating natural mortality rates and simulating fishing scenarios for Gulf of Mexico red grouper (*Epinephelus morio*) using the ecosystem model OSMOSE-WFS. J. Mar. Syst. 154, Part B: 264–279.
Guidetti, P., F. Giardina and E. Azzurro. 2010. A new record of *Cephalopholis taeniops* in the Mediterranean Sea, with considerations on the Sicily channel as a biogeographical crossroad of exotic fish. Mar. Biodivers. Rec. 3: e13.
Guidetti, P. and F. Micheli. 2011. Ancient art serving marine conservation. Front. Ecol. Environ. 9: 374–375.
Gulf of Mexico Fishery Management Council. 2016a. Adjust red grouper allowable harvest. Framework action to the fishery management plan for reef fish resources of the Gulf of Mexico. GMFMC - NOAA, Tampa, Florida: 117 pp.
Gulf of Mexico Fishery Management Council. 2016b. Modifications to Gag minimum size limits, recreational season and black grouper minimum size limits. Framework action for the fishery management plan for the reef fish resources of the Gulf of Mexico. GMFMC - NOAA, Tampa, Florida: 107 pp.
Hackradt, C.W., F.C. Félix-Hackradt and J.A. García-Charton. 2011. Influence of habitat structure on fish assemblage of an artificial reef in southern Brazil. Mar. Environm. Res. 72: 235–247.
Hackradt, C.W., J.A. García-Charton, M. Harmelin-Vivien, Á. Pérez-Ruzafa, L. Le Diréach, J. Bayle-Sempere, E. Charbonnel, D. Ody, O. Reñones, P. Sanchez-Jerez and C. Valle. 2014.

Response of rocky reef top predators (Serranidae: Epinephelinae) in and around marine protected areas in the Western Mediterranean Sea. PLoS ONE 9: e98206.

Hackradt, C.W., F.C. Félix-Hackradt, J. Treviño-Otón, A. Pérez-Ruzafa and J.A. García-Charton. 2020. Density-driven habitat use differences across fishing zones by predator fishes (Serranidae) in south-western Mediterranean rocky reefs. Hydrobiologia 847: 757–770.

Halpern, B.S. 2003. The impact of marine reserves: do reserves work and does reserve size matter? Ecol. Appl. 13: 117–137.

Hamilton, R.J. 2014. Square-tailed coral grouper *Plectropomus areolatus* (grouper)—Melanesia. p. 10. *In*: Russell, M.W., Y. Sadovy de Mitcheson, B. Erisman, R.J. Hamilton, B.E. Luckhurst and R.S. Nemeth (eds.). Status Report—World's Fish Aggregations 2014. International Coral Reef Initiative, California, USA.

Harmelin-Vivien, M., L. Le Diréach, J. Bayle-Sempere, E. Charbonnel, J.A. García-Charton, D. Ody, A. Pérez-Ruzafa, O. Reñones, P. Sánchez-Jerez and C. Valle. 2008. Gradients of abundance and biomass across reserve boundaries in six Mediterranean marine protected areas: Evidence of fish spillover? Biol. Conserv. 141: 1829–1839.

Heemstra, P.C. and J.E. Randall. 1993. FAO species catalogue. Vol. 16. Groupers of the world (Family Serranidae, Subfamily Epinephelinae). An annotated and illustrated catalogue of the grouper, rockcod, hind, coral grouper and lyretail species known to date. FAO Fisheries Synopsis No. 125, FAO, Rome, Italy: 382 pp.

Heithaus, M.R., A. Frid, A.J. Wirsing and B. Worm. 2008. Predicting ecological consequences of marine top predator declines. Trends Ecol. Evol. 23: 202–210.

Heppell, S.S., S.A. Heppell, F.C. Coleman and C.C. Koenig. 2006. Models to compare management options for a protogynous fish. Ecol. Appl. 16: 238–249.

Hereu, B., D. Diaz, J. Pasqual, M. Zabala and E. Sala. 2006. Temporal patterns of spawning of the dusky grouper *Epinephelus marginatus* in relation to environmental factors. Mar. Ecol. Prog. Ser. 325: 187–194.

Hilborn, R. and C.J. Walters. 1992. Quantitative Fisheries Stock Assessment: Choice, Dynamics and Uncertainty. Chapman and Hall, New York.

Hilborn, R., K. Stokes, J.-J. Maguire, T. Smith, L.W. Botsford, M. Mangel, J. Orensanz, A. Parma, J. Rice, J. Bell, K.L. Cochrane, S. Garcia, S.J. Hall, G.P. Kirkwood, K. Sainsbury, G. Stefansson and C. Walters. 2004. When can marine reserves improve fisheries management? Ocean Coast. Manag. 47: 197–205.

Hilborn, R., R.O. Amoroso, C.M. Anderson, J.K. Baum, T.A. Branch, C. Costello, C.L. de Moor, A. Faraj, D. Hively, O.P. Jensen, H. Kurota, L.R. Little, P. Mace, T. McClanahan, M.C. Melnychuk, C. Minto, G.C. Osio, A.M. Parma, M. Pons, S. Segurado, C.S. Szuwalski, J.R. Wilson and Y. Ye. 2020. Effective fisheries management instrumental in improving fish stock status. Proc. Natl. Acad. Sci. 117: 2218-2224.

Hogg, K., P. Noguera-Méndez, M. Semitiel-García and M. Giménez-Casalduero. 2013. Marine protected area governance: Prospects for co-management in the European Mediterranean. Adv. Oceanogr. Limnol. 4: 241–259.

Hoggarth, D.D., S. Abeyasekera, R.I. Arthur, J.R. Beddington, R.W. Burn, A.S. Halls, G.P. Kirkwood, M.K. McAllister, P. Medley, C.C. Mees, G.B. Parkes, G.M. Pilling, R.C. Wakeford and R.L. Welcomme. 2006. Stock Assessment for Fishery Management: A Framework Guide to the Stock Assessment Tools of the Fisheries Management and Science Programme. Food & Agriculture Organization, Rome, Italy: 19 pp.

Honey, K.T., J.H. Moxley and R.M. Fujita. 2010. From rags to fishes: Data-poor methods for fishery managers. Manag. Data-Poor Fish. Case Stud. Models Solut. 1: 155–180.

Hughes, A.T., R.J. Hamilton, J.H. Choat and K.L. Rhodes. 2020. Declining grouper spawning aggregations in Western Province, Solomon Islands, signal the need for a modified management approach. PLoS ONE 15(3): e0230485.

IUCN. 2016. Motion O53: increasing marine protected area coverage for effective marine biodiversity conservation. WCC-2016-Res-050-EN.

IUCN. 2020. The IUCN Red List of Threatened Species. Version 2020–21. https://www.iucnredlist.org.

Johannes, R.E. 1998. The case for data-less marine resource management: Examples from tropical nearshore finfisheries. Trends Ecol. Evol. 13: 243–246.

Johnston, R.J., D.S. Holland, V. Maharaj and T.W. Campson. 2007. Fish harvest tags: An alternative management approach for recreational fisheries in the US Gulf of Mexico. Mar. Policy 31: 505–516.

Jones, P.J.S. and E.M. De Santo. 2016. Is the race for remote, very large marine protected areas (VLMPAs) taking us down the wrong track? Mar. Pol. 73: 231–234.

Kaunda-Arara, B. and G.A. Rose. 2004. Homing and site fidelity in the greasy grouper *Epinephelus tauvina* (Serranidae) within a marine protected area in coastal Kenya. Mar. Ecol. Prog. Ser. 277: 245–251.

Kennelly, S.J. and M.K. Broadhurst. 2002. By-catch begone: Changes in the philosophy of fishing technology. Fish Fish. 3: 340–355.

Kobara, S. and W.D. Heyman. 2010. Sea bottom geomorphology of multi-species spawning aggregation sites in Belize. Mar. Ecol. Prog. Ser. 405: 231–242.

Kobara, S., W.D. Heyman, S.J. Pittman and R.S. Nemeth. 2013. Biogeography of transient reef fish spawning aggregations in the Caribbean: a synthesis for future research and Management. Ocean. Mar. Biol. Ann. Rev. 51: 281–326.

Kockel, A., N.C. Ban, M. Costa and P. Dearden. 2019. Evaluating approaches for scaling-up community-based marine-protected areas into socially equitable and ecologically representative networks. Conserv. Biol. 34: 137–147.

Koeck, B., J. Pastor, G. Saragoni, N. Dalias, J. Payrot and P. Lenfant. 2014. Diel and seasonal movement pattern of the dusky grouper *Epinephelus marginatus* inside a marine reserve. Mar. Environ. Res. 94: 38–47.

Koenig, C.C., F.C. Coleman and K. Kingon. 2011. Pattern of recovery of Goliath grouper *Epinephelus itajara* population in the Southeastern US. Bull. Mar. Sci. 87: 891–911.

Koenigstein, S., F.C. Mark, S. Gößling-Reisemann, H. Reuter and H.-O. Poertner. 2016. Modelling climate change impacts on marine fish populations: Process-based integration of ocean warming, acidification and other environmental drivers. Fish Fish. 17: 972–1004.

Lassen, H. and P. Medley. 2001. Virtual Population Analysis: A practical manual for stock assessment. FAO Fisheries Technical Paper. No. 400. Food and Agriculture Organization of the United Nations, Rome, Italy: 128 pp.

Leigh, G.M., A.B. Campbell, C.P. Lunow and M.F. O'Neill. 2014. Stock assessment of the Queensland east coast common coral trout (*Plectropomus leopardus*) fishery. The State of Queensland, Brisbane, Australia: 115 pp.

Lester, S.E. and B.S. Halpern. 2008. Biological responses in marine no-take reserves versus partially protected areas. Mar. Ecol. Prog. Ser. 367: 49–56.

Lima, E.G., A. Begossi, G. Hallwass and R.A.M. Silvano. 2016. Fishers' knowledge indicates short-term temporal changes in the amount and composition of catches in the southwestern Atlantic. Mar. Policy 71: 111–120.

Lindberg, W.J., T.K. Frazer, K.M. Portier, F. Vose, J. Loftin, D.J. Murie, D.M. Mason, B. Nagy and M.K. Hart. 2006. Density-dependent habitat selection and performance by a large mobile reef fish. Ecol. Appl. Publ. Ecol. Soc. Am. 16: 731–746.

Lindfield, S.J., J.L. McIlwain and E.S. Harvey. 2014. Depth refuge and the impacts of scuba spearfishing on coral reef fishes. PLoS ONE 9: e92628.

Lloret, J., N. Zaragoza, D. Caballero, T. Font, M. Casadevall and V. Riera. 2008. Spearfishing pressure on fish communities in rocky coastal habitats in a Mediterranean marine protected area. Fish. Res. 94: 84–91.

Lloret, J. and T. Font. 2013. A comparative analysis between recreational and artisanal fisheries in a Mediterranean coastal area. Fish. Manag. Ecol. 20: 148–160.

Lubchenco, J., S.R. Palumbi, S.D. Gaines and S. Andelman. 2003. Plugging a hole in the ocean: The emerging science of marine reserves. Ecol. Appl. 13: 3–7.

Lubchenco, J. and K. Grorud-Colvert. 2015. Making waves: The science and politics of ocean protection. Science 350: 382–383.

Luckhurst, B.E. and T.M. Trott. 2008. Seasonally-closed spawning aggregation sites for red hind (*Epinephelus guttatus*): Bermuda's experience over 30 years (1974–2003). pp. 331–336. *In*: CGFI (ed.). Proceedings of the 61st Gulf and Caribbean Fisheries Institute. CGFI, Gosier, Guadeloupe, French West Indies. Vol. 61.

Luckhurst, B.E. and R.S. Nemeth. 2014. Red hind *Epinephelus guttatus* (grouper)—Tropical Western Atlantic. p. 9. *In*: Russell, M.W., Y. Sadovy de Mitcheson, B. Erisman, R.J. Hamilton, B.E. Luckhurst and R.S. Nemeth (eds.). Status Report—World's Fish Aggregations 2014. International Coral Reef Initiative, California, USA.

Luiz, O.J., R.M. Woods, E.M.P. Madin and J.S. Madin. 2016. Predicting IUCN extinction risk categories for the world's Data Deficient groupers (Teleostei: Epinephelidae). Conserv. Lett. 9: 342–350.

Madin, E.M.P., S.D. Gaines, J.S. Madin, A.-K. Link, P.J. Lubchenco, R.L. Selden and R.R. Warner. 2012. Do behavioral foraging responses of prey to predators function similarly in restored and pristine foodwebs? PLoS ONE 7: e32390.

Magris, R.A. and R.L. Pressey. 2018. Marine protected areas: Just for show? Science 360: 723–724.

Mancha-Cisneros, M.M., A.N. Suárez-Castillo, J. Torre, J.M. Anderies and L.R. Gerber. 2018. The role of stakeholder perceptions and institutions for marine reserve efficacy in the Midriff Islands Region, Gulf of California, Mexico. Ocean Cost. Manage. 162: 181–192.

Manojkumar, P.P., L. Ranjith, K. Karuppasamy and K.P. Kanthan. 2019. Fishery and population dynamics of *Epinephelus malabaricus* (Bloch & Schneider, 1801) off Tuticorin, southeast coast of India. J. Mar. Biol. Ass. India 61: 26–30.

Marancik, K.E., D.E. Richardson, J. Lyczkowski-Shultz, R.K. Cowen and M. Konieczna. 2012. Spatial and temporal distribution of grouper larvae (Serranidae: Epinephelinae: Epinephelini) in the Gulf of Mexico and Straits of Florida. Fish. Bull. 110: 1–20.

Marzloff, M., Y.-J. Shin, J. Tam, M. Travers and A. Bertrand. 2009. Trophic structure of the Peruvian marine ecosystem in 2000–2006: Insights on the effects of management scenarios for the hake fishery using the IBM trophic model Osmose. J. Mar. Syst. 75: 290–304.

Mateos-Molina, D., M.T. Schärer-Umpierre, R.S. Appeldoorn and J.A. García-Charton. 2014. Measuring the effectiveness of a Caribbean oceanic island no-take zone with an asymmetrical BACI approach. Fish. Res. 150: 1–10.

Matley, J.K., M.R. Heupel and C.A. Simpfendorfer. 2015. Depth and space use of leopard coral grouper *Plectropomus leopardus* using passive acoustic tracking. Mar. Ecol. Prog. Ser. 521: 201–216.

Maunder, M.N. and K.R. Piner. 2015. Contemporary fisheries stock assessment: many issues still remain. ICES J. Mar. Sci. 72: 7–18.

Mavruk, S., I. Saygu, F. Bengil, V. Alan and E. Azzurro. 2018. Grouper fishery in the Northeastern Mediterranean: An assessment based on interviews on resource users. Mar. Pol. 87: 141–148.

Mavruk, S. 2020. Trends of white grouper landings in the Northeastern Mediterranean: reliability and potential use for monitoring. Med. Mar. Sci. 21: 183–190.

McClanahan, T.R., N.A.J. Graham, J.M. Calnan and M.A. MacNeil. 2007. Toward pristine biomass: Reef fish recovery in coral reef marine protected areas in Kenya. Ecol. Appl. 17: 1055–1067.

McClanahan, T.R. 2019. Coral reef fish community life history traits as potential global indicators of ecological and fisheries status. Ecol. Indic. 96(1): 133–145.

McClenachan, L. 2009. Historical declines of goliath grouper populations in South Florida, USA. Endanger. Species Res. 7: 175–181.

McGoodwin, J.R. 1990. Crisis in the World's Fisheries: People, Problems, and Policies. Stanford University Press, Stanford, California.

McLeod, E., R. Salm, A. Green and J. Almany. 2009. Designing marine protected area networks to address the impacts of climate change. Front. Ecol. Environ. 7: 362–370.

Medeiros, M.C., R.R.D. Barboza, G. Martel and J.S. Mourão. 2018. Combining local fishers' and scientific ecological knowledge: Implications for comanagement. Ocean Coast. Manage. 158: 1–10.

Meissa, B., D. Gascuel and E. Rivot. 2013. Assessing stocks in data-poor African fisheries: A case study on the white grouper *Epinephelus aeneus* of Mauritania. Afr. J. Mar. Sci. 35: 253–267.

Messmer, V., M.S. Pratchett, A.S. Hoey, A.J. Tobin, D.J. Coker, S.J. Cooke and T.D. Clark. 2017. Global warming may disproportionately affect larger adults in a predatory coral reef fish. Global Change Biol. 23: 2230–2240.

Methot Jr., R.D. and C.R. Wetzel. 2013. Stock synthesis: A biological and statistical framework for fish stock assessment and fishery management. Fish. Res. 142: 86–99.

Micheli, F.M., B.S. Halpern, L.W. Botsford and R.R. Warner. 2004. Trajectories and correlates of community change in no-take marine reserves. Ecol. Appl. 14: 1709–1723.

Morales, A. and E. Rosello. 2004. Fishing down the food web in Iberian prehistory ? A new look at the fishes from Cueva de Nerja (Málaga, Spain). Petits Animaux Sociétés Hum. Complements Aliment. Aux Ressour. Util. 111–123.

Morales-Nin, B., J. Moranta, C. García, M.P. Tugores, A.M. Grau, F. Riera and M. Cerdà. 2005. The recreational fishery off Majorca Island (western Mediterranean): Some implications for coastal resource management. ICES J. Mar. Sci. J. Cons. 62: 727–739.

Morato, T., R. Watson, T.J. Pitcher and D. Pauly. 2006. Fishing down the deep. Fish Fish. 7: 24–34.

Morris, A.V., C.M. Roberts and J.P. Hawkins. 2000. The threatened status of groupers (Epinephelinae). Biodivers. Conserv. 9: 919–942.

Mumby, P.J., A.J. Edwards, J. Ernesto Arias-González, K.C. Lindeman, P.G. Blackwell, A. Gall, M.I. Gorczynska, A.R. Harborne, C.L. Pescod, H. Renken, C.C.C. Wabnitz and G. Llewellyn. 2004. Mangroves enhance the biomass of coral reef fish communities in the Caribbean. Nature 427: 533–536.

Mumby, P.J., C.P. Dahlgren, A.R. Harborne, C.V. Kappe, F. Micheli, D.R. Brumbaugh et al. 2006. Fishing, trophic cascades, and the process of grazing on coral reefs. Science 311: 98–101. doi: 10.1126/science.1121129.

Mumby, P.J., A.R. Harborne and D.R. Brumbaugh. 2011. Grouper as a natural biocontrol of invasive lionfish. PLoS ONE 6: e21510.

Muntoni, M., R. Devillers and M. Koen-Alonso. 2019. Science should not be left behind during the design of a marine protected area: meeting conservation priorities while integrating stakeholder interests. FACETS 4: 472–492.

Murawski, S.A. 2000. Definitions of overfishing from an ecosystem perspective. ICES J. Mar. Sci. J. Cons. 57: 649–658.

Myers, R.A. and B. Worm. 2003. Rapid worldwide depletion of predatory fish communities. Nature 423: 280–283.

Ndiaye, W., M. Thiaw, K. Diouf, P. Ndiaye, O.T. Thiaw and J. Panfili. 2013. Changes in population structure of the white grouper *Epinephelus aeneus* as a result of long-term overexploitation in Senegalese waters. Afr. J. Mar. Sci. 35: 465–472.

Nemeth, R.S. 2005. Population characteristics of a recovering US Virgin Islands red hind spawning aggregation following protection. Mar. Ecol. Prog. Ser. 286: 81–97.

Nemeth, R.S., J. Blondeau, S. Herzlieb and E. Kadison. 2007. Spatial and temporal patterns of movement and migration at spawning aggregations of red hind, *Epinephelus guttatus*, in the US Virgin Islands. Environ. Biol. Fishes 78: 365–381. doi: 10.1007/s10641-006-9161-x.

Newman, D., J. Berkson and L. Suatoni. 2015a. Current methods for setting catch limits for data-limited fish stocks in the United States. Fish. Res. 164: 86–93.

Newman, S.J., C.B. Wakefield, A.J. Williams, J.M. O'Malley, S.J. Nicol, E.E. DeMartini, T. Halafihi, J. Kaltavara, R.L. Humphreys, B.M. Taylor, A.H. Andrews and R.S. Nichols. 2015b. International workshop on methodological evolution to improve estimates of life history parameters and fisheries management of data-poor deep-water snappers and groupers. Mar. Policy 60: 182–185.

NOAA – National Oceanic and Atmospheric Administration. 2014. Nassau Grouper, *Epinephelus striatus* (Bloch 1792): Biological Report. 117 pp.

Nowlis, J.S. 2006. Virtual population analysis of the Gulf of Mexico gag grouper (*Mycteroperca microlepis*) stock: The continuity case. NOAA Fisheries - SEDAR. SEDAR10-RW-01, Miami, Florida: 13 pp.

O'Leary, B.C., M. Winther-Janson, J.M. Bainbridge, J. Aitken, J.P. Hawkins and C.M. Roberts. 2016. Effective coverage targets for ocean protection. Conserv. Lett. 9: 398–404.

O'Leary, B.C., N.C. Ban, M. Fernandez, A.M. Friedlander, P. García-Borboroglu, Y. Golbuu, P. Guidetti, J.M. Harris, J.P. Hawkins, T. Langlois, D.J. McCauley, E.K. Pikitch, R.H. Richmond and C.M. Roberts. 2018. Addressing criticisms of large-scale marine protected areas. BioScience 58: 359–370.

Ojeda-Serrano, E., R. Appeldoorn and I. Ruiz-Valentin. 2007. Reef fish spawning aggregations of the Puerto Rican Shelf. CCRI — Caribbean Coral Reef Institute. 31 pp.

Ormerod, S.J. 2003. Current issues with fish and fisheries: Editor's overview and introduction. J. Appl. Ecol. 40: 204–213.

Pace, M.L., S.E. Hampton, K.E. Limburg, E.M. Bennett, E.M. Cook, A.E. Davis, J.M. Grove, K.Y. Kaneshiro, S.L. LaDeau, G.E. Likens, D.M. McKnight, D.C. Richardson and D.L. Strayer. 2010. Communicating with the public: Opportunities and rewards for individual ecologists. Front. Ecol. Environ. 8: 292–298.

Palacios, M.M., S.S. Killen, L.E. Nadler, J.R. White and M.I. McCormick. 2016. Top predators negate the effect of mesopredators on prey physiology. J. Anim. Ecol. 85: 1078–1086.

Pauly, D., V. Christensen, J. Dalsgaard, R. Froese and F. Torres. 1998. Fishing down marine food webs. Science 279: 860–863.

Pauly, D., Christensen and Walters. 2000. Ecopath, Ecosim, and Ecospace as tools for evaluating ecosystem impact of fisheries. ICES J. Mar. Sci. 57: 697.

Pauly, D., V. Christensen, S. Guénette, T.J. Pitcher, U.R. Sumaila, C.J. Walters, R. Watson and D. Zeller. 2002. Towards sustainability in world fisheries. Nature 418: 689–695.

Pauly, D. and D. Zeller. 2015. Sea Around Us concepts, design and data (seaaroundus.org).

Peck, M.A., K.B. Huebert and J.K. Llopiz. 2012. Chapter 3 – Intrinsic and extrinsic factors driving match-mismatch dynamics during the early life history of marine fishes. pp. 177–302. *In*: Woodward, G., U. Jacob and E.J. O'Gorman (eds.). Global Change in Multispecies Systems Part 2. Advances in Ecological Research. Academic Press. Vol. 47.

Pelc, R.A., M.L. Baskett, T. Tanci, S.D. Gaines and R.R. Warner. 2009. Quantifying larval export from South African marine reserves. Mar. Ecol. Prog. Ser. 394: 65–78.

Pelc, R.A., R.R. Warner, S.D. Gaines and C.B. Paris. 2010. Detecting larval export from marine reserves. Proc. Natl. Acad. Sci. 107: 18266–18271.

Pérez-Ruzafa, A., J.A. García-Charton and C. Marcos. 2017. Northeast Atlantic vs. Mediterranean marine protected areas as fisheries management tool. Front. Mar. Sci. 4: 245.

Pikitch, E.K., C. Santora, E.A. Babcock, A. Bakun, R. Bonfil, D.O. Conover, P. Dayton, P. Doukakis, D. Fluharty, B. Heneman, E.D. Houde, J. Link, P.A. Livingston, M. Mangel, M.K. McAllister, J. Pope and K.J. Sainsbury. 2004. Ecosystem-based fishery management. Science 305: 346–347.

Pina-Amargós, F., G. González-Sansón, F. Martín-Blanco and A. Valdivia. 2014. Evidence for protection of targeted reef fish on the largest marine reserve in the Caribbean. PeerJ 2: e274.

Pita, A., J. Casey, S.J. Hawkins, M. Ruiz-Villarreal and M.J. Gutiérrez. 2016. Conceptual and practical advances in fish stock delineation. Fish. Res. 173: 185–193.

Pitt, J., T. Warren and C. Trott. 2017. Managing fish spawning aggregations in a changing climate: A case study of Red hind (*Epinephelus guttatus*) in Bermuda. Proc. 70th Gulf Caribb. Fish. Inst., Nov. 6–10, 2017, Mérida, Mexico: 306–308.

Plagányi, É.E. 2007. Models for an Ecosystem Approach to Fisheries. FAO Fisheries Technical Paper n.477, Food and Agriculture Organization of the United Nations, Rome, Italy: 128 pp.

Planes, S., G.P. Jones and S.R. Thorrold. 2009. Larval dispersal connects fish populations in a network of marine protected areas. Proc. Natl. Acad. Sci. 106: 5693–5697.

Polidoro, B.A., S.R. Livingstone, K.E. Carpenter, B. Hutchinson, R.B. Mast, N. Pilcher, Y. Sadovy de Mitcheson and S. Valenti. 2008. Status of the world's marine species. p. 12. *In*: Vié, J.-C., C. Hilton-Taylor and S.N. Stuart (eds.). The 2008 Review of The IUCN Red List of Threatened Species. IUCN, Gland, Switzerland.

Polunin, N.V.C. and C.M. Roberts. 1993. Greater biomass and value of target coral-reef fishes in two small Caribbean marine reserves. Mar. Ecol.-Prog. Ser. 100: 167–167.

Porch, C.E., A.-M. Eklund and G.P. Scott. 2006. A catch-free stock assessment model with application to goliath grouper (*Epinephelus itajara*) off southern Florida. Fish. Bull. 104: 89–101.

Pörtner, H.O. and A.P. Farrell. 2008. Physiology and climate change. Science 322: 690–692.

Pratchett, M.S., D.S. Cameron, J. Donelson, L. Evans, A.J. Frisch, A.J. Hobday, A.S. Hoey, N.A. Marshall, V. Messmer, P.L. Munday, R. Pears, G. Pecl, A. Reynolds, M. Scott, A. Tobin, R. Tobin, D.J. Welch and D.H. Williamson. 2017. Effects of climate change on coral grouper (*Plectropomus* spp.) and possible adaptation options. Rev. Fish Biol. Fish. 27: 297–316.

Prato, G., D. Gascuel, A. Valls and P. Francour. 2014. Balancing complexity and feasibility in Mediterranean coastal food-web models: Uncertainty and constraints. Mar. Ecol. Prog. Ser. 512: 71–88.

Prato, G. 2016. Stratégie d'échantillonnage et modélisation trophique : Des outils de gestion pour évaluer le fonctionnement des écosystèmes et le statut des prédateurs de haut niveau trophique dans les aires marines protégées méditerranéennes. PhD Thesis, Université de Nice, Nice Sofia-Antipolis, France.

Prato, G., C. Barrier, P. Francour, V. Cappanera, V. Markantonatou, P. Guidetti, L. Mangialajo, R. Cattaneo-Vietti and D. Gascuel. 2016. Assessing interacting impacts of artisanal and recreational fisheries in a small Marine Protected Area (Portofino, NW Mediterranean Sea). Ecosphere 7: e01601.

Préfecture, P.A.C.A. 2013. Arrêté n°2013357-0004 du 23 décembre 2013. Available at: http://www.var.gouv.fr/IMG/pdf/Interdiction_de_chasse_du_Merou_cle691d3e.pdf [Accessed June 15, 2016].

Retnonigntyas, H., I. Yulianto, A. Soemodinoto, Y. Herdiana, T. Kartawijaya, M. Natsir and J.T. Haryanto. 2021. Stakeholder participation in management planning for grouper and snapper fisheries in West Nusa Tenggara Province, Indonesia. Mar. Pol. 128: 104452.

Reum, J.C.P., H. Townsend, S. Gaichas, S. Sagarese, I.C. Kaplan and A. Grüss. 2021. It's not the destination, it's the journey: Multispecies model ensembles for ecosystem approaches to fisheries management. Front. Mar. Sci. 8: 631839.

Rhodes, K.L. and M.H. Tupper. 2007. A preliminary market-based analysis of the Pohnpei, Micronesia, grouper (Serranidae: Epinephelinae) fishery reveals unsustainable fishing practices. Coral Reefs 26: 335–344.

Rhodes, K.L., B.M. Taylor, C.B. Wichilmel, E. Joseph, R.J. Hamilton and G.R. Almany. 2013. Reproductive biology of squaretail coral grouper *Plectropomus areolatus* using age-based techniques. J. Fish Biol. 82: 1333–1350.

Rhodes, K.L., I.E. Baremore, B.M. Taylor, J. Cuetos-Bueno and D. Hernandez. 2020. Aligning fisheries management with life history in two commercially important groupers in Chuuk, Federated States of Micronesia. Aquatic Conserv.: Mar. Freshw. Ecosyst. 2020: 1–15.

Ribeiro, A.R., L.M.A. Damasio and R.A.M. Silvano. 2021. Fishers' ecological knowledge to support conservation of reef fish (groupers) in the tropical Atlantic. Ocean Coast. Manage. 204: 105543.

Richu, A., N. Dahanukar, A. Ali, K. Ranjeet and R. Raghavan. 2018. Population dynamics of a poorly known serranid, the duskytail grouper *Epinephelus bleekeri* in the Arabian Sea. J. Fish Biol. 93: 741–744.

Rizzari, J.R., A.J. Frisch, A.S. Hoey and M.I. McCormick. 2014. Not worth the risk: Apex predators suppress herbivory on coral reefs. Oikos 123: 829–836.

Roberts, C.M. and N.V.C. Polunin. 1993. Marine reserves: Simple solutions to managing complex fisheries? Ambio 22: 363–368.

Roberts, C.M. 2007. The Unnatural History of the Sea. Island Press, Washington, DC.

Roberts, C.M., J.P. Hawkins and W.E.S. Campaign. 2000. Fully-Protected Marine Reserves: A Guide. Vol. 1250. WWF, Washington, DC.

Roberts, C.M., J.A. Bohnsack, F. Gell, J.P. Hawkins and R. Goodridge. 2001. Effects of marine reserves on adjacent fisheries. Science 294: 1920–1923.

Roberts, C.M., B.C. O'Leary and J.P. Hawkins. 2020. Climate change mitigation and nature conservation both require higher protected area targets. Phil. Trans. R. Soc. B 375: 20190121.

Rocha, L.A. 2018. Bigger is not better for ocean conservation. The New York Times, March 21, 2018, Section A: 23.

Rocklin, D., J.-A. Tomasini, J.-M. Culioli, D. Pelletier and D. Mouillot. 2011. Spearfishing regulation benefits artisanal fisheries: The ReGS indicator and its application to a multiple-use Mediterranean marine protected area. PLoS ONE 6: e23820.

Rocklin, D., H. Levrel, M. Drogou, J. Herfaut and G. Veron. 2014. Combining telephone surveys and fishing catches self-report: The French sea bass recreational fishery assessment. PLoS ONE 9: e87271.

Rodwell, L.D., E.B. Barbier, C.M. Roberts and T.R. McClanahan. 2003. The importance of habitat quality for marine reserve fishery linkages. Can. J. Fish. Aquat. Sci. 60: 171–181.

Rojo, I., J. Sánchez-Meca and J.A. García-Charton. 2019. Small-sized and well-enforced Marine Protected Areas provide ecological benefits for piscivorous fish populations worldwide. Mar. Environ. Res. 149: 100–110.

Rojo, I., J.D. Anadón and J.A. García-Charton. 2021. Exceptionally high but still growing predatory reef fish biomass after 23 years of protection in a Marine Protected Area. PLoS One 16(2): e0246335.

Rolim, F.A., T. Langlois, P.F.C. Rodrigues, T. Bond, F.S. Motta, L.M. Neves and O.B.F. Gadig. 2019. Network of small no-take marine reserves reveals greater abundance and body size of fisheries target species. PLoS One 14(1): e0204970.

Rudd, M.A. and M.H. Tupper. 2002. The impact of Nassau grouper size and abundance on scuba diver site selection and MPA economics. Coast. Manag. 30: 133–151.

Runde, B.J. and J.A. Buckel. 2018. Descender device are promising tools for increasing survival in deepwater groupers. Mar. Coast. Fish.: Dyn. Manage. Ecosyst. Sci. 10: 100–117.

Runde, B.J., T. Michelot, N.M. Bacheler, K.W. Shertzer and J.A. Buckel. 2020. Assigning fates in telemetry studies using Hidden Markov Models: an application to deepwater groupers released with descender devices. N. Am. J. Fish. Manage. 40: 1417–1434.

Russ, G.R. and A. Alcala. 1996. Do marine reserves export adult fish biomass? Evidence from Apo Island, central Philippines. Mar. Ecol. Prog. Ser. 132: 1–9.

Russ, G.R. 2002. Yet another review of marine reserves as reef fishery management tools. pp. 421–443. *In*: Sale, P.S. (ed.). Coral Reef Fishes: Dynamics and Diversity in a Complex Ecosystem. Academic Press, San Diego, California, USA.

Russell, M. 2006. Leopard coral grouper (*Plectropomus leopardus*) management in the Great Barrier Reef Marine Park, Australia. SPC Live Reef Fish Inf. Bull. 16: 10–12.

Russell, M.W., B.E. Luckhurst and K.C. Lindeman. 2012. Management of spawning aggregations. pp. 371–404. *In*: de Mitcheson, Y.S. and P.L. Colin (eds.). Reef Fish Spawning Aggregations: Biology, Research and Management. Springer, Netherlands.

Sadovy de Mitcheson, Y., A. Cornish, M. Domeier, P. Colin, M. Russel and K.C. Lindeman. 2008. A global baseline for spawning aggregations of reef fishes. Conserv. Biol. 22, 1233–1244. doi: 10.1111/j.1523-1739.2008.0 1020.x

Sadovy de Mitcheson, Y. and B. Erisman. 2012. Fishery and biological implications of fishing spawning aggregations, and the social and economic importance of aggregating fishes. pp. 225–264. *In*: Sadovy de Mitcheson, Y. and P.L. Colin (eds.). Reef Fish Spawning Aggregations: Biology, Research and Management. Springer, Dortrecht.

Sadovy de Mitcheson, Y. and P.L. Colin. 2012. Reef fish spawning aggregations: biology, research and management. Reef Fish Spawning Aggregations: Biology, Research and Management. https://doi.org/10.1007/978-94-007-1980-4.

Sadovy de Mitcheson, Y., M.T. Craig, A.A. Bertoncini, K.E. Carpenter, W.W.L. Cheung, J.H. Choat, A.S. Cornish, S.T. Fennessy, B.P. Ferreira, P.C. Heemstra, M. Liu, R.F. Myers, D.A. Pollard, K.L. Rhodes, L.A. Rocha, B.C. Russell, M.A. Samoilys and J. Sanciangco. 2012. Fishing groupers towards extinction: A global assessment of threats and extinction risks in a billion dollar fishery. Fish Fish. 14: 119–136.

Sadovy de Mitcheson, Y. and R.S. Nemeth. 2014. Square-tailed coral grouper *Plectropomus areolatus*, camouflage grouper *Epinephelus polyphekadion*, and brown marbled grouper *Epinephelus fuscoguttatus* – Palau and Pohnpei. p. 11. *In*: Russell, M.W., Y. Sadovy de

Mitcheson, B. Erisman, R.J. Hamilton, B.E. Luckhurst and R.S. Nemeth (eds.). Status Report—World's Fish Aggregations 2014. International Coral Reef Initiative, California, USA.

Sadovy de Mitcheson, Y. 2016. Mainstreaming fish spawning aggregations into fishery management calls for a precautionary approach. Bioscience 66: 295–306.

Sadovy de Mitcheson, Y., C. Linardich, J.P. Barreiros, G.M. Ralph, A. Aguilar-Pereira, P. Afonso, B.E. Erisman, D.A. Pollard, S.T. Fennessy, A.A. Bertoncini, R.J. Nair, K.L. Rhodes, P. Francour, T. Brulé, M.A. Samoilys, B.P. Ferreira and M.T. Craig. 2020a. Valuable but vulnerable: Over-fishing and under-management continue to threaten groupers so what now? Mar. Pol. 116: 103909.

Sadovy de Mitcheson, Y., M.C.P. Triana, J.O. Azueta and K.C. Lindeman. 2020b. Regional Fish spawning aggregation fishery management plan: Focus on Nassau grouper and Mutton snapper. Caribbean Fish. Manage. Council. 104 pp.

Sadovy de Mitcheson, Y., P.L. Colin, S.J. Lindfield and A. Bukurrou. 2020c. A decade of monitoring an Indo-Pacific grouper spawning aggregation: Benefits of protection and importance of survey design. Front. Mar. Sci. 7: 571878.

Sadovy, Y.J., T.J. Donaldson, T.R. Graham, F. McGilvray, G.J. Muldoon, M.J. Phillips, M.A. Rimmer, A. Smith and B. Yeeting. 2003. While stocks last: The live reef food fish trade. Asian Development Bank, Manila, Philippines. 169 pp.

Sadovy, Y. 1997. The case of the disappearing grouper: *Epinephelus striatus* (Pisces: Serranidae). J. Fish Biol. 46: 961–976.

Sadovy, Y. and A.M. Eklund. 1999. Synopsis of biological information on the Nassau Grouper, *Epinephelus striatus* (Bloch, 1792), and the Jewfish, *E. itajara* (Lichtenstein, 1822). NOAA Technical Report NMFS 146. Technical Report of the Fishery Bulletin. FAO Fisheries Synopsis 157. US Department of Commerce, Seattle, WA USA, 65 pp.

Sadovy, Y. and M. Domeier. 2005. Are aggregation-fisheries sustainable? Reef fish fisheries as a case study. Coral Reefs 24(2): 254–262.

Sáenz–Arroyo, A., C.M. Roberts, J. Torre and M. Cariño-Olvera. 2005. Using fishers' anecdotes, naturalists' observations and grey literature to reassess marine species at risk: The case of the gulf grouper in the Gulf of California, Mexico. Fish Fish. 6: 121–133.

SAFMC – South Atlantic Fisheries Management Council. 2013. Amendment 22 to the Fishery Management Plan for the snapper grouper fishery of the South Atlantic Region. 11 pp.

SAFMC – South Atlantic Fisheries Management Council. 2015. Content and status of active amendments. 8 pp.

Sagarese, S.R., M.D. Bryan, J.F. Walter, M. Schirripa, A. Grüss and M. Karnauskas. 2015. Incorporating ecosystem considerations within the Stock Synthesis integrated assessment model for Gulf of Mexico Red Grouper (*Epinephelus morio*). SEDAR42-RW-01. 27 pp.

Sagarese, S.R., N.R. Vaughan, J.F. Walter III and M. Karnauskas. 2021. Enhancing single-species stock assessments with diverse ecosystem perspectives: a case study for Gulf of Mexico Red Grouper (*Epinephelus morio*) and red tides. Can. J. Fish. Aquat. Sci. in press.

Sala, E., E. Ballesteros and R.M. Starr. 2001. Rapid decline of Nassau grouper spawning aggregations in belize: fishery management and conservation needs. Fisheries 26: 23–30.

Sala, E., C. Costello, D. Dougherty, G. Heal, K. Kelleher, J.H. Murray, A.A. Rosenberg and R. Sumaila. 2013. A general business model for marine reserves. PLoS ONE 8: e58799.

Sala, E., C. Costello, J. De Bourbon Parme, M. Fiorese, G. Heal, K. Kelleher, R. Moffitt, L. Morgan, J. Plunkett, K.D. Rechberger, A.A. Rosenberg and R. Sumaila. 2016. Fish banks: An economic model to scale marine conservation. Mar. Policy 73: 154–161.

Sala, E., J. Lubchenco, K. Grorud-Colvert, C. Novelli, C. Roberts and U.R. Sumaila. 2018. Assessing real progress towards effective ocean protection. Mar. Pol. 91: 11–13.

Sánchez Lizaso, J.l., R. Goñi, O. Reñones, J.A. Garcia Charton, R. Galzin, J.T. Bayle, P.S. Jerez, A. Pérez Ruzafa and A.A. Ramos. 2000. Density dependence in marine protected populations: A review. Environ. Conserv. 27: 144–158.

Sandin, S.A., J.E. Smith, E.E. DeMartini, E.A. Dinsdale, S.D. Donner, A.M. Friedlander, T. Konotchick, M. Malay, J.E. Maragos, D. Obura, O. Pantos, G. Paulay, M. Richie, F. Rohwer,

R.E. Schroeder, S. Walsh, J.B.C. Jackson, N. Knowlton and E. Sala. 2008. Baselines and degradation of coral reefs in the northern line islands. PLoS ONE 3: e1548.

Sbragaglia, V., S. Coco, R.A. Correia, M. Coll and R. Arlinghaus. 2021. Analyzing publicly available videos about recreational fishing reveals key ecological and social insights: A case study about groupers in the Mediterranean Sea. Sci. Tot. Environm. 765: 142672.

Schärer-Umpierre, M.T., D. Mateos-Molina, R. Appeldoorn, I. Bejarano, E.A. Hernández-Delgado, R.S. Nemethk, M.I. Nemeth, M. Valdés-Pizzini and T.B. Smithk. 2014. Marine managed areas and associated fisheries in the US Caribbean. Mar. Manag. Areas Fish. 129.

Schiller, L., J.J. Alava, J. Grove, G. Reck and D. Pauly. 2015. The demise of Darwin's fishes: Evidence of fishing down and illegal shark finning in the Galápagos Islands. Aquat. Conserv. Mar. Freshw. Ecosyst. 25: 431–446.

Schirripa, M.J. and C.P. Goodyear. 1994. Status of the gag stock of the Gulf of Mexico: Assessment 1.0. NOAA Fisheries - SEFSC. MIA 93/94-61, Miami, Florida.

Schmitt, E.F. and K.M. Sullivan. 1996. Analysis of a volunteer method for collecting fish presence and abundance data in the Florida Keys. Bull. Mar. Sci. 59: 404–416.

Schunter, C., J. Carreras-Carbonell, S. Planes, E. Sala, E. Ballesteros, M. Zabala, J.-G. Harmelin, M. Harmelin-Vivien, E. Macpherson and M. Pascual. 2011. Genetic connectivity patterns in an endangered species: The dusky grouper (*Epinephelus marginatus*). J. Exp. Mar. Biol. Ecol. 401: 126–133.

SEDAR. 2015. Gulf of Mexico Red Grouper. SEDAR 42 Stock Assessment Report. Southeast Data, Assessment, and Review. 612 pp.

SEDAR. 2019. Gulf of Mexico Red Grouper. SEDAR 61 Stock Assessment Report. Southeast Data, Assessment, and Review. 251 pp.

Selig, E.R., K.M. Kleisner, O. Ahoobim, F. Arocha, A. Cruz-Trinidad, R. Fujita, M. Hara, L. Katz, P. McConney, B.D. Ratner, L.M. Saavedra-Díaz, A.M. Schwarz, D. Thiao, E. Torell, S. Troëng and S. Villasante. 2017. A typology of fisheries management tools: using experience to catalyse greater success. Fish Fish. 18: 543–570.

Semitiel-García, M. and P. Noguera-Méndez. 2019. Fishers' participation in small-scale fisheries. A structural nálisis of the Cabo de Palos-Islas Hormigas MPA, Spain. Mar. Pol. 101: 257–267.

Shapiro, D.Y. 1987. Reproduction in groupers. pp. 295–327. *In*: Polovina, J.J. and S. Ralston (eds.). Tropical Snappers and Groupers: Biology and Fisheries Management. Westview Press, London, UK.

Shapiro, D.Y., D.A. Hensley and R.S. Appeldoorn. 1988. Pelagic spawning and egg transport in coral-reef fishes: a skeptical overview. Environ. Biol. Fishes 22(1): 3–14.

Shertzer, K.W., N.M. Bacheler, G.T. Kellison, J. Fieberg and R.K. Wiggers. 2018. Release mortality of endangered Warsaw grouper *Hyporthodus nigritus*: a state-space model applied to capture-recapture data. Endang. Species Res. 35: 15–22.

Shideler, G.S., D.W. Carter, C. Liese and J.E. Serafy. 2015. Lifting the goliath grouper harvest ban: Angler perspectives and willingness to pay. Fish. Res. 161: 156–165.

Shideler, G.S. and B. Pierce. 2016. Recreational diver willingness to pay for goliath grouper encounters during the months of their spawning aggregation off eastern Florida, USA. Ocean Coast. Manag. 129: 36–43.

Shiffman, D.S., A.J. Gallagher, J. Wester, C.C. Macdonald, A.D. Thaler, S.J. Cooke and N. Hammerschlag. 2014. Trophy fishing for species threatened with extinction: A way forward building on a history of conservation. Mar. Policy 50, Part A: 318–322.

Shin, Y.-J. and P. Cury. 2001. Exploring fish community dynamics through size-dependent trophic interactions using a spatialized individual-based model. Aquat. Living Resour. 14: 65–80.

Shin, Y.-J. and P. Cury. 2004. Using an individual-based model of fish assemblages to study the response of size spectra to changes in fishing. Can. J. Fish. Aquat. Sci. 61: 414–431.

Sih, A., J. Cote, M. Evans, S. Fogarty and J. Pruitt. 2012. Ecological implications of behavioural syndromes. Ecol. Lett. 15: 278–289.

Silva, M.R.O. and P.F.M. Lopes. 2015. Each fisherman is different: Taking the environmental perception of small-scale fishermen into account to manage marine protected areas. Mar. Pol. 51: 347–355.

Silvano, R.A.M. and J. Valbo-Jørgensen. 2008. Beyond fishermen's tales: Contributions of fishers' local ecological knowledge to fish ecology and fisheries management. Environ. Dev. Sustain. 10: 657–675.

Silvano, R.A.M., V. Nora, T.B. Andreoli, P.F.M. Lopes and A. Begossi. 2017. The 'ghost of past fishing': Small-scale fisheries and conservation of threatened groupers in subtropical islands. Mar. Pol. 75: 125–132.

Singleton, R.L. and C.M. Roberts. 2014. The contribution of very large marine protected areas to marine conservation: Giant leaps or smoke and mirrors? Mar. Poll. Bull. 87: 7–10.

Sluka, R., M. Chiappone, K.M. Sullivan and R. Wright. 1997. The benefits of a marine fishery reserve for Nassau grouper *Epinephelus striatus* in the central Bahamas. Proc. 8th Int. Coral Reef Symp. Panama 2: 1961–1964.

Sluka, R.D. and K.M. Sullivan. 1998. The influence of spear fishing on species composition and size of groupers on patch reefs in the upper Florida Keys. Fish. Bull. 96: 388–392.

Smith, D., A. Punt, N. Dowling, A. Smith, G. Tuck and I. Knuckey. 2009. Reconciling approaches to the assessment and management of data-poor species and fisheries with Australia's harvest strategy policy. Mar. Coast. Fish. Dyn. Manag. Ecosyst. Sci. 1: 244–254.

Souza, C.H.S. 2000. O homem da ilha e os pioneiros da caça submarina. Dehon, Tubarão.

Sparr, P. and S.C. Venema. 1998. Introduction to tropical fish stock assessment—Part 1. FAO Fisheries Technical Paper 306/1, FAO, Rome, Italy: 407 pp.

Stallings, C.D. 2008. Indirect effects of an exploited predator on recruitment of coral-reef fishes. Ecology 89: 2090–2095.

Sujatha Kandula, K.V.L. Shrikanya and V.A. Iswarya Deepti. 2015. Species diversity and some aspects of reproductive biology and life history of groupers (Pisces: Serranidae: Epinephelinae) off the central eastern coast of India. Mar. Biol. Res. 11: 18–33.

Sweeting, C.J. and N.V.C. Polunin. 2005. Marine protected areas for management of temperate North Atlantic fisheries: Lessons learned in MPA use for sustainable fisheries exploitation and stock recovery. Department of Environment, Food and Rural Affairs, University of Newcastle upon Tyne, UK.

Thorson, J.T., M.D. Scheuerell, B.X. Semmens and C.V. Pattengill-Semmens. 2014. Demographic modeling of citizen science data informs habitat preferences and population dynamics of recovering fishes. Ecology 95: 3251–3258.

Travers-Trolet, M., Y.-J. Shin, L.J. Shannon, C.L. Moloney and J.G. Field. 2014. Combined fishing and climate forcing in the southern Benguela upwelling ecosystem: An end-to-end modelling approach reveals dampened effects. PLoS ONE 9: e94286.

Tsehaye, I. and L.A.J. Nagelkerke. 2008. Exploring optimal fishing scenarios for the multispecies artisanal fisheries of Eritrea using a trophic model. Ecol. Model. 212: 319–333.

Turner, S.C., C.E. Porch, D. Heinemann, G.P. Scott and M. Ortiz. 2001. Status of Gag in the Gulf of Mexico, Assessment 3.0. SFD 01/02-134, NOAA Fisheries - SEFSC. SFD 01/02-134, Miami, Florida: 156 pp.

Usseglio, P., A.M. Friedlander, E.E. DeMartini, A. Schuhbauer, E. Schemmel and P. Salinas-de-Léon. 2015. Improved estimates of age, growth and reproduction for the regionally endemic Galapagos sailfin grouper *Mycteroperca olfax* (Jenyns, 1840). PeerJ 3: e1270.

Usseglio, P., A.M. Friedlander, H. Koike, J. Zimmerhackel, A. Schuhbauer, T. Eddy and P. Salinas-de-León. 2016. So long and thanks for all the fish: Overexploitation of the regionally endemic Galapagos grouper *Mycteroperca olfax* (Jenyns, 1840). PLoS ONE 11: e0165167.

Valdemarsen, J.W. 2001. Technological trends in capture fisheries. Ocean. Coast. Manag. 44: 635–651.

Valdivia, A, C.E. Cox and J.F. Bruno. 2017. Predatory fish depletion and recovery potential on Caribbean reefs. Sci. Adv. 3: e1601303.

Valdivia, A., J.F. Bruno, C.E. Cox, S. Hackerott and S.J. Green. 2014. Re-examining the relationship between invasive lionfish and native grouper in the Caribbean. PeerJ 2: e348.

Voorberg, W. and R. Van der Veer. 2020. Co-management as a successful strategy for marine conservation. J. Mar. Sci. Eng. 8: 491.

Waterhouse, L., S.A. Heppell, C.V. Pattengill-Semmens, C. McCoy, P. Bush, B.C. Johnson and B.X. Semmens. 2020. Recovery of critically endangered Nassau grouper (*Epinephelus striatus*) in the Cayman Islands following targeted conservation actions. Proceedings of the National Academy of Sciences 117(3): 1587–1595.

Weisberg, R.H., L. Zheng and E. Peebles. 2014. Gag grouper larvae pathways on the West Florida Shelf. Cont. Shelf Res. 88: 11–23.

Wielgus, J., F. Ballantyne, E. Sala and L.R. Gerber. 2007. Viability analysis of reef fish populations based on limited demographic information. Conserv. Biol. 21: 447–454.

Wilhem, T., C.R.C. Sheppard, A.L.S. Sheppard, C.F. Gaymer, J. Parks, D. Wagner and N. Lewis. 2014. Large marine protected areas—advantages and challenges of going big. Aquat. Conserv. Mar. Freshw. Ecosyst. 24: 24–30.

Williams, I.D., B.L. Richards, S.A. Sandin, J.K. Baum, R.E. Schroeder, M.O. Nadon, B. Zgliczynski, P. Craig, J.L. McIlwain and R.E. Brainard. 2011. Differences in reef fish assemblages between populated and remote reefs spanning multiple archipelagos across the central and western Pacific. J. Mar. Biol. 2011: e826234.

Williamson, D.H., D.M. Ceccarelli, R.D. Evans, G.P. Jones and G.R. Russ. 2014. Habitat dynamics, marine reserve status, and the decline and recovery of coral reef fish communities. Ecol. Evol. 4: 337–354.

Wilson, E.O. 2016. Half-Earth: Our planet's fight for life. Liveright/W.W. Norton.

Wilson, Jr. R.R. and K.M. Burns. 1996. Potential survival of released groupers caught deeper than 40 m based on shipboard and in-situ observations, and tag-recapture data. Bull. Mar. Sci. 58: 234–247.

Young, J.L., Z.B. Bornik, M.L. Marcotte, K.N. Charlie, G.N. Wagner, S.G. Hinch and S.J. Cooke. 2006. Integrating physiology and life history to improve fisheries management and conservation. Fish Fish. 7: 262–283.

Young, J.W., B.P.V. Hunt, T.R. Cook, J.K. Llopiz, E.L. Hazen, H.R. Pethybridge, D. Ceccarelli, A. Lorrain, R.J. Olson, V. Allain, C. Menkes, T. Patterson, S. Nicol, P. Lehodey, R.J. Kloser, H. Arrizabalaga and C. Anela Choy. 2015. The trophodynamics of marine top predators: Current knowledge, recent advances and challenges. Deep Sea Res. Part II Top. Stud. Oceanogr. 113: 170–187.

Young, M.a.L., S. Foale and D.R. Bellwood. 2016. The last marine wilderness: Spearfishing for trophy fishes in the Coral Sea. Environ. Conserv. 43: 90–95.

Zabala, M., A. García-Rubies, P. Louisy and E. Sala. 1997. Spawning behaviour of the Mediterranean dusky grouper *Epinephelus marginatus* (Lowe, 1834) (Pisces, Serranidae) in the Medes Islands Marine Reserve (NW Mediterranean; Spain). Sci. Mar. 61: 65–77.

Zapelini, C., V.J. Giglio, R.C. Carvalho, M.G. Bender and L.C. Gerhardinger. 2017. Assessing fishing experts' knowledge to improve conservation strategies for an endangered grouper in the Southwestern Atlantic. J. Ethnobiol. 37: 478–493.

Zorrilla-Pujana, J. and S. Rossi. 2014. Integrating environmental education in marine protected areas management in Colombia. Ocean Coast. Manag. 93: 67–75.

CHAPTER 2.2

Grouper Aquaculture World Status and Perspectives

Branko Glamuzina[1,*] and *Michael A. Rimmer*[2,*]

Introduction

Diversification of aquaculture with new species is an important approach to increasing aquaculture production. Unlike terrestrial animal farming, which is characterized by a small number of widely accepted and cultured species, aquaculture is much more diverse. It is estimated that there are 567 aquatic species currently being cultured (FAO 2014), while in the research sector, the number of species being developed for culture is difficult to estimate.

Groupers (class Actinopterygii, order Perciformes, family Serranidae, subfamily Epinephelinae) are classified in 14 genera of the subfamily Epinephelinae, that comprises at least half of the 449 species in the family Serranidae (Tucker 1999). There are 15 major grouper species that are presently cultured. The most commonly cultured species belong to the genus *Epinephelus* and include *E. coioides*, *E. malabaricus*, and *E. fuscoguttatus*. Other important species are *E. bleekeri*, *E. akaara*, *E. awoara*, and *E. areolatus*. Also cultured in smaller amounts are *E. amblycephalus*, *E. lanceolatus*, *E. sexfasciatus*, *Cromileptes altivelis*, *Plectropomus leopardus*, and *P. maculatus* (Ottolenghi et al. 2004). In the last few years, hybrid groupers have become popular in Asia and the most commonly cultured hybrids are *E. lanceolatus* × *E. fuscoguttatus* and *E. polyphekadion* × *E. fuscoguttatus*. Groupers (particularly *Epinephelus* species) are often misidentified and many records of grouper species in

[1] University of Dubrovnik, Department for Aquaculture, ĆiraCarića 4, 20000 Dubrovnik, Croatia.

[2] School of Science, Technology and Engineering, University of the Sunshine Coast, Sippy Downs 4556 Queensland, Australia.

* Corresponding authors: branko.glamuzina@unidu.hr; mrimmer@usc.edu.au

culture are incorrect. For example, much of the Thailand scientific literature deals with research on '*E. malabaricus*', when in fact the species cultured was *E. coioides*. *E. coioides* is often misidentified as '*E. tauvina*' (which is probably not farmed to any significant extent) or as '*E. malabaricus*'. Some research in Indonesia with '*E. corallicola*' was in fact *E. caeruleopunctatus*.

Groupers are widely appreciated by consumers as a good quality fish, which has led to high market demand and high prices compared with most other farmed finfish species. Consequently, groupers are subject to intensive fishing activity and in most regions threatened with over-exploitation. Because of over-exploitation of capture fisheries and high demand, aquaculture of groupers has been practised for at least the last 40 years in different parts of the world.

Grouper aquaculture is supported by seedstock sourced from wild capture as well as from hatcheries. The relative contribution of wild-caught and hatchery-reared seed to grouper aquaculture production is still unclear. Sadovy et al. (2003) estimated that 'more than 60%' of cultured grouper were sourced from wild-caught seed. The proportion of wild-caught grouper seed is difficult to accurately estimate because some species are reared in hatcheries in some countries, but not in others. For example, *E. coioides* is reared in substantial numbers in Taiwan (Su et al. 2008), but not in Indonesia, where most *E. coioides* are collected from the wild (Komarudin et al. 2010). Trying to estimate the relative contribution of wild-caught and hatchery-reared seedstock is not helped by misleading claims as to the low contribution of hatchery-reared groupers, e.g., 'that the typical annual amount of seed produced in the hatcheries in the whole of Southeast Asia (excluding Taiwan Province of China) is 20,000 to 80,000 fry' (Tupper and Sheriff 2008, p.240). In fact, 20,000 to 80,000 fry is equivalent to production from 1–4 10-tonne larval rearing tanks, which could be the production from one cycle from a single small-scale Indonesian hatchery (Sugama et al. 2012). One clue to the contribution of hatchery-reared groupers is provided by the increasing dominance of hybrid groupers, which are only produced in hatcheries. Hybrid groupers are now dominating production systems throughout Southeast Asia, and are increasingly seen in the Hong Kong live fish market (Ferdouse 2014). Taiwan and Indonesia are probably the two main suppliers of hatchery-reared grouper fingerlings, although there is significant hatchery production in Vietnam, Thailand, and the Philippines.

Global production of farmed groupers (FAO 2019) has increased steadily from 78,000 tonnes in 2008 to about 184,000 tonnes in 2017 (Table 1). China (71%), Taiwan (12%), and Indonesia (11%) together make up 95% of reported grouper production (Table 1). Other major producers are Malaysia and Thailand, but production in the Philippines appears to have declined substantially in the last decade (Table 1). Some producer countries, such as Vietnam, do not disaggregate grouper production from aquaculture production of marine finfish, so there are no specific production data available. Grouper is presently also cultured in small quantities in Australia,

Table 1. Aquaculture production of groupers by country, and total value of global production, for the last 10 years. Data from FAO (2019).

Country	**Aquaculture production (tonnes live weight)**									
	2008	**2009**	**2010**	**2011**	**2012**	**2013**	**2014**	**2015**	**2016**	**2017**
Bahrain	1	1	2	1	1	0	1	0	0	0
Brunei Darussalam	3	5	5	5	50	4	28	13	37	65
Cambodia	80	100	120	140	140	150	200	200	300	320
China	45,213	43,645	47,902	56,747	69,747	78,898	84,228	95,790	107,203	131,536
China, Hong Kong SAR	918	475	660	396	286	324	398	206	105	42
Egypt	0	1	2	1	0	0	0	0	0	0
Indonesia	4,641	8,791	10,398	10,580	11,950	18,864	13,346	16,795	11,504	20,600
Korea, Republic of	46	180	270	150	52	56	67	134	217	443
Kuwait	0	0	0	0	0	0	0	0	0	2
Malaysia	4,400	3,806	4,570	6,306	6,009	5,354	7,881	8,003	6,167	6,137
Myanmar	135	45	145	140	140	140	150	13	13	14
Philippines	2,612	921	1,195	1,064	1,290	733	341	337	256	249
Saudi Arabia	50	100	50	105	115	125	140	108	100	100
Singapore	179	184	243	145	131	275	257	238	298	498
Taiwan Province of China	17,042	12,940	11,354	13,456	22,432	25,942	25,682	26,205	20,479	21,785
Thailand	3,179	2,996	2,776	2,726	2,837	2,495	2,586	2,258	2,042	2,007
United Arab Emirates	0	0	0	0	0	0	0	0	0	190
Total	**78,499**	**74,190**	**79,692**	**91,961**	**115,180**	**133,359**	**135,304**	**150,300**	**148,722**	**183,987**
Value (USD millions)	**379**	**311**	**459**	**574**	**674**	**721**	**733**	**714**	**597**	**717**

India, Republic of Korea, and Sri Lanka, but at low levels (FAO 2019). There is no reported commercial production in Europe, Mediterranean, and Africa. There was some limited production reported from the USA in 2003 and 2004 (FAO 2019).

Overview of Grouper Culture in the World

Although groupers are distributed worldwide in tropical and sub-tropical seas (Craig et al. 2011), most aquaculture production is undertaken in Asia. In other regions like the Americas and the Mediterranean, grouper aquaculture is still in research phase. This section updates the status of grouper culture in selected countries.

China

China produces a variety of grouper species—Liu and Sadovy de Mitcheson (2008) list 11 species that are routinely farmed, and this list was compiled before the recent surge in production of hybrid groupers. Hong and Zhang (2003) list six grouper species that are routinely produced in hatcheries, but note that annual production of five species is less than 100,000 fingerlings per annum. Fingerlings are sourced from both hatcheries and from wild-caught juveniles or even sub-adults (Liu and Sadovy de Mitcheson 2008). Both hatchery and wild-caught fingerlings are imported from Taiwan and other South-east Asian countries to meet the substantial demand for fingerlings in China (Kongkeo et al. 2010, Liu and Sadovy de Mitcheson 2008).

Floating cages and ponds are used for grouper grow-out in China, with most production from farms in south-east China (Kongkeo et al. 2010, Li et al. 2011, Liu and Sadovy de Mitcheson 2008). Most floating cages farms are small- or medium-scale, with small (3.5 m × 3.5 m × 4.5 m) cages made from locally available materials (Chen et al. 2007, Kongkeo et al. 2010). Because of constraints to expansion of aquaculture in inshore areas, China has begun to develop offshore farming systems using larger circular cages constructed from HDPE (Chen et al. 2007, Kongkeo et al. 2010). These larger cages are used to culture fast-growing species, such as cobia (*Rachycentron canadum*) and giant grouper *E. lanceolatus* (Kongkeo et al. 2010).

Pond culture of groupers is particularly well developed in Hainan Province and production may be segmented: from eggs/larvae up to 5–10 cm juveniles; from small juveniles to sub-marketable size (< 500 g); and from sub-marketable to marketable size (> 500 g) (Liu and Sadovy de Mitcheson 2008).

There has been limited development of recirculating production systems for grow-out of groupers in Hong Kong. Reportedly, these systems have been used to culture *E. lanceolatus* and *C. altivelis* (Liu and Sadovy de Mitcheson 2008). Production is limited and fish are sold locally (Hong Kong) or exported to mainland China (Liu and Sadovy de Mitcheson 2008).

The bulk of farms use 'trash' fish as the feed source, although farm-made feeds may be used when availability of 'trash' fish is low (Kongkeo et al.

2010, Li et al. 2011, Liu and Sadovy de Mitcheson 2008). Pellet diets have been used in the recirculation production systems in Hong Kong, and an overall FCR of about 1.4:1 is reported for *E. lanceolatus* in these systems (Liu and Sadovy de Mitcheson 2008). In contrast, the use of 'trash' fish provides FCRs in the range 7:1 to 15:1 (Li et al. 2011).

Taiwan

Taiwan is a major regional producer of grouper fingerlings which are supplied to countries throughout Asia. Fry production from Taiwanese hatcheries increased from the mid-1990's to a reported peak of around 280 million fry in 1999 (Chang et al. 2008). From 2001 to 2006, fry production ranged from around 40 to 65 million fingerlings per annum (Chang et al. 2008).

The technology for production of marine finfish, including groupers, has been reviewed by Liao et al. (2001) and Su et al. (2008). Taiwanese larval rearing techniques differ from those used elsewhere in Asia in two regards: the use of 'outdoor' larval rearing systems, and the relatively common use of copepods in larval rearing.

In addition to the commonly-used indoor hatchery system, many Taiwanese farms use 'outdoor' (i.e., concrete or earthen ponds) systems for larval rearing (Liao et al. 2001, Su et al. 2008). Preferred size for outdoor rearing ponds is around 0.3 ha or 3,000–5,000 m^3 (Liao et al. 2001). Su et al. (2008) note that, although the indoor system is more productive in terms of fingerling output, the outdoor system is popular for family-run farms, particularly in the south of Taiwan, where weather conditions are more favourable for outdoor larval rearing. Production from these systems can be substantial: Su et al. (2008) cite an example of commercial production of 3 million *E. coioides* fingerlings from 5–6 crops (April to October) using three 0.3 ha larval rearing ponds, supported by nine 450 m^2 rotifer production ponds and nine copepod production ponds, ranging from 1 to 4 ha each. However, production from outdoor systems is less reliable than from indoor systems (Liao et al. 2001, Su et al. 2008). Water quality can be problematic, resulting in mass mortality or contributing to disease outbreaks (Liao et al. 2001). Su et al. (2008) found that copepods harvested from outdoor ponds tested positive for VNN using PCR, and suggested that copepods may be a major source of infection in outdoor larval rearing systems, and could potentially transfer the virus to indoor rearing systems. Other causes of mortality in these systems include cannibalism and predation by birds (Liao et al. 2001).

Outdoor copepod production ponds may also be used to provide copepods to indoor larval rearing systems (Rayner et al. 2015). Grouper larvae fed copepods generally demonstrate faster growth and higher survival rates (Su et al. 2008, Toledo et al. 2005). In Taiwan, copepods may be used to partially or completely replace *Artemia* as a feed source for grouper larvae from 7 mm to 2–3 cm (Su et al. 2008).

Indonesia

Indonesia has a strong grouper aquaculture industry, and along with Taiwan, is a major source of grouper fingerlings for the Asia-Pacific region. Industry estimates suggest that Indonesia produces around 30–35 million grouper fingerlings per annum (W. Sudja, pers. comm.). Grouper fingerlings are mainly produced from hatcheries in northern Bali, East Java, and at Lampung in Sumatra (Kongkeo et al. 2010, Sugama et al. 2008).

The development of grouper aquaculture in Indonesia has been strongly supported by the Indonesian government, through the Ministry of Marine Affairs and Fisheries (MMAF). The development of hatchery technology for a range of grouper species was undertaken by the Institute for Mariculture Research and Development Gondol, and much of the research and development effort was conducted in collaboration with Japanese and Australian research agencies (Sugama et al. 2008). Uptake of this hatchery technology was rapid in the areas around Gondol, where many 'backyard' hatcheries were already established to culture milkfish *Chanos chanos* (Heerin 2002, Ikenou and Ono 1999). Grouper hatcheries have also proliferated in East Java (around Situbondo) and in southern Sumatra (Lampung) (Sugama et al. 2008). Expansion of grouper hatchery production beyond northern Bali has been effected through a national network of Technical Implementation Units (TIUs) established by MMAF. The TIUs provide local support for aquaculture technologies, including extension and technical support services, and importantly, provide a source of eggs for hatcheries that do not have access to broodstock (Hishamunda et al. 2009, Rimmer et al. 2013a). Development of grouper hatcheries in East Java and in southern Sumatra has been supported by TIUs at Situbondo and Lampung, respectively.

While there are some vertically integrated grouper production systems in Indonesia, much of the industry is segmented. Hatcheries, nurseries, and grow-out phases are in many cases widely separated geographically. For example, the province of Aceh in northern Sumatra was a favoured location for nursing groupers (Komarudin et al. 2010). Nursing is an intermediate stage undertaken after juvenile groupers (2–3 cm in length) leave the hatchery, until they are large enough to be stocked in sea cages (7–10 cm) (Ismi et al. 2012, Komarudin et al. 2010). In other cases, nursing may be done in onshore tanks, in which case the nursery phase is usually located close to the hatchery (Ismi et al. 2012).

Grow-out is generally undertaken in sea cages. Sea cage farms are distributed throughout Indonesia, from Sumatra in the west, as far east as Sumbawa and Flores (Afero et al. 2009, Sugama et al. 2008). Typically, sea cages used for grouper farming range from 3 × 3 m to 5 × 5 m. Kongkeo et al. (2010) categorised sea cage farms as small (4–20 cages), medium (20–100 cages), and large (> 100 cages) scales. Using this definition, Indonesian grouper farms are medium or large scale, most likely because smaller farms are not economically viable. An economic evaluation of grouper aquaculture in Indonesia (Riau Islands, Lampung, East Java, and Bali) showed that, for tiger grouper (*E. fuscoguttatus*), small farms (defined in this study as 7–15 cages) provided negative economic indicators, while medium farms (20–28 cages) provided only marginal positive indicators, and only large (48+ cages) farms culturing tiger grouper provided strongly positive economic indicators (Afero et al. 2010). All farm sizes culturing *C. altivelis* provided positive economic indicators, but these improved as farm sizes increased (Afero et al. 2010).

The tiger grouper has for some years been the most popular species for grow-out culture in Indonesia. More recently, two hybrid groupers have proven to be popular: *E. fuscoguttatus* × *E. polyphekadion* (known in Indonesia as 'kerapu cantik') and *E. fuscoguttatus* × *E. lanceolatus* ('kerapu cantang'). Indonesian grouper farmers report that both hybrids are more robust and less prone to disease than tiger grouper. Fish are generally stocked to reach a final density of 15–20 kg/m^3 (Kongkeo et al. 2010), although this is often much lower during the early stages of grow-out. The duration of culture is about 9–12 months for *Epinephelus* spp., 12–14 months for *Plectropomus leopardus*, and 18–24 months for *Cromileptes altivelis* (Kongkeo et al. 2010).

Although commercial pellet feeds are sometimes used, particularly in western Indonesia, 'trash' fish is the main feed used for grouper aquaculture (Afero et al. 2009, Kongkeo et al. 2010, Sugama et al. 2008). Afero et al. (2009) report FCRs ranging from 3.2 to 4.3, whereas Sugama et al. (2008) report FCRs of 5.8 to 7.8; the latter figures appear more realistic.

Groupers are generally harvested at 500–800 g body weight, and shipped live to Hong Kong. Larger volumes of harvested fish are sold directly to live fish transport vessels that travel from Hong Kong (W. Sudja, pers. comm.). Smaller quantities may be sold to 'collectors' or 'exporters' who aggregate fish (often a combination of wild-caught and cultured fish) for export to Hong Kong (Koeshendrajana and Hartono 2006). Because of the high cost of air freight, only high-value groupers such as *Cromileptes* and *Plectropomus* are shipped by air. *Epinephelus* spp. are generally shipped by live fish transport vessel.

Vietnam

Grouper culture in Vietnam is mainly undertaken in the northern provinces of HaiPhong and QuangNinh, and the south-central provinces of Phu Yen and KhanhHoa (Petersen et al. 2013). Grow-out is undertaken in sea cages and in coastal ponds (Kongkeo et al. 2010, Petersen et al. 2013). As in other parts of Asia, ponds originally culturing shrimp have switched to grouper culture because of recurrent crop losses due to viral disease outbreaks (Petersen et al. 2013). Marine finfish culture in Vietnam is dominated by small-scale (4–20 cages) farms (Kongkeo et al. 2010), and this is particularly the case with grouper farming. It is estimated that there are around 30,000 small cages in Vietnam, mostly in sheltered areas such as HaLong Bay (Kongkeo et al. 2010).

Seedstock are sourced from both wild-caught and hatchery-reared fingerlings, depending on species. Vietnam has government-run and private hatcheries producing grouper fingerlings, but total production is reported to be quite low, at about 1 million fingerlings per annum (Petersen et al. 2013). A number of private sector hatcheries have been established in cooperation with Taiwanese interests (Kongkeo et al. 2010). Eggs and fingerlings are produced locally or imported from Taiwan, China, or Indonesia (Kongkeo et al. 2010, Petersen et al. 2013). However, culture of several species is still dependent on wild-caught fry, including *E. malabaricus*, *E. bleekeri*, *E. areolatus*, *E. akaara*, and *E. awoara* (Kongkeo et al. 2010). The most popular species for grow-out in Vietnam are: *E. coioides*, the hybrid grouper *E. fuscoguttatus* × *E. lanceolatus*, the tiger grouper *E. fuscoguttatus*, followed by the higher value *C. altivelis* and *Plectropomus* spp. (Tran The Muu, pers. comm. 2015).

Average culture periods for grouper in the north are substantially longer (23 months) than in the south-central region (14 months) due to the lower winter water temperatures in the north (Petersen et al. 2013). Stocking densities are generally in the range 10–20 fish/m^3 for sea cages, and 0.5 fish/m^2 for ponds (Petersen et al. 2013). Although pellet diets are available for groupers, most culture still relies heavily on the use of 'trash' fish (Kongkeo et al. 2010, Petersen et al. 2013). FCRs range from 9–12:1 in both cages and ponds (Petersen et al. 2013).

Petersen et al. (2013) note that grouper produced from northern farms are larger (average 3.2 kg) than those produced in the central region (average 0.9–1.0 kg) because the northern farms sell directly to China, where the market prefers larger fish, whereas central region farmers mainly supply local markets. This greater harvest weight and higher farm gate prices for northern farms are factors in the higher profitability of northern Vietnam sea cage farms compared with central region sea cage or pond farms (Petersen et al. 2013).

Thailand

Grouper culture in Thailand uses a combination of wild-caught and hatchery-reared seedstock (Kongkeo et al. 2010, Yashiro 2008). Among the groupers reared in hatcheries are *E. coioides*, *E. malabaricus*, *E. fuscoguttatus*, *C. altivelis*, *P. leopardus*, and *P. maculatus*, while *E. coioides*, *E. malabaricus*, and *E. bleekeri* are collected from the wild (Kongkeo et al. 2010, Yashiro 2008). Most seed production is from the government-run hatcheries located in southern and eastern Thailand (Yashiro 2008). The main seed collection season in southern Thailand is from November to January (Yashiro 2008).

Juvenile groupers are nursed in concrete tanks or in ponds until they are about 7–15 cm in length, which takes 2–3 months (Kongkeo et al. 2010, Yashiro 2008). Fish are stocked at 300–500 fish/m^2 in sea cages, and 25–100 fish/m^2 in ponds or concrete tanks (Kongkeo et al. 2010). Grow-out is undertaken in sea cages and in ponds, although production from ponds is limited (Yashiro

2008). Both *E. coioides* and *E. lanceolatus* are both cultured in ponds (Yashiro 2008). Cage culture is undertaken, using floating and fixed cages ranging in size from 3 × 3 × 2 m to 5 × 5 × 2 m (Kongkeo et al. 2010, Yashiro 2008). The main feed used is 'trash' fish (Kongkeo et al. 2010). Groupers are harvested at 0.6–1.2 kg body weight and sold to domestic markets or exported to Singapore and Malaysia, or sold live into Hong Kong (Yashiro 2008).

Philippines

Most production of groupers in the Philippines is of three species: *E. coioides*, *E. fuscoguttatus*, and *P. leopardus* (Guerrero 2014). Although substantial research and development effort has gone into developing hatchery techniques for *Epinephelus* species (Toledo 2008), the Philippines industry is still strongly reliant on capture of wild seedstock, particularly for *E. fuscoguttatus*, which are captured in bamboo traps or caught using hook and line in tidal rivers, coastal bays, and estuaries from November to June (Guerrero 2014).

Nursing is undertaken in net cages (2 × 2 × 1 m), where the fish are grown from 3–5 cm to 8–10 cm over a period of about two months (Guerrero 2014, Toledo 2008).

Grow-out is undertaken in ponds or in sea cages, the latter up to 6 × 6 × 3 m in size (Guerrero 2014, Pomeroy et al. 2004). 'Trash' fish is the main feed used, although there are reports that pellet feeds are increasingly being used by farmers (Guerrero 2014, Pomeroy et al. 2004, Toledo 2008). FCRs range from 5:1 to 7:1 for 'trash' fish and 2.0:1 to 2.5:1 for pellet feed (Pomeroy et al. 2004). Fish are harvested at 0.4–0.6 kg, and sold live to domestic or export markets (Guerrero 2014). The domestic market is primarily large hotels and seafood restaurants, particularly in Manila (Guerrero 2014).

An economic evaluation of grouper aquaculture in the Philippines concluded that all four scenarios evaluated (broodstock, hatchery/nursery, grow-out, and integrated) were financially viable (Pomeroy et al. 2004). However, the capital costs of broodstock, hatchery/nursery, and integrated systems may be beyond the means of many small producers (Pomeroy et al. 2004).

India

While India has substantial production from freshwater and brackishwater ponds, mariculture is a relatively new development and is still largely in the research and development phase. Grouper aquaculture development is being supported by two ministries: the Ministry of Commerce and Industries, through the Marine Products Export Development Authority (MPEDA) Rajiv Gandhi Centre for Aquaculture (RGCA), and the Ministry of Agriculture's Indian Council of Agricultural Research (ICAR) system, principally through the Central Marine Fisheries Research Institute (CMFRI). The industry development approach for India includes development of grouper aquaculture for sea cage farming (Rimmer et al. 2013b), as well as providing a diversification option for brackishwater pond farmers (Nair et al. 2005, Ranjan et al. 2014).

Research has focused on developing seed production technologies, with the development of captive broodstock populations of *E. fuscoguttatus* and *Plectropomus areolatus* (Rimmer et al. 2013b) and *E. coioides*/*E. tauvina* (Mathew et al. 2002, Ranjan et al. 2014). Grow-out trials have been carried out with both *Epinephelus* species in sea cages, brackishwater ponds, and onshore tanks (Kailasam et al. 2008, Nair et al. 2005). An earlier review of grouper aquaculture in India (Kaipilly and Nandeesha 2008) describes the outcomes of many of these research trials in more detail.

Australia and Pacific Islands

The Australian state of Queensland invested heavily in grouper aquaculture research and development in the 2000s, as described by Rimmer and McBride (2008). However, due to industry development constraints, these research outcomes could not be effectively translated into the development of a grouper aquaculture industry (Rimmer and McBride 2008). During the latter stages of the government-funded research effort, small quantities of groupers were produced in shrimp ponds to demonstrate an alternative to shrimp production (R. Knuckey, pers. comm.).

In 2013, the Queensland Government dismissed research staff and sold off the state research facilities at the Northern Fisheries Centre in Cairns as part of a state-wide cost-cutting strategy. These research facilities were transferred to a private sector company that plans to continue producing grouper fingerlings, focusing on the giant grouper (*E. lanceolatus*) (Knuckey 2014).

Since 2005, the main focus for grouper aquaculture development in Queensland has been on diversification of production in shrimp ponds to respond to fluctuating shrimp prices. However, recently a commercial lease for marine finfish aquaculture has been established in the Torres Strait. Because of the deep water and high current velocities found at this site, it is planned to use submersible cages for farming grouper (M. Jeanneret, pers. comm.).

Although the Pacific Islands region has been a major supplier of wild-caught groupers, there has been little grouper aquaculture undertaken in the region. The New Caledonia Economic Development Agency (ADECAL) has established a hatchery and sea cage farm in North Province which is producing red emperor (*Lutjanus seabae*) and mouse grouper (*Cromileptes altivelis*).

Mediterranean

The Mediterranean finfish mariculture represents the most technologically advanced mariculture industry worldwide, especially in the culture of gilthead sea bream, *Sparus aurata* and European sea bass, *Dicentrarchus labrax*. The yearly production of almost 1.1 billion juveniles of marine species has been recently achieved by Mediterranean hatcheries, with a yearly production of 279,892 tonnes of these two species in 2013 (www.feap.info).

The diversification of production with new fish species has always been a popular activity in scientific and commercial sector, with numerous species under research and at pilot production levels. Although significant funds have been invested in diversification, most of the full cycle production culture is still based on gilthead sea bream and European sea bass. Only a small percentage of other sparid species, such as common dentex *Dentex dentex* or fast growers such as meagre *Agryosomus regius* are now under commercial culture, but with economic results not comparable to the two major species. Despite this, diversification is still a hot issue for Mediterranean finfish culture, but now the focus is mainly on fast growing species. Among these species, yellowfin tuna *Thunus thynus,* has enormous potential, but its culture is still based on capture-based aquaculture principles and as such has an unpredictable future. In addition, the Mediterranean groupers are still of high interest.

The history of Mediterranean research on grouper culture started in the middle 1990s, when results of growth trials of dusky grouper, *Epinephelus marginatus* in Spain (Gracia López and Castelló-Orvay 1995) and spawning in captivity in Italy (Spedicato et al. 1995) were published. These initial activities were followed with research projects in Italy and Croatia (Marino et al. 1998a, b, Glamuzina et al. 1998a, b, Spedicato et al. 1998), which resulted in the first successes of artificial spawning and larval culture of dusky grouper.

Although numerous scientific articles have been published on dusky grouper reproduction and early life history, this has not been followed with transfer into commercial hatcheries and farms. Several other species have also been investigated, including *Epinephelus costae* and *Epinephelus aeneus*. Research on white grouper *Epinephelus aeneus* was undertaken in Israel (Hassin et al. 1997). About 250 fish were captured and maintained in 16 m^3 concrete tanks supplied with sea water in a flow-through system. The captive fish fed readily on dry pellets supplemented with chopped frozen fish, and gained an average of 3.3 g/day during the initial growth phase (0.5–1.5 kg), and 11.3 g/day during the secondary growth phase (1.5–3.0 kg). In adult females held in captivity, the oocytes reached the final stages of vitellogenesis. However, final oocyte maturation, ovulation, and spawning did not occur. Sustained release of [-Ala6,Pro9NEt]-GnRH from implanted devices was highly effective in inducing ovulation, but did not result in natural spawning (Hassin et al. 1997). This research was later followed with growth trials (Cnaani et al. 2012), but there was no adoption by commercial farms. However, long term plans for the domestication of this species in Israel are proposed (Gorskhov 2010). The research on *Epinephelus costae* exhibited a similar pattern as with dusky grouper. Spawning and larval culture (Glamuzina et al. 2000) and growth under ambient conditions (Glamuzina et al. 2003) showed similar characteristics, presenting serious obstacles for future practice, as in trials with dusky grouper.

Early reproduction trials in captivity showed significant problems and reproductive dysfunctions in wild dusky grouper broodstock maintained in captivity. Generally, most of the females fail to complete vitellogenesis in captivity, resulting in the absence of final oocyte maturation, ovulation, and spawning (Marino et al. 2001), with a consequent oocyte atresia problems and mortalities of broodstock fish. If ovulation occurs, spontaneous spawning was also not observed, and the reasons should be only discussed as social factors needed for successful mating (Zabala et al. 1997) or absence of the natural triggers for specific reproduction events, mainly temperature (Glamuzina et al. 1998a). Standard practice in these early days was manual stripping (Glamuzina et al. 1998a), with all problems following, such as over-ripening of eggs, frequent control of fish using anesthetization as common practice, and lack of standardized egg quality (Glamuzina et al. 1998b). This problem was later solved using GnRHa implants (Marino et al. 2003). This treatment used a sustained-release delivery system (implant) loaded with gonadotropin-releasing hormone agonist [D-Ala6, Pro9, NEt]-GnRH (GnRHa). Around 85% females responded positively to the GnRHa implant, and ovulated between 60 and 238 hours after treatment, while none of the control fish showed any signs of maturation. However, no spontaneous spawning was observed, and the eggs were also manually stripped from the females. This treatment enhances number of ovulations per fish from 3.8 to even nine per female (Marino et al. 2003). This method enabled sustainable broodstock management of dusky grouper in captivity, with successful induction and finalization of gametogenesis and prolonged duration of spawning season with more spawning events. However, the need for manual striping is still present, and more investigation is needed to secure spontaneous spawning in captivity.

The second problem of reproduction in captivity is the "male problem". Groupers generally are protogynous hermaphrodites, and change sex in natural Mediterranean conditions only above 10 kg of weight. Because of this, the dusky grouper reproduction trials were faced with lack of male broodstock. This was reported by all research groups in the early days of research (Marino et al. 1998a, Glamuzina et al. 1998a). The solution was found in the earlier developed treatment of masculinization of females with 17-α methyltestosterone added in food or injected in muscles (Kuo et al. 1988). After a 4-month feeding period, with 17-α methyltestosterone combined with frozen fish as feed, all experimental fish gave sperm of varying quality (Glamuzina et al. 1998c). However, the results of the treatment were short-term sex reversal, and after the spawning season the treated fish returned to female (Marino et al. 1998). This was later improved by the use of 17-α methyltestosterone implants, with high efficiency in inducing both sex inversion and complete spermatogenesis in pre-pubertal dusky grouper at the completion of their second year in life (Sarter et al. 2006). Production of male dusky grouper at an early age reduces the time required for the production of sperm by dusky grouper reared under culture conditions.

Based on these achievements, we may conclude that reproduction in captivity for dusky grouper has largely been solved and these approaches could be used in commercial scale production. However, although some developments in reliable larval rearing techniques were developed (Russo et al. 2009, Cunha et al. 2013), commercial larval culture still remains problematic.

Groupers populations in the Mediterranean live at the upper limit of their distribution area, especially in the northern part. It should be expected that growth in these areas is significantly slower than for groupers in tropical waters. The first experiments on growth of young dusky groupers (Gracia López and Castelló-Orvay 2003) showed that mean weight gain was 451 g, and the feed conversion ratio was 1.23, during a 15-month growth experiment at a constant temperature of 26°C. This growth rate is comparable to the growth rate of common cultured fish in Mediterranean, such as European sea bass, in natural environmental conditions. It was documented that groupers (*E. costae* and dusky grouper) started to grow only at temperatures above 20°C, and in natural conditions in the northern Mediterranean, these temperatures occur only in the June-September period (Glamuzina et al. 2003). The growth rate of dusky grouper at lower ambient spring and early summer temperatures (from 16–22°C) was only 0.52 grams per day (Glamuzina et al. 1998a). This slow growth rate could be improved using prolonged growth of juveniles in pre-growout nurseries at a constant temperature of 25°C for 8 months, and then finishing the culture period in cages at ambient summer temperatures. However, this strategy would require investment in new recirculating systems for the early-stage grow-out, and the financial aspect of this approach is not promising based on present prices of groupers in Mediterranean (around US$15/kg).

Americas

For farming in the southeastern U.S. and Caribbean, Nassau grouper (*Epinephelus striatus*), gag grouper (*Mycteroperca microlepis*), black grouper (*Mycteroperca bonaci*), and jewfish (*Epinephelus itajara*) were suggested to have good potential (Tucker 1999). However, of these only *E. striatus* has been investigated by several research groups. The Nassau grouper has historically supported reef fisheries throughout much of the tropical western Atlantic Ocean. However, fisheries have severely declined in most areas because of overfishing. These declines have inspired efforts to develop culture technology for Nassau grouper and to determine the feasibility of stock enhancement (Tucker et al. 1991). Traditional Asian-style grouper farming was not seen as appropriate for the U.S., because inshore areas suitable for cage culture are very rare in the southern U.S., and a collection of wild juveniles for culture activities is considered unethical and illegal (Tucker 1999). Modern hatchery technologies are recommended if grouper culture goes ahead in this area (Tucker 1999).

Research on Nassau grouper started in the 1990s with induced spawning experiments (Tucker et al. 1991, Tucker 1994, Watanabe et al. 1995) and the achievement of voluntary spawning in captivity (Tucker et al. 1996). This in turn allowed the comprehensive description of embryogenesis and larval development (Powell and Tucker 1992). This was followed by research on larval culture and evaluation of first feeding regimes for larvae usings rotifers *Brachionus* sp.-type (Hawaiian strain) and ss-type (Thailand strain) rotifers and cryogenically preserved Pacific oyster trochophores. The advantage of small prey size was documented, and a mixed-prey treatment of 50% oyster trochophores and 50% ss-type rotifers produced higher survival than for those fed rotifers only, or trochophores and rotifers in sequence. Oyster trochophores, although inadequate when used exclusively, enhanced survival when used in combination with rotifers, possibly by improving size selectivity and dietary quality (Watanabe et al. 1996).

Feed utilization and growth of hatchery-reared, post settlement stage Nassau grouper (mean weight = 3.20 g) were compared for 63 days at different temperatures (22, 25, 28, and 31°C), and growth rates were highest at 28 and 31°C than at lower temperature regimes. Higher growth was related to feed consumption, which increased significantly from 1.60% of total fish weight at 22°C to 2.23% at 31°C. The results demonstrate that sea temperature within an ecological range has pronounced and direct effects on feeding and growth of juvenile Nassau grouper, and a temperature range of 28–31°C was recommended for culture of early juveniles (Ellis et al. 1996, 1997).

The successful hatchery production of Nassau grouper led to stock enhancement trials with hatchery-reared juveniles stocked into natural waters. This study showed that hatchery-reared Nassau grouper can tolerate tagging and survive in the ocean if released at a large size, but in general Nassau grouper stock enhancement should be considered only when simpler management methods for attaining wild stock recovery failed (Roberts et al. 1995). Since this work, there have been no reports on commercial stock enhancement projects or any commercial farming activity with any species of groupers in the Caribbean area.

Research on grouper aquaculture in Brazil has been focussed on the dusky grouper *Epinephelus marginatus*, as in the Mediterranean. The larval rearing system chosen was mesocosms with rotifers, *Artemia* sp., and artificial diets as food sources. Newly hatched larvae at an initial density of 10 ind. L^{-1} were stocked in 2 m^3 circular tanks. The survival investigation from 3 to 35 days post hatching revealed a high mortality during the transition from endogenous to exogenous feeding, then to artificial food and during larval metamorphosis, resulting in a low percentage of survivors. The main reason to explain this initial failure suggests that the dusky grouper mouth gape is too small for traditional live feeds and also due to its limited yolk reserves (Kerber et al. 2012). Subsequently, there have been no further reports on development of grouper aquaculture in Brazil.

General Problems and Perspectives

Although grouper culture is well established, particularly in Asia, there remain several problems in development of reliable aquaculture technologies for groupers, including larval culture, suitable preys, food and feeding, diseases and genetic improvements. These issues are discussed in the following text.

Larval Culture

Larval culture is the most sensitive stage in production of all marine species, because of the small larval mouth opening and the difficulty to provide food of satisfactory size and nutritional profile to support larval development. This is even more critical for marine fish species producing small larvae, such as groupers.

Early larval survival of groupers is very low when compared to other finfish (Duray et al. 1996, Duray et al. 1997, Glamuzina et al. 1998a). High mortality during the early life stages has been observed in most grouper species reared in hatcheries, including tiger grouper *Epinephelus fuscoguttatus*, green grouper *Epinephelus coioides* (Duray et al. 1997, Toledo et al. 1999, Kohno et al. 1990), dusky grouper, *Epinephelus marginatus* (Marino et al. 1998, Glamuzina et al. 1998b), Nassau grouper, *Epinephelus striatus* (Watanabe et al. 1996), and other investigated species (Ma et al. 2013).

Problems in larval culture are still the major bottleneck in the development of mass juvenile production (Kohno et al. 1997, Ma et al. 2013). These problems can be briefly summarized as: (1) spawned eggs and larvae are very small and the small functional mouth gape in early larvae limits the choice of initial live feed; (2) grouper larvae are extremely sensitive to mechanical disturbance; and (3) problems in developing of sustainable mass production of appropriate first prey; both in quantity, size, health, and nutritional value.

Most hatchery production in Asia is from low-intensity 'indoor' systems using small ('S-strain') rotifers (*Brachionus rotundiformis*) and brine shrimp (*Artemia franciscana*) as the main food sources. Sometimes, other prey organisms, such as oyster or mussel larvae are also used as initial feed. As noted above, 'outdoor' systems are used in Taiwan for production of lower-value grouper species, such as *E. coioides* and other marine finfish.

Suitable Starter Feeds

Groupers are widely recognised as being more difficult to rear in the hatchery than many other marine finfish. Reasons include: delayed development and small size of the bony elements forming the oral cavity, small mouth and body size, poor reserves of endogenous nutrition, and lower initial feeding rates (Kohno et al. 1997). Significant effort has gone into attempts to circumvent the problem of rearing smaller larvae with small mouths. In many hatcheries, the rotifers fed to early-stage larvae are sieved (using a 50 μm mesh sieve) to maximise the number of small rotifers available to the larvae. In reality, this method selects mainly neonate rotifers, which subsequently will grow to a larger size in the rearing tanks if not devoured (Knuckey et al. 2004). Other approaches to providing smaller prey include the use of oyster trochophores (Su et al. 1997) and copepod nauplii (Toledo et al. 1999). Another approach, investigated by Japanese and Indonesian researchers, is to develop a rotifer smaller than *B. rotundiformis* for hatchery use. Wullur et al. (2009) isolated and cultured a 'minute' rotifer *Proales similis* and assessed its suitability as an initial prey organism for grouper larvae. They found that *P. similis* was 38% smaller and 60% narrower than *B. rotundiformis* (Wullur et al. 2009). In feeding trials with both rotifers, sevenband grouper (*Epinephelus septemfasciatus*) preferentially selected the smaller *Proales* at 4 DAH before shifting to a preference for *Brachionus* at 6 DAH (Wullur et al. 2011). Growth and survival were highest when the larvae were fed a mixed culture of both *Proales* and *Brachionus* (Wullur et al. 2011). Several other minute rotifers belonging to the genera *Colurella* and *Lecane* have been isolated in Indonesia, and limited trials have shown that one species (*Colurella* cf. *adriatica*) is ingested by mouse grouper (*C. altivelis*) larvae (Rimmer et al. 2015). The further development of minute rotifers as initial feed for grouper larvae, as well as larvae of other marine finfish species with small mouth size, appears promising (Hagiwara et al. 2014).

Larval Nutrition

The nutritional requirements of grouper larvae are a topic that has received scant attention, despite recognition of the importance of larval nutrition in successful larviculture. *E. coioides* eggs contained high DHA (22:6n-3), EPA (20:5n-3), and ARA (20:4n-6), indicating the importance of these fatty acids in larval development (Alava et al. 2004). Wild grouper larvae had higher levels of phospholipid than neutral lipid, whereas hatchery-sourced eggs and larvae contained higher levels of neutral lipid than phospholipid (Alava et al. 2004), suggesting that diets high in phospholipid are essential for larval survival and normal development. HUFA-enriched live food organisms (rotifers and brine shrimp) enhanced growth, survival, and pigmentation in *E. coioides* larvae (Alava et al. 2004).

Copepods generally have relatively high levels of the essential fatty acid DHA and have high DHA:EPA ratios (Rayner et al. 2015, Toledo et al. 1999, Toledo et al. 1995, 2005). Despite their acknowledged benefits when used as a live food for larval rearing of many species, including groupers, the use of copepods is uncommon in commercial aquaculture due to limitations in production systems compared with rotifers and brine shrimp (Ajiboye et al. 2011, Støttrup 2000). As noted above, copepods sourced from earthen ponds are often used in Taiwanese grouper hatcheries (Rayner et al. 2015), but this may provide a source of infection for the nervous necrosis virus (Su et al. 2008).

Environmental and Physical Conditions

Grouper larvae are relatively sensitive to environmental and physical conditions, and optimising these would undoubtedly improve the success of larval rearing. However, the optimal conditions, as well as practical limits, have not been established for most species. Toledo et al. (2002) found that different levels of salinity, aeration, and light intensity all affected growth and survival of early-stage *E. coioides* larvae. Adding oil to the surface of larval rearing improves survival of *C. altivelis* (Sugama et al. 2004) by reducing mortality associated with surface-tension death, and this technique is commonly used in Indonesian hatcheries for all grouper species.

Food and Feeding

There is a substantial body of knowledge about grouper nutrition (Williams 2009), and commercial feeds are available in most of the countries where groupers are cultured (Hasan 2012). However, as noted previously, most grouper culture in Asia is supported by the use of 'trash' or low-value fish (Afero et al. 2009, Guerrero 2014, Kongkeo et al. 2010, Li et al. 2011, Liu and Sadovy de Mitcheson 2008, Petersen et al. 2013, Pomeroy et al. 2004, Sugama et al. 2008, Toledo 2008). The use of commercial pellet feeds appears to be limited to Hong Kong (in a recirculating production system) (Liu and Sadovy de Mitcheson 2008), in some western Indonesian grow-out farms (Kongkeo et al. 2010), and in the Philippines where Guerrero (2014) notes that the use of pellet feeds is increasing. An assessment of the reasons for lack of adoption is beyond the scope of this review (see Hasan (2012) for a discussion of the issues driving and constraining uptake of compounded feeds), but for groupers, they include the poor performance of pellet diets (many pellet diets are formulated for 'marine fish' rather than being tailored to specific species or species groups), and the high cost of purchasing small quantities of pellets, as required by the small-scale farming sector.

The use of 'trash' fish as feed has been implicated as a major source of parasites in grouper grow-out culture (Rückert et al. 2009), yet farmers do not consider the health implications of using 'trash' fish as feed.

Genetics

Application of genetics is a crucial part of animals and plant culture development and improvement. However, unlike terrestrial farming, the application of genetics in aquaculture has so far been limited to several species only. Many other aquatic species are still cultured with mostly wild genetic characteristics, and this is the case with most grouper aquaculture. The genetic improvements of cultured species should be developed using two ways: long-term and short-term strategies. Long-term strategies lasted for several decades and are based on different selection methods and strain crossbreeding, and because of this are very costly and difficult to finance over longer periods. The short-term strategies, such as genome manipulation and hybridization, are usually used in the beginning of species cultivation, due to their low cost and immediate results. Interspecific hybridization is frequently used in aquaculture to produce fish with desirable traits from two different species. Many interspecific hybrids have been produced, but most of them are of no applied importance (Hulata 1995). Conversely, a significant number of hybrids are suitable for aquaculture and have desirable characteristics from both parents (Basaravaju et al. 1995, Salami et al. 1993, Harrell 1998). These desirable characteristics were accounted for by heterosis, which depends on two main complementary genetic mechanisms: the combination of useful dominant genes accumulated by both crossed forms, and the increase of the total level of heterozygosity (Kirpichnikov 1981). Hence, identifying the right hybrid combination, which expresses superior growth and development in culture conditions, has been one of the goals of past genetic investigations in the genus *Epinephelus* (Tseng and Poon 1983).

The first reported hybridisation of groupers involved *E. akaara* and *E. amblycephalus*, in an attempt to produce a hybrid with the red colouring of *E. akaara* and the faster growth rates of *E. amblycephalus* (Tseng and Poon 1983). Further early research on hybridisation of groupers was done by Charles James, who crossed the tiger grouper *E. fuscoguttatus* with the camouflage grouper *E. polyphekadion* to produce a hybrid that combined the faster growth of *E. fuscoguttatus* with the robustness and disease resistance of *E. polyphekadion* (James et al. 1999). The resulting hybrid demonstrated growth rates equal to, or slightly higher than, *E. fuscoguttatus* (James et al. 1999). There was little further development of hybrid groupers until Shigeharu Senoo at the Borneo Marine Research Institute in Sabah began developing a series of hybrids: *E. fuscoguttatus* × *E. lanceolatus* (Ch'ng and Senoo 2008), *E. coioides* × *E. fuscoguttatus* (Koh et al. 2008), and *E. coioides* × *E. lanceolatus* (Koh et al. 2010). A major focus of research on hybrids has been to utilise the rapid growth of *E. lanceolatus*. *E. lanceolatus* itself is difficult to breed and rear, so incorporating its genome in a hybrid typically improves growth rate. *E. lanceolatus* hybrids typically use *E. lanceolatus* sperm that has been stripped from male giant grouper and cryopreserved (Fan et al. 2014). Other hybrids that have been developed, but have not been widely adopted to date include: *E. fuscoguttatus* × *E. caeruleopunctatus*, *E. lanceolatus* × *E. polyphekadion*, *E. tukula* × *E. lanceolatus*, and, reportedly, *E. lanceolatus* × *C. altivelis*.

Today, hybrid groupers account for a significant proportion of production. In Indonesia, the two commonly farmed hybrids are *E. fuscoguttatus* × *E. polyphekadion* and *E. fuscoguttatus* × *E. lanceolatus*. Together, these two hybrid groupers probably make up around 70% of total Indonesian grouper production. From 2012 to 2013, the quantity of hybrid groupers imported into Hong Kong SAR increased by about 50 percent (Ferdouse 2014).

Hybridization was used also in the Mediterranean in order to enhance larval culture (Glamuzina et al. 1999, 2001), and the similarity of early development between the hybrid and goldblotch and dusky grouper suggested that these hybrids could be useful in future production investigations, once problems with larval first feeding are solved.

Markets

Grouper aquaculture specifically targets the high-value live fish markets of Hong Kong SAR and China (Chan et al. 2006, Sadovy et al. 2003). Methods of transport vary depending on the value of the product. High-value grouper species (*C. altivelis* and *Plectropomus* spp.) are often shipped by air freight, but *Epinephelus* spp. are usually transported by live fish vessel because the per fish cost of vessel transport is considerably lower than air transport (W. Sudja, pers. comm.). In the last few years, the Hong Kong SAR grouper market has seen substantial decreases in price and demand, due to increasing supply from increased production of farmed groupers in Taiwan and China, coupled with China's drive to cut spending of public funds on banquets (Ferdouse 2014).

Socio-economic Aspects of Grouper Aquaculture

Grouper aquaculture is an important economic activity in coastal communities throughout Asia. In Taiwan, an estimated 10,000 people are involved directly in grouper production or indirectly in marketing or other components of the value chain (Chang et al. 2008). In northern Bali, there are numerous small-scale hatcheries producing seedstock of milkfish, shrimp, and groupers (Heerin 2002, Siar et al. 2002). These hatcheries are important sources of employment for local people, and generate income for local communities (Heerin 2002, Siar et al. 2002). Capital costs are relatively low, and 7 out of 11 hatcheries evaluated by Siar et al. (2002) had capital payback periods of less than one year. This makes them relatively affordable, and consequently large numbers of small hatcheries have been constructed in northern Bali (Heerin 2002, Siar et al. 2002). As much as 20% of the local population is estimated to have some association with these hatcheries, which produce a range of species, including groupers (Heerin 2002). Hatcheries also provide income generation for local women, who are commonly employed to count and grade grouper fingerlings (Siar et al. 2002).

In parts of Indonesia, notably Aceh, North Sumatra, and East Java provinces, a specialised sub-sector of grouper nursing has developed. The nursery phase fills the gap between hatcheries, which harvest fish at 2–3 cm TL, and grow-out farms, which prefer fish 7–10 cm TL for stocking in the sea cages (Ismi et al. 2012, Komarudin et al. 2010). The grouper fingerlings are nursed in brackish water ponds originally constructed for milkfish and shrimp culture, in *hapa* nets where they are fed small shrimp or fish and regularly graded (Komarudin et al. 2010). Income from grouper nursing is relatively high compared with either shrimp or milkfish culture in ponds, which makes it a profitable option for brackish water pond farmers (Komarudin et al. 2010).

A study of grouper grow-out in southern Thailand indicated that grouper culture specifically, was not confined to any wealth category, and was equally incorporated into the livelihood portfolios of wealthy, middle income, and poor households (Sheriff et al. 2008). Although substantial investment is required to establish even a small grouper farm, in the case of poorer households, the investment in fish cages was supported by government agencies (Sheriff et al. 2008).

References

Afero, F., S. Miao and C. Huang. 2009. Bioeconomic analysis of the tiger grouper (*Epinephelus fuscoguttatus*) cage farming in Indonesia. Journal of the Fisheries Society of Taiwan 36: 105–118.

Afero, F., S. Miao and A. Perez. 2010. Economic analysis of tiger grouper *Epinephelus fuscoguttatus* and humpback grouper *Cromileptes altivelis* commercial cage culture in Indonesia. Aquaculture International 18: 725–739.

Ajiboye, O., A. Yakubu, T. Adams, E. Olaji and N. Nwogu. 2011. A review of the use of copepods in marine fish larviculture. Reviews in Fish Biology and Fisheries 21: 225–246.

Alava, V.R., F.M.P. Priolo, J.D. Toledo, J.C. Rodriguez, G.F. Quinitio, A.C. Sa-an, M.R. de la Pena and R.C. Caturao. 2004. Lipid nutrition studies on grouper (*Epinephelus coioides*) larvae. pp. 47–52. *In*: Rimmer, M.A., S. McBride and K.C. Williams (eds.). Advances in Grouper Aquaculture. ACIAR Monograph 110. Australian Centre for International Agricultural Research, Canberra.

Basaravaju, Y., K.V. Devaraj and S.P. Ayyar. 1995. Comparative growth of reciprocal carp hybrids between Catla catla and *Labeo fimbriatus*. Aquaculture 129: 187–193.

Ch'ng, C.L. and S. Senoo. 2008. Egg and larval development of a new hybrid grouper, tiger grouper *Epinephelus fuscoguttatus* × giant grouper *E. lanceolatus*. Aquaculture Science 56: 505–512.

Chan, N.W.W., J. Bennett and B. Johnston. 2006. Consumer demand for sustainable wildcaught and cultured live reef food fish in Hong Kong. Crawford School of Economics and Government, The Australian National University, Canberra.

Chang, C.-Y., C.-C. Chiu and J.A.C. John. 2008. Chapter 12—Prophylaxis for iridovirus and nodavirus infections in cultured grouper in Taiwan. pp. 207–224. *In*: Liao, I.C. and E.M. Leaño (eds.). The Aquaculture of Groupers. Asian Fisheries Society, World Aquaculture Society, Fisheries Society of Taiwan, National Taiwan Ocean University.

Chen, J., C. Guang, H. Xu, Z. Chen, P. Xu, X. Yan, Y. Wang and J. Liu. 2007. A review of cage and pen aquaculture: China. pp. 50–68. *In*: Halwart, M., D. Soto and J.R. Arthur (eds.). Cage Aquaculture—Regional Reviews and Global Overview. FAO Fisheries Technical Paper. No. 498. Food and Agriculture Organisation of the United Nations, Rome.

Cnaani, A., A. Stavi, M. Smirnov and S. Harpaz. 2012. Rearing white grouper (*Epinephelus aeneus*) in low salinity water: effects of dietary salt supplementation. Isr. J. Aquacul-Bamid. 64: 760.

Craig, M.T., Y.J. Sadovy de Mitcheson and P.C. Heemstra. 2011. Groupers of the World—A Field and Market Guide. National Inquiry Services Centre (NISC), Grahamstown, South Africa.

Cunha, M.E., P. Ré, H. Quental-Ferreira, P.J. Gavaia and P. Pousão-Ferreira. 2013. Larval and juvenile development of dusky grouper *Epinephelus marginatus* reared in mesocosms. J. Fish. Biol. 83(3): 448–65.

Duray, M.M., C.B. Estudillo and L.G. Alpasan. 1996. The effect of background colour and rotifer density on rotifer intake, growth and survival of the grouper (*Epinephelus suillus*) larvae. Aquaculture 146: 217–225.

Duray, M.M., C.B. Estudillo and L.G. Alpasan. 1997. Larval rearing of the grouper *Epinephelus suillus* under laboratory conditions. Aquaculture 150: 63–76.

Ellis, S., G. Viala and W.O. Watanabe. 1996. Growth and feed utilization of hatchery-reared juvenile Nassau grouper fed four practical diets. The Progressive Fish-Culturist 58: 167–172.

Ellis, S.C., W.O. Watanabe and E.P. Ellis. 1997. Temperature effects on feed utilization and growth of postsettlement stage Nassau grouper. T. Am. Fish. Soc. 126: 309–315.

Fan, B., X.C. Liu, Z.N. Meng, B.H. Tan, L. Wang, H.F. Zhang, Y. Zhang, Y.X. Wang and H.R. Lin. 2014. Cryopreservation of giant grouper *Epinephelus lanceolatus* (Bloch, 1790) sperm. Journal of Applied Ichthyology 30: 334–339.

FAO. 2014. The State of World Fisheries and Aquaculture 2014. Rome. 223 pp.

FAO. 2019. Fishery and Aquaculture Statistics. Global aquaculture production 1950–2017 (FishstatJ). *In*: FAO Fisheries and Aquaculture Department [online]. Rome. Updated 2019. www.fao.org/fishery/statistics/software/fishstatj/en.

Ferdouse, F. 2014. Live fish trade in Asia in 2013. Aquaculture Asia-Pacific Magazine 10: 44–45.

Glamuzina, B., N. Glavić, B. Skaramuca and V. Kozul. 1998a. Induced sex reversal of the dusky grouper, *Epinephelus marginatus*. Aquac. Res. 29(8): 563–567.

Glamuzina, B., B. Skaramuca, N. Glavic and V. Kožul. 1998b. Preliminary studies on reproduction and early stage rearing trial of dusky grouper, *Epinephelus marginatus* (Lowe, 1834). Aquac. Res. 29(10): 769–771.

Glamuzina, B., V. Kožul, P. Tutman and B. Skaramuca. 1999. Hybridization of Mediterranean groupers: *Epinephelus marginatus* ♀ × *E. aeneus* ♂ and early development. Aquac. Res. 30(8): 625–628.

Glamuzina, B., N. Glavić, P. Tutman, V. Kožul and Skaramuca. 2000. Egg and early larval development of laboratory reared gold blotch grouper, *Epinephelus costae* (Steindachner, 1878) (Pisces, Serranidae). Sci. Mar. 64: 341–345.

Glamuzina, B., N. Glavić, B. Skaramuca, V. Kožul and P. Tutman. 2001. Early development of the hybrid *Epinephelus costae* x *E. marginatus*. Aquaculture 198: 55–61.

Glamuzina, B., P. Tutman, A.J. Conides, V. Kožul, N. Glavić, J. Bolotin, D. Lučić and B. Skaramuca. 2003. Preliminary results on growth and feeding of wild-caught young goldblotch grouper, *Epinephelus costae* in captivity. J. Appl. Ichthyol. 19(4): 209–214.

Gorshkov, S. 2010. Long-term plan for domestication of the white grouper (*Epinephelus aeneus*) in Israel. Isr. J. Aquacul-Bamid. 62(4): 215–224.

Gracia Lopez, V. and F. Castello-Orvay. 1995. Growth of *Epinephelus guaza* under different culture conditions. Cah. Options Mediterr. ISSN 16: 149–155.

Gracia López, V. and F. Castelló-Orvay, 2003. Preliminary data on the culture of juveniles of the dusky grouper, *Epinephelus marginatus* (Lowe, 1834). Hidrobiológica. 13(4): 321–327.

Guerrero, R.D.I. 2014. Philippines target grouper farming for live fish exports. Aquaculture Asia-Pacific, pp. 42–43.

Hagiwara, A., S. Wullur, H.S. Marcial, N. Hirai and Y. Sakakura. 2014. Euryhaline rotifer Proales similis as initial live food for rearing fish with small mouth. Aquaculture 432: 470–474.

Harrell, R.M. 1998. Genetics of striped bass and other Morone. World Aquacult. 29: 56–59.

Hasan, M.R. 2012. Transition from low-value fish to compound feeds in marine cage farming in Asia. FAO Fisheries and Aquaculture Technical Paper No. 573. Food and Agriculture Organisation of the United Nations, Rome.

Hassin, S., D. de Monbrison, Y. Hanin, A. Elizur, Y. Zohar and D.M. Popper. 1997. Domestication of the white grouper, *Epinephelus aeneus* 1. Growth and reproduction. Aquaculture 156: 305–316.

Heerin, S.V. 2002. Technology transfer—backyard hatcheries bring jobs, growth to Bali. Global Aquaculture Advocate 5: 90–92.

Hishamunda, N., N.B. Ridler, P. Bueno and W.G. Yap. 2009. Commercial aquaculture in Southeast Asia: some policy lessons. Food Policy 34: 102–107.

Hong, W. and Q. Zhang. 2003. Review of captive bred species and fry production of marine fish in China. Aquaculture 227: 305–318.

Hulata, G. 1995. A review of genetic improvement of the common carp *Cyprinus carpio* L. and other cyprinids by crossbreeding, hybridization and selection. Aquaculture 129: 143–157.

Ikenou, H. and S. Ono. 1999. The spread of small-scale milkfish hatcheries in Bali, Indonesia—a study on ODA fisheries technical cooperation and technical extension. Tokyo Suisandai Kempo. 86: 41–54.

Ismi, S., T. Sutarmat, N.A. Giri, M.A. Rimmer, R.M.J. Knuckey, A.C. Berding and K. Sugama. 2012. Nursery Management of Grouper: A Best-Practice Manual. ACIAR Monograph No. 150. Australian Centre for International Agricultural Research, Canberra, Australia.

James, C., S. Al-Thobaiti, B. Rasem and M. Carlos. 1999. Potential of grouper hybrid (*Epinephelus fuscoguttatus* x *Epinephelus polyphekadion*) for aquaculture. Naga. 22: 19–23.

Kailasam, M., A.R. Thirunavukkarasu, M. Abraham, G. Thiagarajan, K. Karaiyan, R. Subburaj and S.J. Mohan. 2008. Preliminary report on cage culture of brown spotted grouper *Epinephelus tauvina* (Forsskal). Indian J. Fish. 55: 353–354.

Kaipilly, D. and M.C. Nandeesha. 2008. Chapter 6—Groupers: current status and culture in India. pp. 95–110. *In*: Liao, I.C. and E.M. Leaño (eds.). The Aquaculture of Groupers. Asian Fisheries Society, World Aquaculture Society, Fisheries Society of Taiwan, National Taiwan Ocean University.

Kerber Ehlers, C., S.H. Kerber Azevedo, P.A. dos Santos and E. Gomes Sanches. 2012. Reproduction and larviculture of dusky grouper *Epinephelus marginatus* (Lowe 1834) in Brazil. J. Agr. Sci. Tech. B 2: 229–234.

Kirpichnikov, V.S. 1981. Genetic Bases of Fish Selection. Springer, Berlin, 410 pp.

Knuckey, R. 2014. Commercialisation of grouper aquaculture in Australia—transitioning from government to the private sector, World Aquaculture 2014. World Aquaculture Society, Adelaide, South Australia, pp. 332.

Knuckey, R.M., I. Rumengan and S. Wullur. 2004. SS-strain rotifer culture for finfish larvae with small mouth gape. pp. 21–25. *In*: Rimmer, M.A., S. McBride and K.C. Williams (eds.). Advances in Grouper Aquaculture. ACIAR Monograph 110. Australian Centre for International Agricultural Research, Canberra.

Koeshendrajana, S. and T.T. Hartono. 2006. Indonesian live reef fish industry: status, problems and possible future direction. pp. 74–86. *In*: Johnston, B. and B. Yeeting (eds.). Economics and Marketing of the Live Reef Fish Trade in Asia–Pacific. ACIAR Working Paper No. 60. Australian Centre for International Agricultural Research, Canberra.

Koh, I.C.C. and S.R. Shaleh, Muhd. 2008. Egg and larval development of a new hybrid orange-spotted grouper *Epinephelus coioides* × tiger grouper *E. fuscoguttatus*. Aquaculture Science 56: 441–451.

Koh, I.C.C., S.R. Shaleh, Muhd., N. Akazawa, Y. Oota and S. Senoo. 2010. Egg and larval development of a new hybrid orange-spotted grouper *Epinephelus coioides* × giant grouper *E. lanceolatus*. Aquaculture Science 58: 1–10.

Kohno, H., S. Diani, P. Sunyoto, B. Slamet and P.T. Imanto. 1990. Early developmental events associated with change over of nutrient sources in the grouper, *Epinephelus fuscoguttatus*, larvae. Bulletin Penelitian Perikanan (Fisheries Research Bulletin) Special Edition No. 1: 51–64.

Kohno, H., R. Ordonio-Aguilar, A. Ohno and Y. Taki. 1997. Why is grouper larval rearing difficult?: An approach from the development of the feeding apparatus in early stage larvae of the grouper, *Epinephelus coioides*. Ichthyological Research 44: 267–274.

Komarudin, U., M.A. Rimmer, Islahuttaman, Zaifuddin and S. Bahrawi. 2010. Grouper nursing in Aceh, Indonesia. Aquaculture Asia-Pacific Magazine, pp. 21–25.

Kongkeo, H., C. Wayne, M. Murdjani, P. Bunliptanon and T. Chien. 2010. Current practices of marine finfish cage culture in China, Indonesia, Thailand and Viet Nam. Aquaculture Asia Magazine, pp. 32–40.

Kuo, C.M., Y.Y. Ting and S.L. Yeh. 1988. Induced sex reversal and spawning of blue-spotted grouper, *Epinephelus fario*. Aquaculture 74: 113–126.

Li, X., J. Li, Y. Wang, L. Fu, Y. Fu, B. Li and B. Jiao. 2011. Aquaculture industry in China: current state, challenges, and outlook. Reviews in Fisheries Science 19: 187–200.

Liao, I.C., H.M. Su and E.Y. Chang. 2001. Techniques in finfish larviculture in Taiwan. Aquaculture 200: 1–31.

Liu, M. and Y. Sadovy de Mitcheson. 2008. Chapter 7—Grouper aquaculture in Mainland China and Hong Kong. pp. 111–142. *In*: Liao, I.C. and E.M. Leaño (eds.). The Aquaculture of Groupers. Asian Fisheries Society, World Aquaculture Society, Fisheries Society of Taiwan, National Taiwan Ocean University.

Ma, Z., H. Guo, N. Zhang and Z. Bai. 2013. State of art for larval rearing of grouper. Int. J. Aquacul. 3: 63–72.

Marino, G., G. Maricchiolo, E. Azzurro, A. Massari and A. Mandich. 1998a. Induced sex change of dusky grouper, *Epinephelus marginatus*. In Proceedings of Symposium International sur les Me´rous de Me´diterrane´e, 5–7 November 1998, Ile des Embiez, France. Me´moire de l'Institut Oce´anographique Paul Ricard: 135–137.

Marino, G., E. Azzurro, C. Boglione, A. Massari and A. Mandich. 1998b. Induced spawning and first larval rearing in Epinephelus marginatus. In Proceedings of Symposium International sur les Me´rous de Me´diterrane´e, 5–7 November 1998, at Ile des Embiez, France, Me´moire de l'Institut Oce´anographique Paul Ricard: 139–142.

Marino, G., E. Azzurro, A. Massari, M.G. Finoia and A. Mandich. 2001. Reproduction of dusky grouper from the southern Mediterranean. J. Fish Biol. 58: 909–927.

Marino, G., E. Panini, A. Longobardi, A. Mandich, M.G. Finoia, Y. Zohar and C.C. Mylona. 2003. Induction of ovulation in captive-reared dusky grouper, *Epinephelus marginatus* (Lowe, 1834), with a sustained-release GnRHa implant. Aquaculture 219: 841–858.

Mathew, G., N.K. Sanila, N. Sreedhar, K.S. Leela Bhai, L.R. Kambadkar and N. Palaniswamy. 2002. Experiments on broodstock development and spawning of *Epinephelus tauvina* (Forskal). Indian J. Fish. 49: 135–139.

Nair, K.V.S., P.P. Manojkumar, K.P. Said Koya and V.K. Suresh. 2005. Experiments on grow out culture of groupers, *Epinephelus malabaricus* (Schneider) and *Epinephelus tauvina* (Forskal). Indian J. Fish. 52: 469–475.

Ottolenghi, F., C. Silvestri, P. Giordano, A. Lovatelli and M.B. New. 2004. Capture-based Aquaculture. The Fattening of Eels, Groupers, Tunas and Yellowtails. Rome, FAO. 2004. 308p.

Petersen, E.H., D.T. My Chinh, N.T. Diu, V.V. Phuoc, T.H. Phuong, N.V. Dung, N.K. Dat, P.T. Giang and B.D. Glencross. 2013. Bioeconomics of grouper, Serranidae: Epinephelinae, culture in Vietnam. Reviews in Fisheries Science 21: 49–57.

Pomeroy, R.S., R. Agbayani, M. Duray, J. Toledo and G. Quinito. 2004. The financial feasibility of small-scale grouper aquaculture in the Philippines. Aquacult. Econ. Manage. 8: 61–83.

Powell, A.B. and J.W. Tucker, Jr. 1992. Egg and larval development of laboratory-reared Nassau grouper, *Epinephelus striatus* (Pisces, Serranidae). B. Mar. Sci. 50: 171–185.

Ranjan, R., B. Xavier, B. Dash, L.l. Edward, G. Maheswarudu and G. Syda Rao. 2014. Domestication and brood stock development of the orange spotted grouper, *Epinephelus coioides* (Hamilton, 1822) in open sea cage off Visakhapatnam coast Indian J. Fish. 61: 21–25.

Rayner, T.A., N.O.G. Jørgensen, E. Blanda, C.-H. Wu, C.-C. Huang, J. Mortensen, J.-S. Hwang and B.W. Hansen. 2015. Biochemical composition of the promising live feed tropical calanoid copepod *Pseudodiaptomus annandalei* (Sewell 1919) cultured in Taiwanese outdoor aquaculture ponds. Aquaculture 441: 25–34.

Rimmer, M., R. Whittington, J. Becker, N. Dhand, S. Wullur and S. Raharjo. 2015. Scoping study for fish health-mariculture in Indonesia, and rabbitfish aquaculture development. Final Report to ACIAR. Australian Centre for International Agricultural Research (ACIAR), Canberra, pp. 29.

Rimmer, M.A. and S. McBride. 2008. Chapter 10—Grouper aquaculture in Australia. pp. 177–188. *In*: Liao, I.C. and E.M. Leaño (eds.). The Aquaculture of Groupers. Asian Fisheries Society, World Aquaculture Society, Fisheries Society of Taiwan, National Taiwan Ocean University.

Rimmer, M.A., K. Sugama, D. Rakhmawati, R. Rofiq and R.H. Habgood. 2013a. A review and SWOT analysis of aquaculture development in Indonesia. Reviews in Aquaculture 5: 255–279.

Rimmer, M.A., Y.C. Thampisamraj, P. Jayagopal, D. Thineshsanthar, P.N. Damodar and J.D. Toledo. 2013b. Spawning of tiger grouper *Epinephelus fuscoguttatus* and squaretail coralgrouper *Plectropomus areolatus* in sea cages and onshore tanks in Andaman and Nicobar Islands, India. Aquaculture 410–411: 197–202.

Roberts, C.M., N. Quinn, J.W. Tucker Jr. and P.N. Woodward. 1995. Introduction of hatchery reared Nassau groupers to a coral reef environment. N. Am. J. Fish. Manage. 15: 159–164.

Russo, T., C. Boglione, P. De Marzi and S. Cataudella. 2009. Feeding preferences of the dusky grouper (*Epinephelus marginatus*, Lowe 1834) larvae reared in semi-intensive conditions: a contribution addressing the domestication of this species. Aquaculture 289: 289–296.

Rückert, S., S. Klimpel, S. Al-Quraishy, H. Mehlhorn and H. Palm. 2009. Transmission of fish parasites into grouper mariculture (Serranidae: *Epinephelus coioides* (Hamilton, 1822)) in Lampung Bay, Indonesia. Parasitology Research 104: 523–532.

Sadovy, Y.J., T.J. Donaldson, T.R. Graham, F. McGilvray, G.J. Muldoon, M.J. Phillips, M.A. Rimmer, A. Smith and B. Yeeting. 2003. While Stocks Last: The Live Reef Food Fish Trade. Asian Development Bank, Manila, Philippines.

Salami, A.A., O.A. Fagbenro and D.H.J. Sydenham. 1993. The production and growth of clariid catfish hybrids in concrete tanks. Isr. J. Aquacult. Bamidgeh 45: 18–25.

Sarter, K., M. Papadaki, S. Zanuy and C.C. Mylonas. 2006. Permanent sex inversion in 1-year-old juveniles of the protogynous dusky grouper (*Epinephelus marginatus*) using controlled-release 17α-methyltestosterone implants. Aquaculture 256: 443–456.

Sheriff, N., D.C. Little and K. Tantikamton. 2008. Aquaculture and the poor—is the culture of high-value fish a viable livelihood option for the poor? Marine Policy 32: 1094–1102.

Siar, S.V., W.L. Johnston and S.Y. Sim. 2002. Study on Economics and Socio-economics of Small-scale Marine Fish Hatcheries and Nurseries, with Special Reference to Grouper Systems in Bali, Indonesia. 36 pp. Report prepared under APEC Project FWG 01/2001 - 'Collaborative APEC Grouper Research and Development Network'. Asia-Pacific Marine Finfish Aquaculture Network Publication 2/2002. Network of Aquaculture Centres in Asia-Pacific, Bangkok, Thailand.

Spedicato, M.T., G. Lembo, P. Di Marco and G. Marino. 1995. Preliminary results in the breeding of dusky grouper *Epinephelus marginatus* (Lowe, 1834). Cah. Options Mediterr. 16: 131–148.

Spedicato, M.T., M. Contegiacomo, P. Carbonara, G. Lembo and C. Boglione. 1998. Artificial reproduction of *Epinephelus marginatus* aimed at the development of restocking techniques. Bio. Mar. Medit. 5: 1248–1257.

Støttrup, J.G. 2000. The elusive copepods: their production and suitability in marine aquaculture. Aquacul. Res. 31: 703–711.

Su, H.M., M.S. Su and I.C. Liao. 1997. Preliminary results of providing various combinations of live foods to grouper (*Epinephelus coioides*) larvae. Hydrobiol. 358: 301–304.

Su, H.M., M.S. Su, K.F. Tseng and I.C. Liao. 2008. Chapter 2—Development of techniques for enhancing seed production of *Epinephelus coioides* in Taiwan. pp. 29–48. *In*: Liao, I.C. and E.M. Leaño (eds.). The Aquaculture of Groupers. Asian Fisheries Society, World Aquaculture Society, Fisheries Society of Taiwan, National Taiwan Ocean University.

Sugama, K., Trijoko, S. Ismi and K.M. Setiawati. 2004. Larval rearing tank management to improve survival of early stage humpback grouper (*Cromileptes altivelis*) larvae. pp. 67–70. *In*: Rimmer, M.A., S. McBride and K.C. Williams (eds.). Advances in Grouper Aquaculture. ACIAR Monograph 110. Australian Centre for International Agricultural Research, Canberra.

Sugama, K., I. Insan, I. Koesharyani and K. Suwirya. 2008. Chapter 4—Hatchery and grow-out technology of groupers in Indonesia. pp. 61–78. *In*: Liao, I.C. and E.M. Leaño (eds.). The Aquaculture of Groupers. Asian Fisheries Society, World Aquaculture Society, Fisheries Society of Taiwan, National Taiwan Ocean University.

Sugama, K., M.A. Rimmer, S. Ismi, I. Koesharyani, K. Suwirya, N.A. Giri and V.R. Alava. 2012. Hatchery Management of Tiger Grouper (*Epinephelus fuscoguttatus*): A Best-Practice Manual. ACIAR Monograph No. 149. Australian Centre for International Agricultural Research, Canberra, Australia.

Toledo, J.D., M.S. Golez, M. Doi and A. Ohno. 1999. Use of copepod nauplii during early feeding stage of grouper *Epinephelus coioides*. Fisheries Science 65: 390–397.

Toledo, J.D., N.B. Caberoy, G.F. Quinitio, C.H. Choresca and H. Nakagawa. 2002. Effects of salinity, aeration and light intensity on oil globule absorption, feeding incidence, growth and survival of early-stage grouper *Epinephelus coioides* larvae. Fisheries Science 68: 478–483.

Toledo, J.D., M. Golez and A. Ohno. 2005. Studies on the use of copepods in the semi-intensive seed production of grouper *Epinephelus coioides*. pp. 169–182. *In*: Lee, C.-S., P.J. O'Bryen and N.H. Marcus (eds.). Copepods in Aquaculture. Proceedings of a Workshop on Culture of Copepods and Applications to Marine Finfish Larval Rearing, Honolulu, HI (USA), 5–8 May 2003. Blackwell, Oxford.

Toledo, J.D. 2008. Chapter 5—Grouper aquaculture R&D in the Philippines. pp. 79–93. *In*: Liao, I.C. and E.M. Leaño (eds.). The Aquaculture of Groupers. Asian Fisheries Society, World Aquaculture Society, Fisheries Society of Taiwan, National Taiwan Ocean University.

Tseng, W.Y. and C.T. Poon. 1983. Hybridization of *Epinephelus* species. Aquaculture 34: 177–182.

Tupper, M. and N. Sheriff. 2008. Capture-based aquaculture of groupers. pp. 217–253. *In*: Lovatelli, A. and P.F. Holthus (eds.). Capture-based Aquaculture. Global Overview. FAO Fisheries Technical Paper No. 508. Food and Agriculture Organisation of the United Nations, Rome.

Tucker, J.W. Jr., J.E. Parsons, G.C. Ebanks and P.G. Bush. 1991. Induced spawning of Nassau grouper *Epinephelus striatus*. J. World Aquacult. Soc. 22: 187–191.

Tucker, J.W. Jr. 1994. Spawning by captive serranid fishes: a review. J. World Aquacult. Soc. 25: 345–359.

Tucker, J.W. Jr. 1999. Species profile: grouper aquaculture. SRAC Publication No. 721. Southern Regional Aquaculture Center (SRAC), Fort Pierce, Florida, USA.

Tucker, J.W., Jr., P.N. Woodward and D.G. Sennett. 1996. Voluntary spawning of captive Nassau groupers *Epinephelus striatus* in a concrete raceway. J. World Aquacult. Soc. 27: 373–383.

Williams, K.C. 2009. A review of feeding practices and nutritional requirements of postlarval groupers. Aquaculture 292: 141–152.

Wullur, S., Y. Sakakura and A. Hagiwara. 2009. The minute monogonont rotifer Proales similis de Beauchamp: culture and feeding to small mouth marine fish larvae. Aquaculture 293: 62–67.

Wullur, S., Y. Sakakura and A. Hagiwara. 2011. Application of the minute monogonont rotifer Proales similis de Beauchamp in larval rearing of seven-band grouper *Epinephelus septemfasciatus*. Aquaculture 314: 355–360.

Yashiro, R. 2008. Chapter 8—An overview of grouper aquaculture in Thailand. pp. 143–154. *In*: Liao, I.C. and E.M. Leaño (eds.). The Aquaculture of Groupers. Asian Fisheries Society, World Aquaculture Society, Fisheries Society of Taiwan, National Taiwan Ocean University.

Watanabe, W.O., S.C. Ellis, E.P. Ellis, W.D. Head, C.D. Kelley, A. Moriwake, C.-S. Lee and P.K. Bienfang. 1995. Progress in controlled breeding of Nassau grouper (*Epinephelus striatus*) broodstock by hormone induction. Aquaculture 138: 205–219.

Watanabe, W.O., S.C. Ellis, E.P. Ellis, V.G. Lopez, P. Bass, J. Ginoza and A. Moriwake. 1996. Evaluation of first-feeding regimens for larval Nassau grouper *Epinephelus striatus* and preliminary, pilot-scale culture through metamorphosis. J. World Aquacult. Soc. 27: 323–331.

Zabala, M., A. Garcia-Rubieas, P. Louisy and E. Sala. 1997. Spawning behaviour of the Mediterranean dusky grouper *Epinephelus marginatus* (Lowe, 1834) (Pisces, Serranidae) in the Medes Islands Marine Reserve (NW Mediterranean, Spain). Sci. Mar. 61: 65–77.

CHAPTER 2.3

The Importance of Groupers and Threats to Their Future

Yvonne Sadovy de Mitcheson[1,2,4,*] and *Min Liu*[3,4]

Introduction

In this chapter, following a brief background on taxonomy, biology, and ecology (covered in more detail in other chapters), we highlight the importance of groupers and consider their roles in reef ecosystems and benefits for humans, examine the growing threats that they face and the difficulties associated with their monitoring, population status evaluation, conservation, and management. We recognise that overexploitation is the major threat to this taxon. We also briefly consider habitat degradation, climate change and, most recently, covid-19, and the possible implications of these in addition to overexploitation for the taxon. Our overview is not intended to be a comprehensive coverage of all the topics, most of which are addressed more fully in other chapters, but seeks to identify and explore the importance and major uses of, and issues and challenges faced by, groupers and the implications should current trends continue. We consider possible solutions and future steps.

Taxonomy

The groupers (family Epinephelidae, formerly tribe Epinephelini under subfamily Epinephelinae and family Serranidae) form a distinctive assemblage

[1] Swire Institute of Marine Science, School of Biological Sciences, The University of Hong Kong, Hong Kong Special Administrative Region.
[2] Science and Conservation of Fish Aggregations (www.SCRFA.org).
[3] State Key Laboratory of Marine Environmental Science, College of Ocean and Earth Sciences, Xiamen University, Xiamen, Fujian, China. Email: minliuxm@xmu.edu.cn
[4] IUCN Groupers & Wrasses Specialist Group.
* Corresponding author: yjsadovy@hku.hk

of almost 169 species, described to date, of coral and rocky reef fishes found predominantly in tropical and sub-tropical areas of the Atlantic and Indo-Pacific regions (Smith and Craig 2007, Craig et al. 2011, Ma and Craig 2018, see Chapter 1.1). Most species occur in shallow coastal waters, often extending into offshore and deeper reefs, with a few species limited to deep shelf slopes down to > 300 m. The taxon includes the largest of all reef fishes (among teleosts), the Giant (*Epinephelus lanceolatus*), the Pacific Goliath (*E. quinquefasciatus*), and the Atlantic Goliath (*E. itajara*) groupers that can exceed some 2 m in total length (TL), although few exceed 1 m (Craig et al. 2011).

While a small number of groupers is well-studied because they are relatively large and conspicuous, particularly accessible and convenient for field work, or because of their relatively high economic importance, we still know little about the majority of species in this taxon. This situation is somewhat surprising given their commercial and food importance, and is reflected in the number of species (15% of all groupers) that are classified as Data Deficient (DD) in the IUCN Red List, which means that there is insufficient information available to enable them to be assessed regarding their conservation status (Sadovy de Mitcheson et al. 2020a). Another challenge is that identification is unresolved for several species, due to morphological similarities, because body colour can vary markedly with location, size, and reproductive state, or because some species can look very different when dead versus alive, which can lead to uncertain identifications and confusion for workers (Craig et al. 2011). With the advent of molecular data, many novel relationships among grouper species have emerged, but the taxon as defined today continues to challenge taxonomists.

The term 'grouper' typically applies to the species of three genera, *Epinephelus*, *Hyporthodus* (formerly most species were placed in the genus *Epinephelus*), and *Mycteroperca*, with species of *Plectropomus* also commonly referred to as coral grouper or coral trout (Craig et al. 2011). Also included among the groupers are a number of smaller species in several other genera, such as *Alphestes*, *Cephalopholis*, *Cromileptes*, *Dermatolepis*, and *Variola*. All these genera are considered groupers in this chapter.

Biology and Ecology—A Brief Primer

This diverse taxon exhibits an interesting array of biological and ecological characteristics, including sexual patterns ranging from gonochorism to different forms of functional hermaphroditism, complex and diverse social structures and mating systems, long distance migrations, and interspecific relationships (see Chapters 1.3 and 1.5). Most species are geographically widespread and are variously distributed, from shallow to deep waters along continental and insular shelves of the tropics and sub-tropics. Groupers include some of the longest living reef fishes yet described, and some of the most threatened of all reef fishes due to uncontrolled fishing activities. Some species mate in small social groups over many months each year, and others in brief and sometimes spectacular spawning aggregations. Some species are bold active mid-water or demersal predators, and others extremely shy. Certain groupers are particularly attractive to divers and many are highly coveted (and hence economically valued) food fishes targeted both chilled and live, for subsistence as well as for domestic and international trade. A few species are popular in aquaria and recreational fisheries (Sadovy de Mitcheson et al. 2013) (Fig. 1).

Our understanding of grouper biology is best for reproduction, sexual pattern, habitat, maximum size, social structure, with some work on age and growth, mostly for commercial species (see Chapters 1.2–1.4). Some medium to larger species may not become sexually mature until at least 5–10 years and have longevities exceeding 30 or 40 years (the Giant Grouper, Bullock et al. 1992; the Nassau Grouper *E. striatus*, Bush et al. 1996; the Brown-marbled Grouper *E. fuscoguttatus*, Pears et al. 2006; the Dusky Grouper *E. marginatus*, Condini et al. 2014). Many smaller species may not live much more than 10 years and mature at about a year (e.g., the Coney *Cephalopholis fulva* [maximum size 21 cm TL], Potts and Manooch 1995; the Chocolate Hind

Fig. 1 Attractive groupers, such as the Humpback Grouper *Cromileptes altivelis* (a) and the Giant Grouper *Epinephelus lanceolatus* (b) are popular in aquaria. The two species are also highly valued for their flesh quality in the live food fish trade. [(Photos: Michael Patrick O'Neill/SeaPics (a), Bai-an Lin (b)].

C. boenak [maximum size 26 cm TL], Chan and Sadovy 2002). However, some smaller species can be surprisingly long-lived with quite variable longevities. For example, longevities of the Peacock Grouper *C. argus*, which reaches about 50 cm, vary widely throughout its geographic range, from 14–40 years (Pears 2005, J.H. Choat, James Cook University, Australia, unpublished data).

Groupers are particularly intriguing for their reproductive biology. Sexual patterns range from gonochorism to several forms of hermaphroditism (although neither protandry nor simultaneous hermaphroditism) and can be quite labile; diagnosis of sexual pattern in this group, however, is particularly challenging, with even congeners differing in this aspect of life history (Sadovy de Mitcheson and Liu 2008a). Mating can be in groups or in pairs and the relative sizes of ripe testes may reflect the mating system and presence of sperm competition (e.g., Erisman et al. 2013, Ohta et al. 2017).

Many medium to larger groupers form distinctive reproductive 'aggregations' for short periods (a few weeks to a few months) each year and migrate (sometimes tens to hundreds of kilometres) to spawning sites, predictable spatially and temporally, to gather briefly and spawn in large numbers each year (Sadovy de Mitcheson and Colin 2012). These aggregations can consist of tens to tens of thousands of fishes concentrated in small areas, and are often a target of fishing due to high fish catchabilities while aggregated. Recruitment can be sporadic, with a few year classes sometimes dominating the age structure of a fishery (Russ et al. 1998, papers in Sadovy de Mitcheson and Colin 2012).

Importance of Groupers

Ecosystem Role

In terms of biomass, groupers count among the larger species in reef fish assemblages and, at natural levels, can occur in significant numbers and biomass in reef environments, with important implications for the reef ecosystem (Craig et al. 2011, Sadovy de Mitcheson et al. 2013). Although we have much to learn about the role of groupers, particularly as apex predators, in marine ecosystems, their loss in significant numbers could be important in terms of shaping reef assemblages, because the top-down effect of grouper predation could influence community structure in highly diverse systems (Bellwood et al. 2004, Boaden and Kingsford 2015). Predators can influence lower trophic levels through their interactions with prey, and this could be a significant factor in reef communities over both short term and evolutionary time scales. Predators can play an important role in regulating prey communities, while key traits, from age, size and growth, to colour patterns and reproduction in prey species may be influenced by local predator species and their densities (e.g., Ruttenberg et al. 2011, Walsh et al. 2012, Boaden and Kingsford 2015). In the Bahamas, the Nassau Grouper, as an example of a predator, along with high reef complexity, had a positive and additive effect on total fish abundance in study areas; the study concluded that management of large-bodied piscivores and reef habitat are critical for management and conservation of reef ecosystems (Hensel et al. 2019).

As adults, top marine predators such as sharks and larger groupers seldom have natural predators (Essington et al. 2006). Their presence in an ecosystem is usually associated with top-down control of prey abundance, behaviour and habitat structure, and their loss or reduction in their biomass can affect reef ecosystems, both directly and indirectly (Thrush 1999, Heithaus et al. 2008, McCauley et al. 2010) (see Chapter 1.4). A specific example is an inverse relation between grouper density and that of the coral-eating, Crown-of-thorns Starfish *Acanthaster planci*, with reduction of predators by fishing correlated to higher densities of the starfish, which in unchecked numbers can damage living coral (Fiji, Dulvy et al. 2004) (see Chapter 1.4).

Groupers may be important predators of the invasive Red Lionfish *Pterois volitans* in the Caribbean and tropical western Atlantic. In these regions, the invasive fish is rapidly spreading and has been recorded in the stomachs of large-bodied Caribbean groupers. There are concerns that the lionfish is a voracious predator on many native species, hence its control is of much interest (Albins and Hixon 2008). Whether grouper predation of the lionfish is sufficient to act as a biocontrol of this invasive species is unknown, but pest biocontrol by predatory fishes has been reported in other ecosystems, and there is certainly concern that chronic overfishing of groupers on reefs of the region could compromise an important component of natural biocontrol (Maljkovic et al. 2008, Mumby et al. 2011).

The possible ecosystem implications of the massive seasonal movements of biomass associated with large numbers of migrating groupers as they move, sometimes for hundreds of kilometres away from their home reefs to spawning sites, have been little explored. These movements of hundreds to tens of thousands to fish, sometimes in groups (depending on the species and involving many tonnes of fish), during particular periods each year to aggregate to spawn in small areas are a little understood ecological component of adult connectivity within marine ecosystems (Nemeth 2012). It is reasonable to expect that such movements can involve considerable energy transfer due to the activities of a large biomass of moving fish and their feeding and defecation activities (Nemeth 2012). Large numbers of aggregated grouper adults, predictably concentrated, can be an important source of food for larger predators, such as sharks (Fig. 2). In one small French Polynesia atoll pass, for example, about 17 tonnes of fish (approximately 20,000 fish) gather over a small area, and tracking results suggest that sharks may overcome low local energy availability by feeding on this unfished highly seasonal aggregation, which effectively concentrates energy from other local trophic pyramids in the lagoon (Mourier et al. 2016). Hence, fish spawning aggregations can play a significant role in the maintenance of local inverted pyramids in pristine marine areas.

Several groupers influence or even shape the structure of the habitat or community they occupy by leveraging their ecological influence as ecosystem

Fig. 2 Sharks waiting to prey on aggregating Camouflage Groupers *Epinephelus polyphekadion* (a) which aggregate to spawn each year at a predictable time and place (French Polynesia) (a, b). Egg (caesionids) and grouper (sharks) predators seem to know when the groupers will assemble (Photos: Yvonne Sadovy de Mitcheson).

engineers and architects, or exhibit other interspecific relationships (see Chapter 1.5). By their activities of excavating sandy substrate to expose rock, for example, the Yellowedge *E. flavolimbatus* and the Red *E. morio* groupers create habitat structure by clearing away sediment, thereby supporting other fishes and invertebrates, by providing refuge from predation in an otherwise less complex habitat. This habitat may also serve as a cleaning station for the Yellowedge Grouper (Coleman and Williams 2002). The excavated areas provide important habitats for the Spiny Lobster *Panulirus argus* and the Vermilion Snapper *Rhomboplites aurorubens* in the Gulf of Mexico, while the cleaned substrate allows for the attachment of sessile invertebrates and refuge for a range of mobile species. Coleman et al. (2010) demonstrated increased biodiversity and abundance associated with habitat structured by the Red Grouper, and speculated on its importance as habitat for other economically important species. Some groupers are also known for their interspecific relationships, including collaborative hunting with moray eel,

Fig. 3 Interspecific relationship. The Nassau Grouper *Epinephelus striatus* and trumpetfish *Aulostomus* species sometimes hunt together (Photo: Doug Perrine/SeaPics).

trumpetfish, and octopus (Diamant and Shpigel 1985, Bshary et al. 2006) (Fig. 3) (see Chapter 1.5).

Commercial Significance

While representing but a tiny percentage of fisheries globally by weight, groupers are nonetheless disproportionately valuable components (i.e., relatively high value per kilogram) of coastal rocky and coral reef fisheries for many countries and communities in the tropics and sub-tropics. They are variously important for subsistence and for domestic and international trade in small- and medium-scale fisheries, and highly appreciated for their firm white flesh which can be prepared in many ways; they are also well-suited for freezing and processing, while some species are quite hardy and able to withstand long-distance transport alive. The production of a few species (and hybrids) by mariculture (marine aquaculture or fish farming) is also growing, and that importance has increased substantially over the last decade (e.g., Dennis et al. 2020) (see Chapter 2.2). Grouper is often featured on menus as a fish of choice and is the dominant type of fish, traded mainly live, in the lucrative, luxury, Chinese seafood market. They are marketed in multiple forms, including whole fish and fillets (fresh, chilled, and frozen) and live. They are also often valued recreationally.

Unfortunately, although groupers are highly valued food fishes, their fisheries are not particularly productive (i.e., in terms of tonnes produced per year, which can range from hundreds to thousands of tonnes per fishery) relative to the widely recognized major global fisheries, like an Atlantic Cod, *Gadus morhua*, which produce hundreds of thousands of tonnes per year (Sadovy de Mitcheson 2016). Hence, both their landings and economic value tend to be poorly documented in many source countries. This in turn obscures the overall social and economic importance of, and trends in, their fisheries, often resulting in little government interest in their management (Sadovy 1994, Sadovy de Mitcheson et al. 2020a).

The major global source of fisheries catch and trade data is that by the United Nations Food and Agriculture Organization (FAO) capture fisheries statistics (landings), which has been in place for over 60 years and summarizes national capture fisheries and aquaculture statistics annually. This database depends on each country reporting its landings. In this database, two categories of 'Groupers nei (*Epinephelus* spp.; nei means 'not elsewhere included')' and 'Groupers, seabasses nei (Serranidae)' are recorded, in addition to a third category of about 40 grouper species from six genera (*Aethaloperca, Cephalopholis, Cromileptes, Epinephelus, Mycteroperca,* and *Plectropomus*) noted to species level; these include many species of high interest for subsistence and commercial fisheries (http://www.fao.org/fishery/statistics/global-capture-production/en). While several *Hyporthodus* species are also important to local fisheries, they are not distinguished from the genus *Epinephelus* in FAO records because of the relatively recent taxonomic revision of the genus (Craig et al. 2011, Ma and Craig 2018).

However, many grouper landings go unrecorded at the national level, exports are poorly documented or not indicated at all, and multiple species are often consolidated into a simple 'Groupers nei' category obscuring any species-specific trends (Sadovy de Mitcheson and Yin 2015). Hence, it is difficult to understand global volumes of grouper catches by species and trends in these over time and, as a result, their economic contributions are not well recognized. Even if there are national databases for groupers these usually do not include data on fishing effort making it hard to use landings data alone to assess population status (e.g., Craig et al. 2011).

Fresh and Frozen Groupers

Data on global grouper capture production suggest a steady overall increase of about 30 times (not correcting for under- or over-reporting-see below or including live groupers) from approximately 15,700 tonnes (t) in 1950 to 442,400 t in 2018, with trends differing substantially by geographic region (Fig. 4). In the 1960s the major areas of capture fisheries were Asia and the Americas in approximately equal quantities. Thereafter, Asian production continued to increase until in 2018 this region accounted for most reported global production, more than 80%, while that in the Americas stagnated, from nearly 50% to less than 10% in recent years. Oceania generally has the lowest reported production, despite massive coral reef areas for groupers, and their regular appearance in local markets, with Kiribati, Fiji, and Australia reporting the most groupers in this region (Figs. 4, 5, 6).

Concerning the extent to which FAO data represent grouper landings globally, we are aware of both under-reporting and over-reporting by different countries, as for reef fisheries generally. With respect to under-reporting, important producers in Asia, such as India and Vietnam, evidently do not report grouper capture production to FAO (http://www.fao.org/fi/oldsite/FCP/en/VNM/profile.htm; http://www.exportersindia.com/indian-suppliers/grouper-fish.htm). Indonesia produces the highest recorded

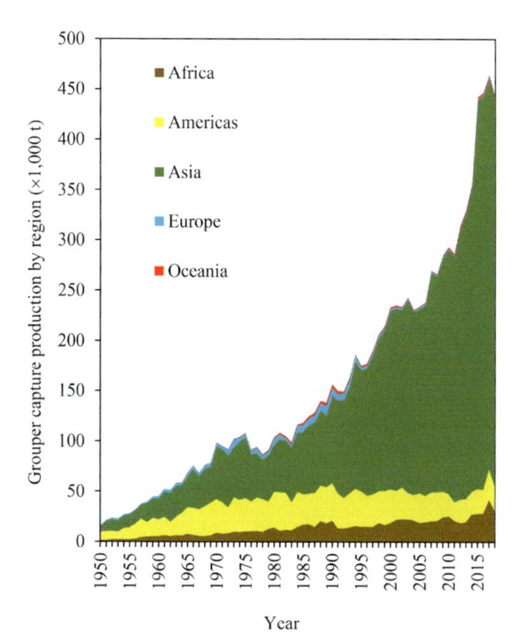

Fig. 4 Global annual grouper capture production (t) from FAO dataset from 1950–2018 (http://www.fao.org/fishery/statistics/global-capture-production/en).

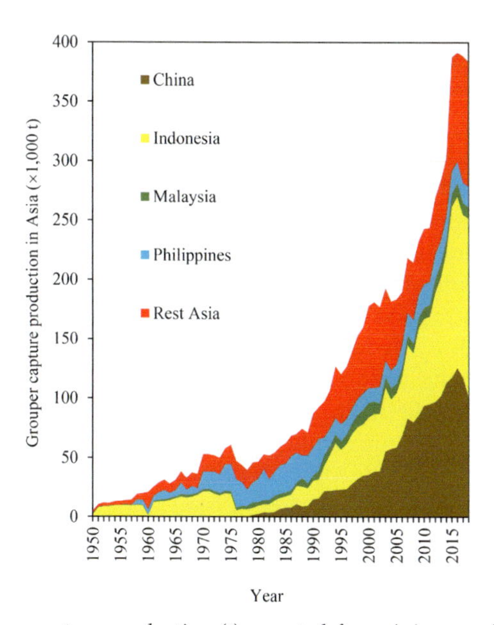

Fig. 5 Annual grouper capture production (t) reported from Asia according to FAO dataset from 1950–2018, with the top four countries indicated being China, Indonesia, Malaysia, and the Philippines (http://www.fao.org/fishery/statistics/global-capture-production/en). Additional information in 1978–1989 for China extracted from China Fishery Statistical Yearbooks is excluded in FAO dataset.

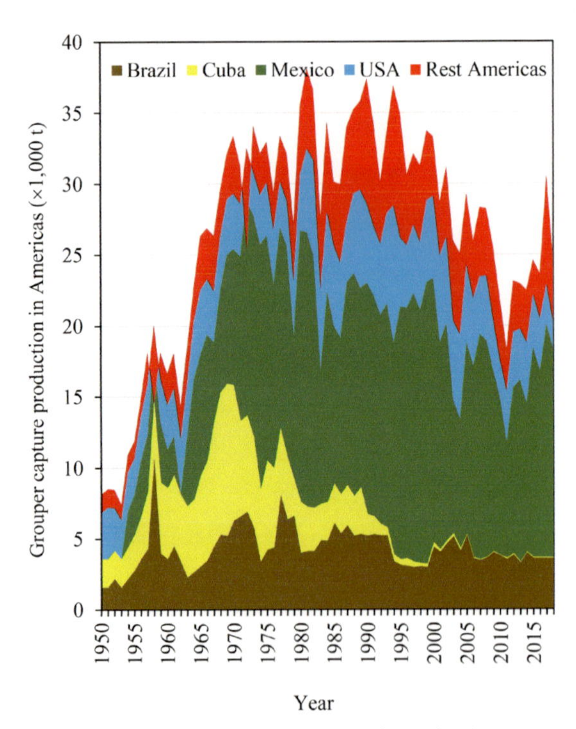

Fig. 6 Annual grouper capture production (t) reported from the Americas according to FAO dataset from 1950–2018, with the top four producing countries, Brazil, Cuba, Mexico, and USA (http://www.fao.org/fishery/statistics/global-capture-production/en).

grouper catches, which is perhaps unsurprising given its large reef area (Fig. 5). Under-reporting may apply in the case of deepwater/shelf edge fisheries for groupers in some locations of Indonesia (J. Pet, The Nature Conservancy, Indonesia, personal communication 2016). In the Maldives, grouper fishery under-reporting was estimated to be as much as 92–99% from 1994–2001 (Sattar and Adam 2005) and in Fiji under-reporting is severe (Lee et al. 2018, Sadovy de Mitcheson et al. 2018). Some Caribbean countries do not report their landings to FAO (e.g., Honduras) despite known landings and international trade (e.g., Box and Canty 2010). Moreover, there is extensive wild capture of juvenile groupers in Asia as seed for mariculture grow-out [a practice called 'capture-based aquaculture' (CBA), Lovatelli and Holthus (2008)] which is effectively reported under mariculture production, rather than capture fisheries (Sadovy 2000, Sadovy de Mitcheson and Liu 2008b). Groupers in the substantial live fish trade are also largely unreported to FAO (see below) (Sadovy et al. 2003, ADMCF 2015).

Conversely, the high wild capture of groupers indicated for domestic landings is over-reported by China (mainland, excluding Hong Kong, Macau, and Taiwan unless otherwise specified) (Fig. 5). Considering the limited available natural habitats in good condition suited to groupers in Chinese waters, overfishing, and severely degraded coastal habitats, these figures cannot represent domestic production. The large grouper landings reported (between a quarter and a third of Asia's total landings) may be accounted for by one or a combination of: (1) non-groupers reported as groupers (i.e., misidentification); (2) imports into China from other areas or countries; (3) fish brought in on Chinese distant water vessels from non-Chinese waters, such as East Africa, South China Sea, Pacific Ocean, or the Andaman Sea, included under domestic landings; (4) inclusion of some local grouper mariculture production in capture fishery reporting; and/or (5) general over-reporting of national statistics data (e.g., as previously reported for China national fishery data by Watson and Pauly 2001). These high reported grouper landings are also subject to question because they were reported from the Northwest Pacific FAO region, not a significant area for grouper production.

In the Americas, overall, landings were almost an order of magnitude less than in Asia in recent decades. Landings steadily grew until the 1970s then declined after about 2000; reported annual production for the whole region barely exceeded 25,000 t after 2012 (Fig. 6). Production in Cuba, which has particularly good national data on fisheries and once had a productive grouper fishery, declined much earlier, in the 1990s (Claro et al. 2009). Today, Mexico reports the highest production in the Americas, followed by Brazil and USA (Fig. 6), while the USA is now a net importer of groupers (Sadovy de Mitcheson and Yin 2015).

Live Groupers

A significant sector of apparent under-reporting to FAO is the substantial, highly valuable, and somewhat secretive live grouper trade, predominantly destined for Hong Kong and mainland China markets; Hong Kong is a major trade hub for mainland China (Sadovy de Mitcheson et al. 2017, 2019). Live groupers are transported to Hong Kong from source countries, mainly Indonesia and the Philippines, but also in smaller quantities from other countries, such as Australia, India (especially the Andaman Islands), the Maldives, and sporadically from Pacific Island Countries (PICs). Most PICs, however, pulled out of the trade due to unsustainable practices, social concerns (e.g., corruption and illegal activities) and ciguatera, among other concerns (Sadovy de Mitcheson et al. 2017).

Groupers are exported after capture at the preferred trade market-size (typically around 400–800 g), caught as juveniles or produced by hatcheries, and then grown out to market size (i.e., for CBA) (Sadovy de Mitcheson et al. 2017). From what we can determine, much of the live grouper trade that comes from the wild is not reflected in FAO capture statistics because the live fish are not recorded as part of national fishery production. Even the international trade in much of this sector is poorly understood with the exception of the customs database import records in Hong Kong, the trade hub for this trade, which runs a well-developed import documenting system

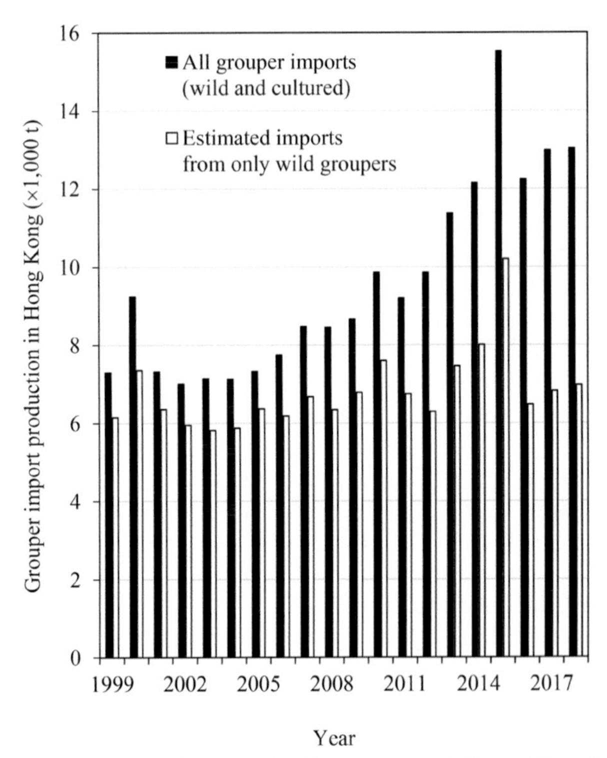

Fig. 7 Total reported import productions (t) of live groupers in Hong Kong from 1999–2018. Data Source: Hong Kong Census and Statistics Department (CSD) (by air and foreign-registered vessels), and Agriculture, Fisheries, and Conservation Department (AFCD) voluntary interviews (by Hong Kong registered vessels).

for several live fish trade categories and a voluntary interview system. As currently understood, a minimal estimate (reported volumes which under-estimate the totals) of this international trade was 13,000 t in 2018 into Hong Kong (Sadovy de Mitcheson et al. 2017), with wild-caught fish making up at least 50% of the volume; the rest comes from hatchery and CBA production (Fig. 7).

Despite the low overall volume of the global live grouper trade (relative to chilled/fresh grouper), its retail value is considerable, estimated as at least US$1 billion at retail, bringing substantial benefits to the value chain, with particularly high profits downstream (Sadovy de Mitcheson et al. 2017). Certain groupers fetch particularly high retail prices; in January 2016 in Hong Kong the Humpback Grouper *Cromileptes altivelis* was retailing at about 150 US$/kg, the Leopard Coral grouper *P. leopardus* at up to 75 US$/kg, with higher prices for wild compared to cultured fish (note that wholesale prices are only available from 2005 to January 2016; https://fish.net/english/fisheries_information/wholesale_prices.php?year=2016). Although significant amounts of this trade are re-exported from Hong Kong into mainland China, little of this trade is documented and much is part of a long-standing seafood smuggling arrangement, between the city and the mainland, which seeks to minimize the tariffs payable on mainland luxury seafood imports (Sadovy de Mitcheson et al. 2017).

There are concerns about depletions in some species, and widely reported illegal, unregulated, and unmonitored trade in some areas (Scales et al. 2006, Sadovy de Mitcheson et al. 2017), leading to actual and potential threats to certain species as indicated by changes in production patterns. One pattern is illustrated by a shifting relationship from the capture of market-sized fish to a predominance in its production by CBA in the Leopard Coral grouper as overfishing increased and market-sized fish declined. This is one of the most highly valued species (due to a combination of volume and unit value) in the live grouper trade, mostly sourced from the Philippines, Indonesia, and Australia. In Australia, this fishery is managed and only market-sized fish are caught and exported. In the Philippines (particularly Palawan), after declines in catches of market-sized fish, CBA was encouraged by the government, and most of this species exported (Russell 2006) (all of it illegally) today come from CBA. This has created concerns over the future of the fishery because of the negative impact that this can have on the reproductive potential of the population, and lack of management (Salao et al. 2013, Sadovy de Mitcheson et al. 2017, Sadovy de Mitcheson 2019). In Indonesia, which appears to fall between the Australia and the Philippines in the condition of the fishery of this species, declines in larger fish in some areas may lead to similar erosion of reproductive capacity and there is no management (e.g., Khasanah et al. 2019).

Mariculture

Global grouper mariculture production has grown substantially since the early 2000s and is dominated by Asia, with the main producers being mainland China, Taiwan and Indonesia (data available since 1970;http://www.fao.org/fishery/statistics/global-aquaculture-production/en, Rimmer and Glamuzina 2019). The Sabah Grouper hybrid, *E. fuscoguttatus* female x *E. lanceolatus* male (Dennis et al. 2020), production data were first included in FAO dataset in 2018. While the focus of this chapter is on wild populations (see Chapter 2.2) and threats to these, and while we recognize the role of mariculture for increasing grouper supply, fish farming also poses additional threats to wild populations. Moreover, increased mariculture does not reduce fishing pressure on groupers, because fishing continues as mariculture develops. Millions of fishers globally (Teh et al. 2013) will continue to depend on wild capture, which often includes groupers among many other species, for food and livelihoods.

Grouper mariculture, both CBA and hatchery-based aquaculture (HBA), impacts wild populations in several ways. In the case of HBA, threats can come from the ongoing need to source wild broodstock to replenish genetic diversity and reproductive output, while the risks from juvenile capture fisheries to obtain 'seed' for CBA grow out have already been discussed (FAO 2011, Rimmer and Glamuzina 2017). There is also at least one example of juveniles being taken for feed. In India, large volumes of trawl-caught juvenile Spinycheek Grouper, *E. diacanthus*, are used as fish feed (Zacharia et al. 1995, R. Nair, ICAR-Central Marine Fisheries Research Institute, India personal communication 2020).

More than any other country, China has dedicated large coastal areas to grouper mariculture production from south (Hainan Province) to north (Tianjin Municipality and Shandong Province), building on its long history of aquaculture and with strong government support (http://www.shuichan.cc/news_view-318052.html, http://www.mingbo-aquatic.com/, http://www.tianjinhaifa.com/). Ongoing problems faced in seeking to increase production in China, or elsewhere, however, are the poor condition of coastal waters in some locations, shortage of broodstock for some species (which need regular replacement to maintain genetic diversity, and for sperm and egg quality) and (as an impact on other species) the high demand for fish feed (the supply of which is largely unsustainable), since groupers are carnivorous and still require a diet rich in fish protein (Naylor et al. 2000, Pierre et al. 2008, Cao et al. 2015, Zhang et al. 2020, Sumaila et al. 2021).

To improve profitability and add diversity to the market, hybrids were first developed in 2007 (http://www.thefishsite.com/fishnews/3629/researchers-breed-first-hybrid-grouper/) in Malaysia and further received great interests in China (Fig. 8). Escapes from culture, or releases, into the wild are not uncommon and, in other species have resulted in problems (e.g., Wringe et al. 2018, Rimmer and Glamuzina 2019). Sabah Grouper hybrids have been found with ripe eggs in the wild (X. Zhang and J.L. Li, Hainan University, China, unpublished data); it is not known if they reproduce in the wild, or what the possible impact would be on native populations if they did.

Fig. 8 Hybrid between the Brown-marbled Grouper *Epinephelus fuscoguttatus* and Humpback Grouper *Cromileptes altivelis* on sale in Hong Kong [Photo: Yvonne Sadovy de Mitcheson].

Recreational, Tourism, and Other Leisure Industries

In addition to their considerable value as food in both subsistence and commercial (domestic and international trade) sectors, groupers are prized in the leisure sector, including for recreational fishing, dive tourism, and in public and private aquaria. Although our understanding of such uses is sketchy, several examples are illustrative. People love to see or spear large groupers. Among divers, particularly large or 'friendly' groupers are a popular attraction, for example in the Cod Hole of Australia's Great Barrier Reef (http://www.youdive.tv/Australia-Dive-with-the-biggest-groupers-of-the-world-at-Cod-Hole_v81.html) (Fig. 9) and in east Africa (Lukasik 2016), while recreational divers are willing to pay US$100–200 to view Goliath Groupers (Shideler and Pierce 2016). Sports spearfishers often like to catch groupers which can rapidly lead to depletions of large size individuals (e.g., Giglio et al. 2017). The spectacular large Potato *E. tukula* and Giant Groupers survive well in public aquaria and are commonly included for display, for example, in mainland China, Hong Kong, and Taiwan (e.g., Fig. 1, S. Gendron, Ocean Park, Hong Kong, personal communication 2016). During the covid-19 lockdowns many aquaria were closed; one large grouper made the news when it was reported to be lonely and received many supportive letters from children (ABC 2020).

Grouper spawning aggregations are a strong draw for divers in Palau, Belize, and French Polynesia, among other locations (Y. Sadovy de Mitcheson, The University of Hong Kong, Hong Kong, personal observation, multiple years). In a novel economic analysis, the Nassau Grouper was assessed to have 20 times the economic value alive for tourism, and for its ongoing reproductive contribution to the fishery, compared to its immediate landed value (Sala et al. 2001).

Recreational fishing for groupers is lucrative, although generally not well documented in the scientific literature. For example, in the United States, both shallow and deepwater groupers are important in the highly lucrative

Fig. 9 Diver interacts with the Potato Grouper *Epinephelus tukula* in Australia (Photo: Nigel Marsh/SeaPics).

recreational sector and are targeted in several states (http://www.nmfs.noaa.gov/). Groupers are worth hundreds of millions of US$ for recreational fishing in the Gulf of Mexico (Gentner 2009). In parts of the Caribbean, groupers are one of the species of recreational interest (e.g., Southwick et al. 2016).

Threats to Groupers

Extrinsic Factors and Vulnerabilities

Multiple threats to groupers occur due to high and increasing consumer demand, anthropogenic impacts on habitat, overfishing associated with lack of management of most exploited populations, poor oversight of domestic and international trade in the taxon, possibly also from aspects of climate change. For a few species, their small areas of geographic distribution could make them particularly vulnerable to local impacts, such as for the vulnerable White-edged Grouper, *E. albomarginatus*, and the near-threatened Catface Grouper, *E. andersoni*. These species are endemic to a limited area extending from eastern South Africa to Mozambique, which is widely affected by increasing fishing pressure. Water temperature changes associated with climate change could have a significant effect on those species for which temperature is likely a major determining factor in reproduction (Asch and Erisman 2018). Moreover, there are few alternatives to fishing in many fishing communities across vast areas where many grouper species occur, which makes their management particularly challenging.

These actual and potential, direct and indirect, extrinsic threat factors, combined with vulnerable life histories, and the lack of monitoring or management at the species level in general mean that many populations are now much reduced from natural and even biologically sustainable levels, and likely to become more so if nothing changes. Many of these problems and challenges were recognized for the Caribbean region over two decades ago (Sadovy 1994) and have emerged elsewhere. Nineteen species, 11.4% of all groupers, are considered to be in one of the IUCN Red List globally threatened categories, with a further eight species considered to be near-threatened (IUCN Red List Assessments, updated November 2018) (Table 1). The small absolute size of individual fisheries (in terms of weight produced per year, for example), the substantial numbers of fishermen involved, highly dispersed fishing centres, and overall poor appreciation of grouper economic and local food value, act in combination to obscure their importance and condition. This situation makes assessment, monitoring, and the introduction of management a major challenge (e.g., Sadovy 1994, Sadovy de Mitcheson et al. 2013, Lee et al. 2018).

Today, many grouper populations are overfished, and many of the former refuges in remote or deep areas, formerly little fished, are disappearing, often driven by offshore fishery expansions as inshore areas become overfished (Sadovy de Mitcheson et al. 2013). Non-selective fishing gear in multi-species fisheries makes protected species difficult to avoid catching even

Table 1. List of threatened grouper species including critically endangered (N = 1), endangered (N = 2) and vulnerable (N = 16), and near threatened (N = 8) (IUCN Red List November 2018 update: www.iucnredlist.org).

Species	Common name	Threat category/criteria	Habitat
Epinephelus striatus	Nassau Grouper	Critically Endangered (A2bd)	Coral Reefs
Epinephelus akaara	Hong Kong Grouper	Endangered (A2bcd)	Coral and Rocky Reefs
Mycteroperca jordani	Gulf Grouper	Endangered (A2bd)	Shallow Rocky Reefs
Mycteroperca fusca	Island Grouper	Vulnerable (B2ab(v))	Rocky Reefs
Epinephelus albomarginatus	White-edged Grouper	Vulnerable (A4bd)	Coral and Rocky Reefs
Epinephelus bruneus	Longtooth Grouper	Vulnerable (A2bd)	Deep Reefs
Hyporthodus (=Epinephelus) flavolimbatus	Yellowedge Grouper	Vulnerable (A2bd+3bd)	Deep Reefs and Soft Bottom
Epinephelus fuscoguttatus	Brown-marbled Grouper	Vulnerable (A2bd+4bd)	Coral Reefs
Epinephelus itajara	Atlantic Goliath Grouper	Vulnerable (A2bcd)	Coral Reefs and Mangroves
Epinephelus marginatus	Dusky Grouper	Vulnerable (A2bd+4bd)	Unknown?
Epinephelus morio	Red Grouper	Vulnerable (A2bd)	Coral and Rocky Reefs
Epinephelus polyphekadion	Camouflage Grouper	Vulnerable (A2bd)	Coral Reefs
Hyporthodus (=Epinephelus) niveatus	Snowy Grouper	Vulnerable (A2bd+4bd)	Deep Reefs
Hyporthodus (=Epinephelus) acanthistius	Gulf Coney	Vulnerable (A2bd)	Rocky Reefs
Mycteroperca interstitialis	Yellowmouth Grouper	Vulnerable (A4bd)	Deep Reefs
Mycteroperca microlepis	Gag Grouper	Vulnerable (A4bd)	Hard bottom and rocky ledges
Mycteroperca olfax	Sailfin Grouper	Vulnerable (A2bd)	Rocky Reefs
Plectropomus areolatus	Squaretail Coral grouper	Vulnerable (A2bd)	Shallow Coral Reefs
Plectropomus marisrubri		Vulnerable (A4bd)	Unknown
Epinephelus aeneus	White Grouper	Near Threatened	Deep Reefs
Epinephelus andersoni	Catface Grouper	Near Threatened	Rocky Reefs

Table 1 Contd. ...

...Table 1 Contd.

Species	Common name	Threat category/criteria	Habitat
Epinephelus daemelii	Saddletail Grouper	Near Threatened	Rocky Reefs
Epinephelus goreensis	Dungat Grouper	Near Threatened	Deep, soft and hard bottom habitats
Epinephelus nigritus	Warsaw Grouper	Near Threatened	Deep, hard substrate
Hyporthodus (=*Epinephelus*) *egastularius*		Near Threatened	Unknown
Mycteroperca bonaci	Black Grouper	Near Threatened	Shallow Coral and Rocky Reefs
Mycteroperca venenosa	Yellowfin Grouper	Near Threatened	Coral and Rocky Reefs

with good management. Coastal development can damage nearshore areas, often important for early life history phases, while coral reef habitat is being degraded by human activities, including from changes due to climate change (Hughes et al. 2003). Key biological processes (e.g., spawning aggregations) are eroded by uncontrolled fishing, while juvenile fisheries for 'seed' to supply CBA operations are not monitored or managed; CBA will continue even if species can be hatchery-produced if wild capture is easier and cheaper (Sadovy de Mitcheson and Liu 2008b). Fish being taken in fisheries increasingly span multiple size classes and are regularly sold below the size of sexual maturation (e.g., Hong Kong, To and Sadovy de Mitcheson 2009; Pohnpei, Rhodes and Tupper 2007; USA, New York Times 2015; Fiji, Mavruk et al. 2018; the South China Sea, Min Liu, personal observation 2017–2020).

Intrinsic Factors and Vulnerabilities

Groupers are inherently more vulnerable to anthropogenic impacts, particularly to overfishing, than are many other fish taxa due to several aspects of their biology (e.g., Levin and Grimes 2002, Young et al. 2006). The longevity and relatively large and late maximum and sexual maturation sizes and ages of many species (see Chapter 1.3) mean that many must survive multiple years before they can reproduce, and this is particularly a problem in heavily exploited multi-species fisheries, where groupers are often larger than the average size of fish being caught. For example, when mean size of catch in a multi-species fishery is smaller than the size of maturation of a grouper species, many juvenile groupers are taken amongst other species (Sadovy 1994). Even for species that are relatively (for the taxon) fast-growing and productive, heavy fishing pressure can rapidly deplete populations, as has been shown for *Plectropomus* spp. in some locations (e.g., Frisch et al. 2016).

Complex reproductive systems make some groupers particularly prone to overfishing, and for others may confer some protection. An increasingly female-biased sex ratio could potentially lead to sperm limitation if sex change does not occur quickly enough for females to replace males. This can happen when too many males (the larger sex in a protogynous species) are removed, for example from spawning aggregations, as for the Gag Grouper *M. microlepis* (McGovern et al. 1998, Heppell et al. 2006). Nonetheless, some of the more threatened groupers are thought to be functionally gonochoristic (e.g., Nassau, Camouflage, and Leopard *Mycteroperca rosacea* Groupers) (Sadovy and Colin 1995, Rhodes and Sadovy 2002, Erisman et al. 2008). On the other hand, labile sex change may buffer hermaphroditic species from overfishing to some extent (Bannerot et al. 1987). For example, the Chocolate Hind, a relatively small species with bi-directional sex change, is now one of the most commonly reported groupers in catches (http://www.fao.org/fishery/statistics/global-capture-production/en). The existence of two male developmental pathways (i.e., primary and secondary males) and male to female sex change in this species, may confer greater resilience to fishing pressure through sexual lability (Chan and Sadovy 2002, Liu and Sadovy 2004, Sadovy de Mitcheson and Liu 2008a).

The aggregation-spawning reproductive mode of many mid- to large-sized species makes them particularly prone to overfishing in the absence of management. As a result, and because few are managed, many exploited aggregations of groupers are declining or have even disappeared. For species in which aggregation is their only known mode of reproduction, this can seriously erode reproductive output. Spawning sites and times remain similar each year, and once these sites are discovered, they understandably become a focus for fishing. While the largest known aggregations contain, or once contained, tens of thousands of fishes (the largest known being those of the Camouflage and Nassau Groupers), most are much smaller or have declined, sometimes by ten-fold or more, with a few hundred or few thousands of individuals being more common nowadays (Sadovy de Mitcheson and Colin 2012). Few aggregations are managed sustainably and catches from many are declining, even under relatively low levels of fishing pressure in artisanal fisheries. Of 509 aggregations for which there is information on changes in catches over time, about one third of those of known status are declining or no longer form and some no longer form at all; those that show increasing catches were more recently discovered (Fig. 10). Comparative work among different grouper species shows that those that aggregate are more susceptible to being overfished (Coleman et al. 1996).

Fishing Methods and Sustainability

How are Groupers Caught?

Groupers of a wide range of body sizes are caught by a diversity of fishing gears and methods for the chilled/frozen/live fish markets, and for CBA. While some groupers are specifically targeted by species or size, many are taken by relatively unselective gears in multi-species fisheries. Some are

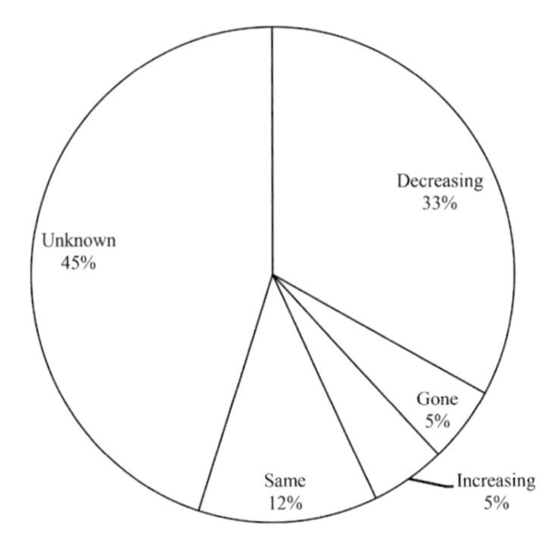

Fig. 10 Current known status reflecting changes of exploited grouper aggregations globally, as noted by fisher interviews, monitoring, or underwater surveys (N = 509) (www.SCRFA.org, accessed 23 June 2016).

particularly easy to catch, such as many shallower water species readily accessible to spear fishing and popular with recreational spearfishers. Some species can be challenging to catch efficiently by conventional gears, such as Leopard Coral grouper for the live export market, with fishers in parts of the Philippines and Indonesia regularly using the poison cyanide (Padilla et al. 2003). In the case of larger groupers included in the live trade, preferred 'plate-sized' fish may be still in their juvenile size range, as in the case of the Brown-marbled Grouper (Kindsvater et al. 2017). In the Pacific region, fish numbers declined as nighttime fishing with spears increased, with groupers being particularly easy to catch at night, especially while aggregated (Gillett and Moy 2006). Hooks take a wide range of body sizes of groupers, given their large mouth extension capability. Fish traps are heavily used in some areas, particularly across the Caribbean region. Large nets may be deployed to capture fish migrating to aggregation sites along predictable routes in some locations, such as in Cuba (Claro et al. 2009). Some species, mainly smaller individuals, are taken incidentally by bottom trawls, including the Hong Kong, Yellow *E. awoara*, Longtooth *E. bruneus*, Rock *E. fasciatomaculosus*, and Longfin *E. quoyanus* Groupers in southern China waters, including Hong Kong (M. Liu, personal observations 2006–2021). The Spinycheek Grouper is heavily taken in trawl fisheries and used as fish feed for mariculture operations and terrestrial farming (see mariculture section).

Juveniles for CBA are caught in large numbers with a wide range of gears, many of them specially adapted to take small individuals of particular species or sizes. These gears range from modified and small mesh fish traps and fyke nets, to specially developed devices including small artificial reefs, or 'gangos', in the Philippines, scissor push nets and twig baskets or 'pums' from Thailand for post-larval settlers (Sadovy 2000, Mous et al. 2006, Sadovy de Mitcheson and Liu 2008b). Once estimated to involve catches of millions of grouper juveniles each year in Southeast Asia for CBA, some of these fisheries continue although the extent of fishing activities for juveniles for CBA currently is unknown, and is likely to have declined substantially since the 1990s, given the increasing success of HBA (Sadovy 2000, Rimmer and Glamuzima 2019).

Challenges for Sustaining Grouper Fisheries

Little is known of the condition of most grouper fisheries, or of the population parameters of exploited species (e.g., mortalities and growth rates). While there are exceptions (e.g., the Red and the Atlantic Goliath Groupers, USA National Marine Fisheries Service; the Coral grouper, Government of Australia; Areolate Grouper, *E. areolatus*, Indonesia, Amorim et al. 2020), formal stock assessments are few because most fisheries are perceived to be too small to attract government interest or funding to conduct assessments. In the absence of formal stock assessments, general indicators of fishery condition, as for other taxa, can include catch trends (ideally measured as catch per unit effort-CPUE), trends in sizes of capture and incidence of juveniles in catches, length-based analyses, and numbers of fish in spawning aggregations. Using various assessment methods, signs of heavy to possible overexploitation are especially marked in longer-lived, larger species, with several considered to be threatened according to IUCN Red List categories and criteria (Table 1). The major threat factor, according to these IUCN listings, is overfishing and this is closely associated with lack of fishery management or controls on fishing levels or trade volumes (Sadovy de Mitcheson et al. 2020a).

In a few cases species may be particularly threatened or vulnerable to declines because they occur over relatively small areas or because their behaviour makes them highly susceptible to certain gear types or technologies. For example, the Saddletail Grouper *E. daemelii*, which only occurs in parts of Australia and New Zealand, is territorial and curious by nature, making it easy to overfish by line and spear fishers; numbers are markedly reduced from recreational as well as commercial fishing (Pogonoski et al. 2002, Francis et al. 2015). Large individuals of the Goliath Grouper are often readily accessible on wrecks and in shallow water and are heavily overfished by spearfishing in Brazil (Giglio et al. 2017). The Island Grouper *M. fusca* is known with certainty only from the Azores and Madeira (Portugal), Cape Verde, and the Canary Islands (Spain), and is classified as Endangered due to fishing pressure throughout much of its small geographic range (http://www.iucnredlist.org/species/64409/42691809).

In terms of behaviours that increase susceptibility, those associated with aggregation-spawning are of particular interest for both monitoring and management. For example, the Nassau Grouper, at other times solitary, aggregates predictably to spawn. Traps set at its aggregation sites fill quickly because individuals appear to be strongly drawn to conspecifics at this time and will squeeze into a full trap even if there is hardly any space left, making the ripe adults very easy and efficient to catch (Y. Sadovy de Mitcheson, personal observation 1988). Several species produce sounds on their spawning aggregations, making them susceptible to new technologies in sound detection. Examples include the Yellowfin Grouper *M. venenosa* (Rowell et al. 2015), the Goliath Grouper (Mann et al. 2009), and the Red Hind *E. guttatus* (Appeldoorn et al. 2013). On the other hand, sound production and its correlation with fish density makes this a valuable monitoring tool (e.g., Appeldoorn et al. 2013).

If aggregation-focused fisheries are not managed, populations can quickly decline, along with the fisheries they support. While the remote locations of many aggregation sites are a particular challenge for enforcement, their brief duration can allow for efficient and highly targeted management action focused on the spawning period. Assessing the condition of aggregation-fisheries can be challenging due to 'hyperstability' whereby CPUE, a common proxy for abundance used in fisheries assessments, becomes delinked from fish abundance, remaining high even as populations decline; this relationship can also vary across species (Sadovy and Domeier 2008, Erisman et al. 2011, Robinson et al. 2015). Aggregating species may need management by multiple measures (e.g., Grüss et al. 2014).

The overall outcome of these various challenges to sustaining fisheries, not surprisingly, is that the more resilient species, those better able to sustain fishing due to their biology, increasingly dominate catches. In some markets for example, surveys note that smaller species, formerly not considered of much commercial significance, are becoming more important in sales. Heemstra and Randall (1993) wrote of the Chocolate Hind several decades ago: "*C. boenak* is too small to be of commercial importance." Yet this species is now the most heavily reported single grouper species in FAO data. Smaller species and smaller sized individuals of larger species are now increasingly common in markets in Asia, such as the Honeycomb Grouper *E. merra*, and the Tomato Hind, *C. sonnerati*, among others (To and Sadovy de Mitcheson 2009, Sadovy de Mitcheson 2019, Kandula et al. 2015). Relatively high natural productivity and larger capture size has enabled a productive fishery of *E. areolatus* (Boddington et al. 2021).

Of particular concern are some of the fisheries that supply high-value groupers and where trader pressure and consumer demand are particularly intense. In the luxury live fish trade, for example, overfishing and boom-and-bust grouper fisheries, use of poison (e.g., cyanide), and heavy take of juveniles for grouper CBA are not uncommon (Sadovy de Mitcheson et al. 2017). Very few of these fisheries are managed (Padilla et al. 2003, Scales et al. 2006, Sadovy de Mitcheson and Liu 2008b). Although it is a valuable trade that often provides good income to source communities and along the trade chain, the paucity of oversight and management, and high pressure for export (most of which is not taxed or is unrecorded) make it particularly challenging to control and to safeguard (Fabinyi and Liu 2014, Sadovy de Mitcheson et al. 2018).

Conflicts, heavy and shifting market forces, and practical constraints can challenge governments, both directly and indirectly, where groupers are in high and increasing demand from different sectors. For example, in Palau the rapid growth in seafood tourism in recent years has put heavy strain on coastal resources and drawn attention to the need to balance domestic food needs for Palauans and those of the tourism sector. Unlike most other PICs, reef fishes, including groupers, are no longer exported for commercial purposes from Palau due to such concerns; minimum sizes and other protective measures are also in force, but implementing management measures effectively continues to be challenging (Wabnitz et al. 2018). Fiji is struggling to maintain affordable domestic market supply for tourism and the local population against pressures to export and recently introduced protection of groupers during the spawning season; this was then relaxed due to hardships associated with the covid-19 virus (Rawalai 2020). In Pohnpei, seasonal protection of groupers led to increased fishing pressures on other taxa, highlighting the need to consider wider issues around grouper management and in relation to multi-species fisheries as a whole (Rhodes et al. 2008). In West Africa, Thiao et al. (2012) identified the role of the booming small-scale fisheries in the collapse of the populations of White Grouper *E. aeneus*, an emblematic species. The price to fishers for the White Grouper in Senegal, highly valued by northern hemisphere markets, increased by almost an order of magnitude between 1993 and 2006, much more than for other species caught in the artisanal fishery there, and management is needed to prevent further declines (Thiao et al. 2012). Protected deepwater species may be difficult to avoid taking in deepwater multi-species fisheries (e.g., the Speckled Hind *E. drummondhayi*, the Warsaw *H. nigritus*, and the Black *M. bonaci* Groupers) even if fishers respect size regulations by releasing undersized fish (IUCN Red List assessments for these species, www.iucnredlist.org). Few of these fish are likely to survive release, due to barotrauma, once they have reached the surface should they be accidently hooked and then released (e.g., Rudershausen et al. 2007).

Other Anthropogenic Impacts

Other, non-fishery, anthropogenic impacts can result in lethal or sub-lethal effects, or potentially reduce reproductive capacity in groupers, with long-term implications for their populations. These range from habitat damage and loss, to pollution, increased sediment loads, and factors related to climate change, particularly temperature.

Increases in contaminated suspended sediment damaged grouper gill structure and caused sub-lethal effects in the Orange-spotted Grouper *E. coioides*, for which suspended sediment was determined to increase hyperplasia and reduce the epithelium, resulting in impaired oxygen uptake and osmoregulatory stress (Wong et al. 2013). In French Polynesia the Honeycomb Grouper was found to have pesticide accumulation with unknown implications for survivorship (Wenger et al. 2015).

Certain species may be particularly prone to exposure to contaminants or other environmental impacts because of aspects of their life history. In one example, mean mercury concentrations observed in Goliath Grouper from U.S. waters were within the range known to cause direct health effects in fish after long-term exposure. Adams and Sonne (2013) suggest that exposure to mercury and other environmental influences, such as pathogens and reduced temperatures in near-shore environments, where the Goliath Groupers spend much of their life, could be stressful and act as co-factors in the observed population declines in the species. Both Goliath and Red Groupers are susceptible to red-tide blooms, with scientists attributing a marked dip in the biomass of Red Grouper to an uncharacteristically persistent (year-long) and large red tide in 2005 (Coleman et al. 2010).

Loss of reef and adjacent habitats from bombing, pollution, construction, and reclamation can reduce habitat for shelter, foraging, or nursery grounds. The endangered Nassau Grouper appears to have quite specific habitat associations at different life history phases (Eggleston et al. 1998), while the threatened Atlantic Goliath Grouper depends heavily on mangroves as juveniles and in the young adult stages (Koenig et al. 2007). A significant proportion of China's coastline has been developed or impacted due to economic development in recent decades, which is likely among the multiple reasons that groupers (among other marine taxa) appear to have declined so substantially in the country (Liu and Sadovy de Mitcheson 2009, He et al. 2014).

Finally, little is understood about the potential effects of climate change on grouper populations, but these could be substantial due to projected loss of coral reef habitat and due to changes in water temperature regimes, which could affect reproductive patterns. Temperature changes could potentially affect the timing of spawning, the formation of reproductive aggregations, and the resulting survivorship of the young produced. For example, the time of spawning is closely temperature-associated in the Nassau Grouper, which spawns within a very well-defined temperature range very close to 25°C; changes in water temperature could substantially alter its reproductive capacity (Colin 1992, Asch and Erisman 2018). Inter-annual variation is the degree of aggregation correlated with annual variation of mean seawater temperature in the 40 days prior to spawning (Nanami et al. 2017). In experimental studies, the Leopard Coral grouper exhibited declines in survivorship, aerobic scope, and activity with relatively moderate increases in temperature. Given that a significant portion of populations of the species is already exposed to temperatures (≥ 30°C), such responses signal the need for more precautionary management for the species (Pratchett et al. 2017). By contrast, temperatures of the Camouflage Grouper at spawning are far more varied, which may confer some resilience under climate change (Y. Sadovy de Mitcheson, unpublished data). Finally, little is known about the effect of ocean acidification on grouper populations; while indications are that these may not be substantial for reef fishes, work is needed on the taxon to evaluate this (Clark et al. 2020).

Conservation and Management of Groupers

Global Conservation Status of Groupers

Of nine coral reef fish species identified by IUCN Red List criteria to be Critically Endangered globally, three are groupers, threatened directly or indirectly principally by fishing (McClenachan 2015). An assessment of all groupers that occur globally, using IUCN criteria (IUCN 2012), determined 27 species (16.2% of groupers) to be Threatened or Near Threatened (NT) (Table 1, Fig. 11). An additional 15.0% of the 167 species were considered DD, which means that there were insufficient data to assess their conservation status, a reflection of our poor understanding of many species; it is possible that some of these DD species are also threatened. The Caribbean Sea, coastal Brazil, and Southeast Asia contain a disproportionate (even after factoring in species diversity) number of Threatened species, as well as numerous poorly documented DD and NT species (IUCN Red List Assessments, updated November 2018; Sadovy de Mitcheson et al. 2020a).

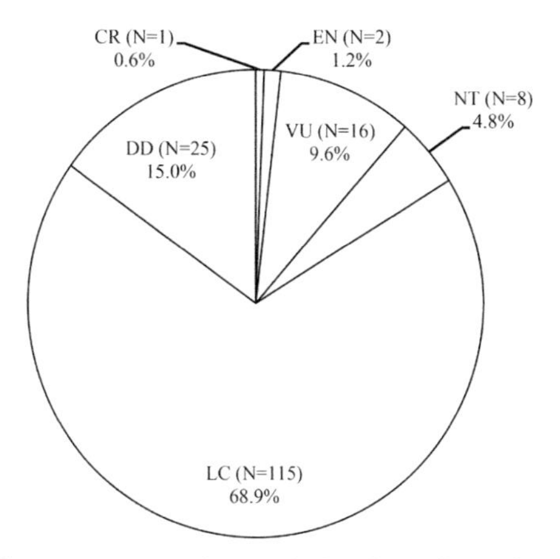

Fig. 11 Categories of all grouper species (N = 167). CR, Critically Endangered; EN, Endangered; VU, Vulnerable; NT, Near Threatened; LC, Least Concern; DD, Data Deficient according to the IUCN Red List (IUCN Red List Assessments, updated November 2018).

As fishing pressures increase and fishing expands deeper and further from urban centres, there remain progressively fewer refuges for shallow water species, while deeper water species, which seem to be particularly long-lived and relatively low in abundance, are increasingly exposed to exploitation. In Southeast Asia, extraction levels of groupers in many areas two decades ago already exceeded the estimated annual rate of sustainable grouper production of reefs of the region (e.g., Warren-Rhodes et al. 2003). Demand for and prices of live/chilled fish and fillets are high as demand for grouper exports grows and becomes a substantial additional pressure to grouper populations on top of domestic needs for local consumption and tourism (see above).

Conservation Measures in Place and Their Effectiveness

While there are many management challenges, management can bring increases in grouper numbers, catches and sizes, and associated benefits. A range of conventional management tools has been applied to groupers, including minimum sizes, gear, and sales controls, among other conventional fishery measures. When effectively protected, groupers show significant increases in size and biomass within no-take Marine Protected Areas (MPAs) especially for smaller and medium-sized species and those that do not migrate (e.g., Edgar et al. 2014, Howlett et al. 2016). MPAs are particularly appropriate for sedentary groupers (i.e., those that do not migrate or only migrate short distances to spawning sites), such as the Leopard Coral grouper, the Dusky Grouper, among other species (Russ 2002, Anderson et al. 2014). Properly sited shallow and deepwater MPAs, and the protection of outer reef passages can be important for maintaining refuges for shallow-water species, and for deepwater species, including certain groupers (Craig et al. 2011, Olavo et al. 2011). Biomass of groupers of the genus *Mycteroperca*, for example, increased over a decade in one MPA, and there are multiple reports of the benefits of MPAs to small- to medium-sized sedentary groupers (e.g., Roberts 1995, Chiappone et al. 2000, Aburto-Oropeza et al. 2011). For example, reproductive potential is influenced by multiple factors and can increase in some groupers with appropriate conditions and siting of MPAs (Carter et al. 2017, Bucol et al. 2021).

For species that migrate long distances to spawn and are not fully protected by the typically small sizes of MPAs or their often near-shore locations, protection of specific spawning migration routes could be implemented at small-enough spatial scales to be practical (e.g., Waldie et al. 2016). Effective protection of spawning sites or spawning seasons can support recovery, as the following examples illustrate. Although regional (i.e., across national boundaries) protection may often be preferable to national-only level protection for managing grouper populations, due to larval dispersal potential and seasonal migrations, localized protection can also be effective.

In the Indo-Pacific, community protection of several spawning aggregations led to increases in aggregation densities at sites for the Camouflage and the Brown-marbled Groupers, while localized recruitment was determined for the Squaretail Coral Grouper *P. areolatus*; both examples, from Papua New Guinea, illustrate the relevance of localized action (Almany et al. 2013, Hamilton et al. 2011). In Palau, a decade of successful protection has started the recovery of the Squaretail Coral and the Camouflage Groupers (Sadovy de Mitcheson et al. 2020b). Studies of groupers that exhibit differences in reproductive mode (i.e., the intensity of aggregation and duration of spawning) suggest that this mode could influence population structure (e.g., Ma et al. 2018), further highlighting the need to tailor management to species biology and to consider reproductive mode when looking at population structure.

In the Caribbean and tropical Atlantic regions, protection of grouper aggregations or complete cessation of fishing have variously resulted in increases in catches, numbers aggregating, and in fish sizes. Examples of increased catches and/or sizes were noted for the Red Hind (Nemeth 2005,

Luckhurst and Trott 2009). The Atlantic Goliath Grouper shows promising signs of recovery in Florida (USA) following a 1990 moratorium on its capture in US waters (Koenig et al. 2011). While genetic studies on population structure suggest that regional level protection is important for the Nassau Grouper (Jackson et al. 2014), local protection has proven productive when enforced. Fifteen years of protection of an aggregation site in the reproduction season in the Cayman Islands led to a marked recovery in the Nassau Grouper at a protected site (Waterhouse et al. 2020).

Inter-annual variation in the degree of aggregation correlated with local annual variation of mean seawater temperature in the 40 days prior to spawning (Nanami et al. 2017).

Management and Monitoring of Groupers

The Challenges of Management

The combination of low volumes and undervalued, poorly documented, and biologically vulnerable species in reef fisheries creates a vicious cycle, whereby the incentive to manage is progressively undermined as populations decline, while the ability to manage becomes more difficult. This happened with the Nassau Grouper which, unusually for a commercial reef fish, is now considered to be critically endangered on the IUCN Red List (Table 1), included as a species of concern (U.S. Endangered Species Act), and in 2017 was added to the Specially Protected Species and Wildlife (SPAW, http://www.cep.unep.org/content/about-cep/spaw) protocol. As populations drop to very low levels, natural phenomena like the 'Allee Effect', whereby heavily reduced populations find it particularly difficult to recover, may occur (Courchamp et al. 2008). This situation is exacerbated when there is a positive correlation between rarity and value (Courchamp et al. 2006); in the case of groupers, some of the more desired species can be among the least common, as in the case of live wild-caught Humpback Grouper, which are highly valuable and heavily sought (Sadovy de Mitcheson et al. 2018).

Management Options

While conventional management methods can be applied to grouper fisheries, such as catch and size limits and quotas, these are unlikely to be effective in many cases unless combined with other measures (e.g., Giglio et al. 2017, see Chapter 2.1). Size limits alone will be ineffective, for example, when species recruit to a multi-species fishery before they are sexually mature, for deepwater species (release mortalities of deep or undersize fish can be high, e.g., Bartholomew and Bohnsack (2005)) or where 'plate-size' market demand, or CBA, focuses fishing on juvenile sizes. Size limits are also likely to be insufficient if most annual catches come from spawning aggregations. Gear restrictions, such as increased hook size, may protect immature fish but promote the capture of larger and older fish, including megaspawner females which are the most fecund (Usseglio et al. 2016). Gear modifications can improve fishing practices for species complexes (e.g., proposed rule Federal Register 2020).

Spawning aggregations can be managed by spatial and/or seasonal protection, ideally in combination with other measures, such as seasonal sales. MPAs have shown positive outcomes for groupers, but few such areas are fully protected or big enough for the less sedentary species (see above). The best management approach is one that seeks to safeguard sufficient reproductive potential by ensuring not only that enough fish grow large enough mature, but also that adults have sufficient opportunity to mate, including in spawning aggregations (Russell et al. 2012).

Monitoring and Assessment

Species-specific monitoring systems for landings (rather than a generic combined species category of 'groupers'), and for exports and imports (for example by adding species-level detail to the international trade Harmonized Code system) are needed for groupers. This would enable identification of trends, volumes, and major international trading partners, and may help to flag particularly vulnerable species or populations, or key opportunities for strategic action. For example, findings of high levels of fishing and serial overexploitation associated with the live fish trade drew attention to its operations and enabled an understanding of its dynamics (e.g., Scales et al. 2006, Sadovy de Mitcheson et al. 2017).

To assess fishery status in data-poor contexts, several approaches are available. For example, vulnerability assessment frameworks combine measures of species productivity and susceptibility to fishery assessments (e.g., Patrick et al. 2010, Hobday et al. 2011). Length-frequency data can be applied to poorly monitored and data-poor fisheries like groupers (e.g., Indonesia, Amorim et al. 2020). Data-limited methods have been widely used in Caribbean and Pacific reef fisheries, including for groupers (e.g., Meissa et al. 2013, Cummings et al. 2014, Prince et al. 2015). Fishery histories can be reconstructed from limited available data (e.g., Léopold et al. 2017).

Understanding fisheries from the perspective of the fisher is key to effective management and fisher interviews are important for understanding socio-economic contexts, fishing effort (needed to interpret landings trends over time) and fisher perspectives, and traditional knowledge on trends in their fisheries (e.g., Hamilton et al. 2012, Mavruk et al. 2018, Khasanah et al. 2019). Bio-economic models are important for evaluating yield and economic benefits. For example, Najamuddin et al. (2016) determined both biological and economic overfishing was occurring, and highlighted the need to reduce fishing effort for Leopard Coral Grouper in Spermonde Island waters of Indonesia; they made a compelling case for reduction in effort.

Other Considerations and Opportunities

Several broader issues, in addition to their role in small-scale fisheries, should be considered in relation to the future of groupers, including the growing economic importance of ecotourism and export demand for wild-caught groupers, options for managing fisheries, sustainable aquaculture, and the contribution of national, regional, and international policies, agreements, accords, and conventions. Moreover, given that species in tropical regions are predicted to become negatively affected by temperature increases linked to climate change, projected to substantially reduce catch potential in the tropics (Cheung et al. 2010), impacts on groupers and their fisheries could be considerable; this situation calls for decidedly precautionary management.

In addition to improving the contribution of groupers directly to fishers in source countries, there are other economic considerations for managing these fisheries well. For example, the economic value to tourism of divers, able to witness spectacular wildlife events such as spawning aggregations or to see large animals in the wild could be substantial and should be evaluated. With growing demand for wild grouper and limited natural supplies, source countries need to consider if exports can be sustained, or are competing excessively with domestic demand, or further compromising fisheries. The development of mariculture needs to be sustainable, ideally to include the phasing out of CBA to focus on HBA production systems to reduce dependence on wild capture, although wild fish feed supply for grouper culture is still a substantial fishing pressure.

With successful management, grouper populations can recover and thrive, which calls for planning around how to sustain the recovery. The importance of involving major resource users in management planning has also become apparent (e.g., Papua New Guinea, Almany et al. 2013; Turkey, Mavruk et al. 2020; Indonesia, Yulianto et al. 2015). For example, one challenge to managers that emerged after Goliath Grouper numbers increased following protection was the perception that they had become a nuisance species. In Florida, anglers became frustrated with Goliath Grouper individuals grabbing their catches, which resulted in poaching and discarding of the protected species. The situation highlighted the need to proactively plan for a situation when recovery occurs (Shideler et al. 2015). Post-recovery planning may also be necessary when spawning aggregations increase after protection. For example, should they be reopened after recovery, or kept closed permanently to continue producing eggs to populate the rest of the fishery into the future, or, in some cases, for eco-tourism possibilities?

Beyond the development of further species-specific management and conservation measures, existing national policies, and regional and international agreements and instruments, codes, and standards, can guide, and steer towards more sustainable practices, more transparent trade, more responsible consumption, and inter-country collaboration. For example, at the national level, in late 2012, a frugality campaign introduced by China's new leadership sought to cut corruption by banning lavish banqueting which was often done by officials; this greatly reduced demand for luxury seafood dining, including for groupers (Godfrey 2013). Many regional accords in Southeast Asia have sustainable fisheries and mariculture management in their core agendas, although little has yet been enacted (Sadovy de Mitcheson 2019). Several international agreements, guidelines, and instruments support conservation of biodiversity and biologically sustainable international trade in natural resources (FAO Code of Conduct for Responsible Fisheries, Convention of Biological Diversity (CBD), Convention on International Trade in Endangered Flora and Fauna (CITES)).

Concluding Comments

Groupers are amongst the most valuable fishes taken in reef fisheries. They are important for coastal communities, traders, culturists, the service sector, leisure activities, tradition, eco-tourism, and consumers alike. They are also fascinating species in their own right, core elements of reef ecosystems. They include some of the largest of all reef fish species; some are iconic and of cultural importance. They are also particularly susceptible to overexploitation, with at least 16% of species already Threatened or Near-Threatened. The situation is particularly bad in Asia, which now accounts for the majority of global grouper landings and has the highest grouper species diversity. So what is the future for groupers?

Strategic Development Goal 14 (SDG 14) of the United Nations 2030 Agenda for Sustainable Development is an overarching commitment, made by many countries where groupers occur, to "Conserve and sustainably use the oceans, seas, and marine resources for sustainable development", while a recent overview called for paying special attention to small-scale fisheries in low-income countries (Costello et al. 2019). We recognize at least 10 measures and opportunities for achieving this goal and for sustaining groupers. (1) Monitoring of capture volumes (with correction of both under- and over- reporting, and training in species identification) and assessment of exploited populations, for which multiple data-poor methods and models can be applied; (2) national and/or regional transboundary management to address the geographically broad distribution of some populations; (3) reducing juvenile capture to safeguard reproductive capacity using minimum size restrictions and by phasing out CBA in favour of HBA; (4) expanding the use of MPAs, in addition to conventional management, especially for smaller to medium-sized species, and those that don't migrate to spawn; (5) management of aggregation spawning species including by spatial and/or seasonal measures to protect aggregations; (6) precautionary management that factors in the possibility of hyperstability and Allee effects; (7) evaluating the ecotourism value of groupers; (8) respecting the traditional and cultural value of groupers (see, for example, in Fiji, http://www.4fj.org.fj/); (9) improving trade documentation of groupers including through the use of harmonized codes in international trade, and (10) conducting value chain and bio-economic analyses to recognize the importance of groupers for fishing communities for food and livelihoods and to identify opportunities for improving income of fishers and other sectors in source countries.

Finally, the potential direct and indirect impacts of climate change will need to be factored into population assessments and evaluations of the future of groupers, valuable but vulnerable fisheries, and also fascinating and ecologically important fish species.

Acknowledgements

We are grateful to Rachel Wong, Baian Lin, and Weidi Yang for assistance with the preparation of this manuscript. We acknowledge comments by George Shideler that improved the chapter.

References

Aburto-Oropeza, O., B. Erisman, G.R. Galland, L. Mascarenas-Osorio, E. Sala and E. Ezcurra. 2011. Large recovery of fish biomass in a no-take marine reserve. PLOS ONE 6: e23601.

Adams, D.H. and C. Sonne. 2013. Mercury and histopathology of the vulnerable goliath grouper, *Epinephelus itajara*, in U.S. waters: a multi-tissue approach. Environmental Research 126: 254–63.

ADMCF. 2015. Mostly legal but not sustainable. ADM Capital Foundation Report (http://admcf.org/wordpress/wp-content/uploads/2015/11/LRFFFINAL_FINAL-NOV61.pdf).

Albins, M.A. and M.A. Hixon. 2008. Invasive Indo-Pacific lionfish *Pterois volitans* reduce recruitment of Atlantic coral-reef fishes. Marine Ecology Progress Series 367: 233–238.

Almany, G.R., R.J. Hamilton, M. Bode, M. Matawai, T. Potuku, P. Saenz-Agudelo, S. Planes, M.L. Berumen, K.L. Rhodes, S.R. Thorrold, G.R. Russ and G.P. Jones. 2013. Dispersal of grouper larvae drives local resource sharing in a coral reef fishery. Current Biology 23: 626–623.

Amorim, P., P. Sousa, E. Jardim, M. Azevedo and G.M. Menezes. 2020. Length-frequency data approaches to evaluate snapper and grouper fisheries in the Java Sea, Indonesia. Fisheries Research 229: 105576.

Anderson, A.B., R.M. Bonaldo, D.R. Barneche, C.W. Hackradt, F.C. Félix-Hackradt, J.A. García-Charton and S.R. Floeter. 2014. Recovery of grouper assemblages indicates effectiveness in a marine protected area in southern Brazil. Marine Ecology Progress Series 514: 207–215.

Appeldoorn, R.S., M.T. Schärer-Umpierre, T.J. Rowell and R.S. Nemeth. 2013. Measuring relative density of spawning red hind (*Epinephelus guttatus*) from sound production: consistency within and among sites. Proceedings of Gulf and Caribbean Fisheries Institute 65: 284–286.

Asch, R.G. and B. Erisman. 2018. Spawning aggregations act as a bottleneck influencing climate change impacts on a critically endangered reef fish. Diversity and Distributions 24: 1712–1728.

Bannerot, S.P., W.W. Fox and J.E. Powers. 1987. Reproductive strategies and the management of snappers and groupers in the Gulf of Mexico and Caribbean. pp. 295–327. *In*: Polovina, J.J. and S. Ralston (eds.). Tropical Snappers and Groupers Biology and Fisheries Management. Westview Press, Boulder.

Bartholomew, A. and J.A. Bohnsack. 2005. A review of catch-and-release angling mortality with implications for no-take reserves. Reviews in Fish Biology and Fisheries 15: 129–154.

Bellwood, D.R., T.P. Hughes, C. Folke and M. Nystrom. 2004. Confronting the coral reef crisis. Nature 429: 827–833.

Boaden, A.E. and M.J. Kingsford. 2015. Predators drive community structure in coral reef fish assemblages. Ecosphere 6: 1–33.

Boddington, D.K., B. Corey, C.B. Wakefield, E.A. Fisher, D.V. Fairclough, D.V. Fairclough and S.J. Newman. 2021. Age, growth and reproductive life-history characteristics infer a high population productivity for the sustainably fished protogynous hermaphroditic yellow spotted rockcod (*Epinephelus areolatus*) in north-western Australia. Journal of Fish Biology. doi: 10.1111/jfb.14889.

Box, S.J. and S.W. Canty. 2010. The long and short-term economic drivers of overexploitation in Honduran coral reef fisheries due to their dependence on export markets. Proceedings of the Gulf and Caribbean Fisheries Institute 63: 43–51.

Bshary, R., A. Hohner, K. Ait-El-Djoudi and H. Fricke. 2006. Interspecific communicative and coordinated hunting between groupers and giant moray eels in the Red Sea. PLOS Biology 4: e431.

Bucol, A.A., R.A. Abesamis, B.L. Stockwell, J.R. Lowe and G.R. Russ. 2021. Development of reproductive potential in protogynous coral reef fishes within Philippine no-take marine reserves. Journal of Fish Biology 99(5): 1561–1575.

Bullock, L.H., M.D. Murphy, M.F. Godcharles and M.E. Mitchell. 1992. Age, growth and reproduction of jewfish *Epinephelus itajara* in the eastern Gulf of Mexico. Fisheries Bulletin 90: 243–249.

Bush, P.G., E.D. Lane and G.C. Ebanks. 1996. Validation of ageing technique for Nassau Grouper (*Epinephelus striatus*) in the Cayman Islands. pp. 150–157. *In*: Arrequin-Sanchez, F.A., J.L. Munro, M.C. Balgos and D. Pauly (eds.). Biology, Fisheries and Culture of Tropical Snappers and Groupers. Proceedings EPOMEX/ICLARM International Workshop on Tropical Snappers and Groupers. October 1993.

Cao, L., R. Naylor, P. Henriksson, D. Leadbitter, M. Metian, M. Troell and W.B. Zhang. 2015. China's aquaculture and the world's wild fisheries. Science 347: 113–135.

Carter, A.B., C.R. Davies, M.J. Emslie, B.D. Mapstone, G.R. Russ, A.J. Tobin and A.J. Williams. 2017. Reproductive benefits of no-take marine reserves vary with region for an exploited coral reef fish. Scientific Reports 7(1): 9693.

Chan, T.C and Y. Sadovy. 2002. Reproductive biology, age and growth in the chocolate hind, *Cephalopholis boenak* (Bloch, 1790), in Hong Kong. Marine and Freshwater Research 53: 791–803.

Cheung, W.W.L., V.W.Y. Lam, J.L. Sarmiento, K. Kearney, R. Watson, D. Zeller and D. Pauly. 2010. Large-scale redistribution of maximum fisheries catch potential in the global ocean under climate change. Global Change Biology 1: 24–35.

Chiappone, M., R. Sluka and S.K. Sullivan. 2000. Groupers (Pisces: Serranidae) in fished and protected areas of the Florida Keys, Bahamas and northern Caribbean. Marine Ecology Progress Series 198: 261–272.

China Fishery Statistical Yearbooks. 1978–1989. China Agriculture Press, Beijing.

Clark, T.D., G.D. Raby, D.G. Roche, S.A. Binning, B. Speers-Roesch, F. Jutfelt and J. Sundin. 2020. Ocean acidification does not impair the behaviour of coral reef fishes. Nature 577: 370–375.

Claro, R., Y. Sadovy de Mitcheson, K.C. Lindeman and A.R. Garcia-Cagide. 2009. Historical analysis of Cuban commercial fishing effort and the effects of management interventions on important reef fishes from 1960–2005. Fisheries Research 99: 7–16.

Coleman, F.C., C.C. Koenig and L.A. Collins. 1996. Reproductive styles of shallow-water groupers (Pisces: Serranidae) in the eastern Gulf of Mexico and the consequences of fishing spawning aggregations. Environmental Biology of Fishes 47: 129–141.

Coleman, F.C. and S.L. Williams. 2002. Overexploiting marine ecosystem engineers: potential consequences for biodiversity. Trends in Ecology and Evolution 17: 40–44.

Coleman, F.C., C.C. Koenig, K.M. Scanlon, S. Heppell, S. Heppell and M.W. Miller. 2010. Benthic habitat modification through excavation by red grouper, *Epinephelus morio*, in the northeastern Gulf of Mexico. The Open Fish Science Journal 3: 1–15.

Colin, P. 1992. Reproduction of the Nassau grouper, *Epinephelus striatus* (Pisces: Serranidae) and its relationship to environmental conditions. Environmental Biology of Fishes 34: 357–377.

Condini, M.V., C.Q. Albuquerque and A.M. Garcia. 2014. Age and growth of dusky grouper (*Epinephelus marginatus*) (Perciformes: Epinephelidae) in the southwestern Atlantic, with a size comparison of offshore and littoral habitats. Fishery Bulletin 112(4): 311–321.

Costello, C., L. Cao, S. Gelcich et al. 2019. The Future of Food from the Sea. Washington, DC: World Resources Institute. Available online at www.oceanpanel.org/future-food-sea.

Courchamp, F., W. Angulo, P. Rivalan, R.J. Hall, L. Signoret, L. Bull and Y. Meinard. 2006. Rarity value and species extinction: The anthropogenic Allee effect. PLOS Biology 4: e415.

Courchamp, F., J. Berec and J. Gascoigne. 2008. Allee Effects in Ecology and Conservation. New York: Oxford University Press. ISBN 978-0-19-956755-3.

Craig, M.T., Y.J. Sadovy de Mitcheson and P.C. Heemstra. 2011. Groupers of the World: A Field and Market Guide. Grahamstown: National Inquiry Services Center (NISC) Ltd.
Cummings, N.J., M. Karnauskas, W.L. Michaels and A. Ascota (eds.). 2014. Report of a GCFI Workshop. Evaluation of Current Status and Application of Data Limited Stock Assessment Methods in the Larger Caribbean Region. Gulf and Caribbean Fisheries Institute Conference, Corpus Christi, Texas.
Dennis, L.P., G. Ashford, T.Q. Thai, V.V. In, N.H. Ninh and A. Elizur. 2020. Hybrid grouper in Vietnamese aquaculture: Production approaches and profitability of a promising new crop. Aquaculture 522: 735108.
Diamant, A. and M. Shpigel. 1985. Interspecific feeding associations of grouper (Teleostei, Serranidae) with octopuses and moray eels in the Gulf of Eilat (Aquaba). Environmental Biology of Fishes 13: 153–159.
Dulvy, N.K., R.P. Freckleton and N.V.C. Polunin. 2004. Coral reef cascades and the indirect effects of predator removal by exploitation. Ecology Letters 7: 410–416.
Edgar, G.J., R.D. Stuart-Smith, T.J. Willis, K. Stuart, S.C. Baker, S.N.S. Barrett, M.A. Becerro, A.T.F. Bernard, J. Berkhout, C.D. Buxton, S.J. Campbell, A.T. Cooper, M. Davey, S.C. Edgar, G. Forsterra, D.E. Galvan, A. Irigoyen, D.J. Kushner and R. Moura. 2014. Global conservation outcomes depend on marine protected areas with five key features. Nature 506: 216–220.
Eggleston, D.B., J.J. Grover and R.N. Lipcius. 1998. Ontogenetic diet shifts in Nassau grouper: trophic linkages and predatory impact. Bulletin of Marine Science 63(1): 111–126.
Erisman, B.E., J.A. Rosales-Casián and P.A. Hastings. 2008. Evidence of gonochorism in a grouper, *Mycteroperca rosacea*, from the Gulf of California, Mexico. Environmental Biology of Fishes 82: 23–33.
Erisman, B.E., L.G. Allen, H.T. Claisse, D.J. Pondella II, E.F. Miller and J.H. Murray. 2011. The illusion of plenty: Hyperstability masks collapses in two recreational fisheries that target fish spawning aggregations. Canadian Journal of Fisheries and Aquatic Sciences 68: 1705–1716.
Erisman, B.E., C.W. Petersen, P.A. Hastings and R.R. Warner. 2013. Phylogenetic perspectives on the evolution of functional hermaphroditism in teleost fishes. Integrative and Comparative Biology 53: 736–754.
Essington, T.E., A.H. Beaudreau and J. Wiedenmann. 2006. Fishing through marine food webs. Proceedings of the National Academy of Science 103: 3171–3175.
Fabinyi, M. and N. Liu. 2014. Seafood banquets in Beijing: consumer perspectives and implications for environmental sustainability. Conservation and Society 12: 218–228.
FAO. 2011. Aquaculture development. 6. Use of wild fishery resources for capturebased aquaculture. FAO Technical Guidelines for Responsible Fisheries. No. 5, Suppl. 6. Rome, FAO. 2011. 81 pp.
Federal Register. 2020. www.federalregister.gov/documents/2020/04/21/2020-08093/fisheries-of-the-caribbean-gulf-of-mexico-and-south-atlantic-snapper-grouper-fishery-of-the-south.
Francis, M.P., D. Harasti and H.A. Malcolm. 2015. Surviving under pressure and protection: a review of the biology, ecology and population status and of the highly vulnerable grouper, *Epinephelus daemelii*. Marine and Freshwater Research 67: 1215–1228.
Frisch, A.J., D.S. Cammeron, M.S. Pratchett, D.H. Williamson, A.J. Williams, A.D. Reynolds, A.S. Hoey, J.R. Rizzari, L. Eyans, B. Kerrigan, G. Muldoon, D.J. Welch and J.A. Hobbs. 2016. Key aspects of the biology, fisheries and management of Coral grouper. Reviews in Fish Biology and Fisheries 26(3): 303–325.
Gentner, B. 2009. Allocation Analysis of the Gulf of Mexico Gag and Red Grouper Fisheries. Gentner Consulting Group.
Giglio, V.J., M.G. Bender, C. Zapelini and C.E.L. Ferreira. 2017. The end of the line? Rapid depletion of a large-sized grouper through spearfishing in a subtropical marginal reef. Perspectives in Ecology and Conservation 15: 115–118.
Gillett, R. and W. Moy. 2006. Spearfishing in the Pacific Islands: Current Status and Management Issues. FAO/Fishcode Review No. 19. Rome, FAO.

Godfrey, M. 2013. Frugality campaign hits China's grouper market. Seafood Source, April 8. www.seafoodsource.com/news/supply-trade/frugality-campaign-hits-china-s-grouper-market.

Grüss, A., J. Robinson, S.S. Heppell, S.A. Heppell and B.X. Semmens. 2014. Conservation and fisheries effects of spawning aggregation marine protected areas: What we know, where we should go, and what we need to get there. ICES Journal of Marine Science 71(7): 1515–1534.

Hamilton, R.J., T. Potuku and J.R. Montambault. 2011. Community-based conservation results in the recovery of reef fish spawning aggregations in the Coral Triangle. Biological Conservation 144: 1850–1858.

Hamilton, R., Y. Sadovy de Mitcheson and A. Aguilar-Perer. 2012. The role of local ecological knowledge in the conservation and management of reef fish spawning aggregations. pp. 331–370. *In*: Sadovy de Mitcheson, Y. and P.L. Colin (eds.). Reef Fish Spawning Aggregations: Biology, Research and Management. Fish & Fisheries Series 35, Springer Science+Business Media B.V.

He, Q., M.D. Bertness, J.F. Bruno, B. Li, G.Q. Chen, T.C. Coverdale, A.H. Altieri, J.H. Bai, T. Sun, S.C. Pennings, J.G. Liu, P.R. Ehrlich and B.S. Cui. 2014. Economic development and coastal ecosystem change in China. Scientific Reports 4: 5995.

Heemstra, P.C. and J.E. Randall. 1993. Groupers of the World. FAO Species Catalogue Vol. 16. FAO, Rome.

Heithaus, M.R., A. Frid, A.J. Wirsing and B. Worm. 2008. Predicting ecological consequences of marine top predator declines. Trends in Ecology and Evolution 23: 202–210.

Hensel, E., J.E. Allgeier and C.A. Layman. 2019. Effects of predator presence and habitat complexity on reef fish communities in the Bahamas. Marine Biology 166: 136.

Heppell, S.S., S.A. Heppell, F.C. Coleman and C.C. Koenig. 2006. Models to compare management options for a protogynous fish. Ecological Applications 16(1): 238–249.

Hobday, A.J., A.D.M. Smith, I.C. Stobutzki, C. Bulman, R. Daley, J.M. Dambacher, R.A. Deng, J. Dowdney, M. Fuller, D. Furlani, S.P. Griffiths, D. Johnson, R. Kenyon, I.A. Knuckey, S.D. Ling, R. Pitcher, K.J. Sainsbury, M. Sporcic, T. Smith, C. Turnbull, T.I. Walker, S.E. Wayte, H. Webb, A. Williams, B.S. Wise and S. Zhou. 2011. Ecological risk assessment for the effects of fishing. Fisheries Research 108: 372–384.

Howlett, S.J., R. Stafford, M. Waller, S. Antha and C. Mason-Parker. 2016. Linking protection with the distribution of grouper and habitat quality in Seychelles. Journal of Marine Biology 2016: 7851425.

Hughes, T.P., A.H. Baird, Y.D.R. Bellwood, M. Card, S.R. Connolly, C. Folke, R. Grosberg, O. Hoegh-Guldberg, J.B.C. Jackson, J. Kleypass, J.M. Lough, P. Marshall, M. Nystrom, S.R. Palumbi, B. Rosen and J. Roughgarden. 2003. Climate change, human impacts, and the resilience of coral reefs. Science 301: 929–933.

IUCN. 2012. IUCN Red List Categories and Criteria: Version 3.1 (2nd edition). International Union for Conservation of Nature, Gland, Switzerland and Cambridge, UK.

Jackson, A.M., B.X. Semmens, Y.J. Sadovy de Mitcheson, R.S. Nemeth, S.A. Heppell, P.G. Bush, A. Aguilar-Perera, J.A.B. Claydon, M.C. Calosso, K.S. Sealey, M.T. Scharer and G. Bernardi. 2014. Population structure and phylogeography in Nassau grouper (*Epinephelus striatus*), a mass-aggregating marine fish. PLOS ONE 9: 1–11.

Kandula, S. Kantimahanti, V.L. Shrikanya and V.A. Iswarya Deepti. 2015. Species diversity and some aspects of reproductive biology and life history of groupers (Pisces: Serranidae: Epinephelinae) off the central eastern coast of India. Marine Biology Research 11: 18–33.

Khasanah, M., N. Nrudin, Y. Sadovy de Mitcheson and J. Jamaluddin. 2019. Management of grouper export trade in Indonesia. Reviews in Fisheries Science & Aquaculture 28(1): 1–15.

Kindsvater, H., J. Reynolds, Y. Sadovy de Mitcheson and M. Mange. 2017. Selectivity matters: rules of thumb for management of plate-sized, sex-changing fish in the live reef food fish trade. Fish and Fisheries 18: 821–836.

Koenig, C.C., F.C. Coleman, A.M. Eklund, J. Schull and J. Ueland. 2007. Mangrove as essential nursery habitat for goliath grouper (*Epinephelus itajara*). Bulletin of Marine Science 80: 567–586.

Koenig, C.C., F.C. Coleman and K. Kingon. 2011. Pattern of recovery of the goliath grouper *Epinephelus itajara* population of the southeastern US. Bulletin of Marine Science 87: 891–911.

Lee, S., A. Lewis, R. Gillett, M. Fox, N. Tuqiri, Y. Sadovy, A. Batibasaga, W. Lalavanua and E. Lovell. 2018. Chapter 23 Groupers. pp. 119–127. *In*: Fiji Fishery Resource Profiles: Information for Management on 44 of the Most Important Species Groups. Gillett, Preston and Associates and the Wildlife Conservation Society, Suva, Fiji.

Léopold, M., G. David, J. Raubani, J. Kaltavara, L. Hood and D. Zeller. 2017. An improved reconstruction total marine fisheries catches for the New Hebrides and the Republic of Vanuatu, 1950–2014. Frontiers in Marine Science 4: 306.

Levin, P.S. and C.B. Grimes. 2002. Reef fish ecology and grouper conservation and management. pp. 377–389. *In*: Sales, P.F. (ed.). Coral Reef Fishes: Dynamics and Diversity in a Complex Ecosystem. Academic Press, San Diego.

Liu, M. and Y. Sadovy. 2004. The influence of social factors on juvenile sexual differentiation and adult sex change in a diandric, protogynous epinepheline, *Cephalopholis boenak*. Journal of Zoology London 264: 239–248.

Liu, M. and Y. Sadovy de Mitcheson. 2009. Exploitation history, mariculture and trade status of the threatened Hong Kong grouper (*Epinephelus akaara*) throughout its geographic range. Ocean Park Conservation Foundation of Hong Kong, Hong Kong.

Lovatelli, A. and P.F. Holthus (eds.). 2008. Capture-Based Aquaculture. Global Overview. FAO Fisheries Technical Paper No. 508. Rome, FAO.

Lukasik, B. 2016. Diving with potato bass. https://africageographic.com/stories/diving-potato-bass/.

Luckhurst, B.E. and T.M. Trott. 2009. Seasonally-closed spawning aggregation sites for red hind (*Epinephelus guttatus*): Bermuda's experience over 30 years (1974–2003). Proceedings of Gulf and Caribbean Fisheries Institute 61: 331–336.

Ma, K.Y. and M.T. Craig. 2018. An inconvenient monophyly: an update on the taxonomy of the groupers (Epinephelidae). Copeia 106(3): 443–456.

Ma, K.Y., L. van Herwerden, S.J. Newman, M.L. Berumen, J.H. Choat, K.H. Chu and Y. Sadovy de Mitcheson. 2018. Contrasting population genetic structure in three aggregating groupers (Percoidei: Epinephelidae) in the Indo-West Pacific: the importance of reproductive mode. BMC Evolutionary Biology 18: 180.

Maljkovic, A., T.E. van Leeuwen and S.N. Cove. 2008. Predation on the invasive red lionfish, *Pterois volitans* (Pisces: Scorpaenidae), by native groupers in the Bahamas. Coral Reefs 27(3): 501.

Mann, D.A., J.V. Locascio, F.C. Coleman and C.C. Koenig. 2009. Goliath grouper *Epinephelus itajara* sound production and movement patterns on aggregation sites. Endanger Species Research 7: 229–236.

Mavruk, S., I. Saygu, F. Bengil, V. Alan and E. Azzurro. 2018. Grouper fishery in the Northeastern Mediterranean: an assessment based on interviews on resource users. Marine Policy 87: 141–148.

Mavruk, S., I. Saygu and F. Bengil. 2020. Fishers' responses towards the banning white grouper fishery in Turkey. Journal of Wildlife and Biodiversity 4: 50–57.

McCauley, D.J., F. Micheli, H.S. Young, D.P. Tittensor, D.R. Brumbaugh, E.M.P. Madin, K.E. Holmes, J.E. Smith, H.K. Lotze, P.A. DeSalles, S.N. Arnold and B. Worm. 2010. Acute effects of removing large fish from a near-pristine coral reef. Marine Biology 157: 2739–2750.

McClenachan, L. 2015. Extinction risk in reef fishes. pp. 199–207. *In*: Mora, C. (ed.). Ecology of Fishes on Coral Reefs. Cambridge University Press, Cambridge.

McGovern, J.C., D.M. Wyanski, O. Pashuk, C.S. II Manooch and G.R. Sedberry. 1998. Changes in the sex ratio and size at maturity of gag, *Mycteroperca microlepis*, from the Atlantic coast of the southeastern United States during 1976–1995. Fishery Bulletin 96: 797–807.

Meissa, B., D. Gascuel and E. Rivot. 2013. Assessing stocks in data-poor African fisheries: A case study on the white grouper *Epinephelus aeneus* of Mauritania. African Journal of Marine Science 35: 253–267.

Mourier, J., J. Maynard, V. Parravicini, L. Ballesta, E. Clua, M.L. Domeier and S. Planes. 2016. Extreme inverted trophic pyramid of reef sharks supported by spawning groupers. Current Biology 26: 2011–2016.

Mous, P.J., Y. Sadovy, A. Halim and J.S. Pet. 2006. Capture for culture: artificial shelters for grouper collection in SE Asia. Fish and Fisheries 7: 58–72.

Mumby, P.J., A.R. Harborne and D.R. Brumbaugh. 2011. Grouper as a natural biocontrol of invasive lionfish. PLOS ONE 6(6): e21510.

Najamuddin, N., A. Baso and R. Arfiansyah. 2016. Bio-economic analyses of coral trout grouper fish in Spermonde Archipelago, Makassar. International Journal of Oceans and Oceanography 10(3): 247–264.

Nanami, A., T. Sato, Y. Kawabata and J. Okuyama. 2017. Spawning aggregation of white-streaked grouper *Epinephelus ongus*: spatial distribution and annual variation in the fish density within a spawning ground. PeerJ. 5: e3000.

Naylor, R.L., R.J. Goldburg, J.H. Primavera, N. Kautsky, M.C.M. Beveridge, J. Clay, C. Folke, J. Lubchence, H. Mooney and M. Troell. 2000. Effect of aquaculture on world fish supplies. Nature 405: 1017–1024.

Nemeth, R.S. 2005. Population characteristics of a recovering US Virgin Islands red hind spawning aggregation following protection. Marine Ecology Progress Series 286: 81–97.

Nemeth, R.S. 2012. Ecosystem aspects of species that aggregate to spawn. pp. 21–55. *In*: Sadovy de Mitcheson, Y. and P.L. Colin (eds.). Reef Fish Spawning Aggregations: Biology, Research and Management. Fish & Fisheries Series 35, Springer Science+Business Media B.V.

New York Times. 2015. In overturning conviction, supreme court says fish are not always tangible (reported by Liptak A.). 25 February 2015.

Ohta, I., Y. Akita, M. Uehara and A. Ebisawa. 2017. Age-based demography and reproductive biology of three *Epinephelus groupers, E. polyphekadion, E. tauvina*, and *E. howlandi* (Serranidae), inhabiting coral reefs in Okinawa. Environmental Biology of Fishes 100: 1451–1467.

Olavo, G., P.A.S. Costa, A.S. Martins and B.P. Ferriera. 2011. Shelf-edge reefs as priority areas for conservation of reef fish diversity in the tropical Atlantic. Aquatic Conservation Marine and Freshwater Ecosystem 21: 199–209.

Padilla, J.E., S. Mamauag, N. Brucal, D. Yu and A. Morales. 2003. Sustainability Assessment of the Live Reef-Fish for Food Industry in Palawan, Philippines. WWF, Philippines.

Patrick, W.S., P. Spencer, J. Link, J. Cope, J. Fields, D. Kobayashi, P. Lawson, T. Gedamke, E. Cortes, O. Ormseth, K. Bigelow and W. Overholtz. 2010. Using productivity and susceptibility indices to assess the vulnerability of United States fish stocks to overfishing. Fishery Bulletin 108: 305–322.

Pears, R.J. 2005. Comparative demography and assemblage structure of serranid fishes: implications for conservation and fisheries management. PhD thesis, James Cook University, Townsville, Australia.

Pears, R.J., J.H. Choat, B.D. Mapstone and G.A. Begg. 2006. Demography of a large grouper, *Epinephelus fuscoguttatus*, from Australia's Great Barrier Reef: implications for fishery management. Marine Ecology Progress Series 307: 259–272.

Pierre, S., S. Gaillard, N. Prévot-D'Alvise, J. Aubert, O. Rostaing-Capaillon, D. Leung-Tack and J. Grillasca. 2008. Grouper aquaculture: Asian success and Mediterranean trials. Aquatic Conservation: Marine and Freshwater Ecosystem 18: 297–308.

Pogonoski, J.J., D.A. Pollard and J.R. Paxton. 2002. Conservation Overview and Action Plan for Australian Threatened and Potentially Threatened Marine and Estuarine Fishes. Environment Australia, Canberra, Australia.

Potts, J.C. and C.S. Manooch. 1995. Age and growth of red hind and rock hind collected from North Carolina through the Dry Tortugas, Florida. Bulletin of Marine Science 56(3): 784–794.

Pratchett, M.S., D.S. Cameron, J. Donelson, L. Evans, A.J. Frisch, A.J. Hobday, A.S. Hoey, N.A. Marshall, V. Massmer, P.L. Munday, R. Pears, G. Pecl, A. Reynolds, M. Scott, A. Tobin, R. Tobin, D.J. Welch and D.H. Williamson. 2017. Effects of climate change on coral grouper

(*Plectropomus* spp.) and possible adaptation options. Reviews in Fish Biology and Fisheries 27: 297–316.

Prince, J., S. Victor, V. Kloulchad and A. Hordyk. 2015. Length based SPR assessment of eleven Indo-Pacific coral reef fish populations in Palau. Fisheries Research 171: 42–58.

Rawalai, L. 2020. Fiji shortens protection due to covid. State lifts grouper ban earlier to help coastal communities. www.fijitimes.com/state-lifts-grouper-ban-earlier-to-help-coastal-communities/. 14 August, 2020.

Rhodes, K.L. and Y. Sadovy. 2002. Reproduction in the camouflage grouper (Pisces: Serranidae) in Pohnpei, Federated States of Micronesia. Bulletin of Marine Science 70(3): 851–869.

Rhodes, K.L. and M.H. Tupper. 2007. A preliminary market-based analysis of the Pohnpei Micronesia, grouper (Serranidae: Epinephelinae) fishery reveals unsustainable fishing practices. Coral Reefs 26: 335–344.

Rhodes, K.L., M.H. Tupper and C.B. Wichilme. 2008. Characterization and management of the commercial sector of the Pohnpei coral reef fishery, Micronesia. Coral Reefs 27: 443–454.

Rimmer, M.A. and B. Glamuzina. 2019. A review of grouper (Family Serranidae: Subfamily Epinephelinae) aquaculture from a sustainability science perspective. Reviews in Aquaculture 11: 58–87.

Roberts, C.M. 1995. Rapid buildup of fish biomass in a Caribbean marine reserve. Conservation Biology 9: 815–826.

Robinson, J., N.A.J. Graham, J.E. Cinner, G.R. Almany and P. Waldie. 2015. Fish and fisher behaviour influence the vulnerability of groupers (Epinephelidae) to fishing at a multispecies spawning aggregation site. Coral Reefs 34: 371–382.

Rowell, T.J., R.S. Nemeth, M.T. Schärer and R.S. Appeldoorn. 2015. Fish sound production and acoustic telemetry reveal behaviors and spatial patterns associated with spawning aggregations of two Caribbean groupers. Marine Ecology Progress Series 518: 239–254.

Rudershausen, P.J., J.A. Buckel and E.H. Williams. 2007. Discard composition and release fate in the snapper and grouper commercial hook-and-line fishery in North Carolina, USA. Fisheries Management and Ecology 14: 103–113.

Russ, G.R., D.C. Lou, J.B. Higgs and B.P. Ferreira. 1998. Mortality rate of a cohort of the coral trout, *Plectropomus leopardus*, in zones of the Great Barrier Reef Marine Park closed to fishing. Marine and Freshwater Research 49: 507–511.

Russ, G.R. 2002. Yet another review of marine reserves as reef fishery management tools. pp. 421–443. *In*: Sale, P. (ed.). Coral Reef Fishes: Dynamics and Diversity in a Complex Ecosystem. Academic Press, San Diego.

Russell, M. 2006. Leopard coral grouper (*Plectropomus leopardus*) management in the Great Barrier Reef Marine Park, Australia. SPC Live Reef Fish Information Bulletin 16: 10–12.

Russell, M., B.C. Luckhurst and K.C. Lindeman. 2012. Management of spawning aggregations. pp. 371–404. *In*: Sadovy de Mitcheson, Y. and P.L. Colin (eds.). Reef Fish Spawning Aggregations: Biology, Research and Management. Fish & Fisheries Series 35, Springer Science+Business Media B.V.

Ruttenberg, B.I., S.L. Hamilton, S.M. Walsh, M.K. Donovan, A. Friedlander, E. DeMartini, E. Sale and S.A. Sandin. 2011. Predator-induced demographic shifts in coral reef fish assemblages. PLOS ONE 6(6): e21062.

Sadovy, Y. 1994. Grouper stocks of the western central Atlantic: the need for management and management needs. Proceedings of Gulf and Caribbean Fisheries Institute 43: 43–64.

Sadovy, Y. and P.L. Colin. 1995. Sexual development and sexuality in the Nassau grouper, *Epinephelus striatus* (Bloch) (Pisces: Serranidae). Journal of Fish Biology 46: 511–516.

Sadovy, Y. 2000. Regional Survey for Fry/Fingerling Supply and Current Practices for Grouper Mariculture: Evaluating Current Status and Long-term Prospects for Grouper Mariculture in South East Asia. Final Report to the Collaborative APEC Grouper Research and Development Network (FWG 01/99 revised). Singapore: APEC Secretariat.

Sadovy, Y. and M. Domeier. 2008. Are aggregation-fisheries sustainable? Reef fish fisheries as a case. Coral Reefs 24: 254–262.

Sadovy, Y.J., T.J. Donaldson, T.R. Graham, F. McGilvray, G.J. Muldoon, M.J. Phillips, M.A. Rimmer, A. Smith and B. Yeeting. 2003. The Live Reef Food Fish Trade While Stocks Last. Manila: Asian Development Bank.

Sadovy de Mitcheson, Y. and M. Liu. 2008a. Functional hermaproditism in teleosts. Fish and Fisheries 9: 1–43.

Sadovy de Mitcheson, Y. and M. Liu. 2008b. Environment and biodiversity impacts of captured-based aquaculture. pp. 5–39. *In*: Lovatelli, A. and P.F. Holthus (eds.). Capture-Based Aquaculture. Global Overview. FAO Fisheries Technical Paper No. 508. Rome, FAO.

Sadovy de Mitcheson, Y. and P.L. Colin (eds.). 2012. Reef Fish Spawning Aggregations: Biology, Research and Management. Fish & Fisheries Series 35, Springer Science+Business Media B.V.

Sadovy de Mitcheson, Y., M.T. Craig, A.A. Bertoncini, K.E. Carpente, W.W.L. Cheung, J.H. Choat, A.S. Cornish, S.T. Fennessy, B.P. Ferreira, P.C. Heemstra, M. Liu, R.F. Meyers, D.A. Pollard, K.L. Rhodes, L.A. Rocha, R.B. Russell, M.A. Samoilys and J. Sanciangco. 2013. Fishing groupers towards extinction: a global assessment of threats and extinction risk in a billion dollar fishery. Fish and Fisheries 14: 119–136.

Sadovy de Mitcheson, Y. and X. Yin. 2015. Cashing in on coral reefs: the implications of exporting reef fishes. pp. 166–179. *In*: Mora, C. (ed.). Ecology of Fishes on Coral Reefs. Cambridge University Press.

Sadovy de Mitcheson, Y. 2016. Mainstreaming fish spawning aggregations into fishery management calls for truly precautionary approach. BioScience 66: 295–306.

Sadovy de Mitcheson, Y., I. Tam, G. Muldoon, S. le Clue, E. Botsford and S. Shea. 2017. The Trade in Live Reef Food Fish—Going, Going, Gone. Volume 1: Main Report. Parts I, II & III. ADM Capital Foundation and The University of Hong Kong, Hong Kong Special Administrative Region. 288 pp.

Sadovy de Mitcheson, Y., S. Mangubhai, A. Witter, N. Kuridrani, A. Batibasaga, P. Waqainabete and R. Sumaila. 2018. Value Chain Analysis of the Fiji Grouper Fishery. Report of Science and Conservation of Fish Aggregations (SCRFA), United States. 57 pp. https://fiji.wcs.org/Portals/82/reports/WCS%20Grouper%20VCA%20Report%20081018%20WEB.pdf?ver=2018-10-31-023525-590.

Sadovy de Mitcheson, Y. 2019. The Live Reef Food Fish Trade: Undervalued, Overfished and Opportunities for Change. International Coral Reef Initiative. pp. 44. www.icriforum.org/wp-content/uploads/2020/05/ICRI%20Live%20Reef%20Food%20Fish%20Report-44p-double_0.pdf.

Sadovy de Mitcheson, Y.J., C. Linardich, J.P. Barreiros, G.M. Ralph, A. Aguilar-Perera, P. Afonso, B.E. Erisman, D.A. Pollard, S.T. Fennessy, A.A. Bertoncini, R.J. Nair, K.L. Rhodes, P. Francour, T. Brulé, M.A. Samoilys, B.P. Ferreira and M.T. Craig. 2020a. Valuable but vulnerable: Over-fishing and under-management continue to threaten groupers so what now? Marine Policy 116: 103909.

Sadovy de Mitcheson, Y.J., P.J. Colin, S.J. Lindfield and A. Bukurrou. 2020b. A decade of monitoring an Indo-Pacific grouper spawning aggregation: Benefits of protection and importance of survey design. Frontiers in Marine Science 7: 853.

Sala, E., E. Ballesteros and R.M. Starr. 2001. Rapid decline of Nassau grouper spawning aggregations in Belize: Fishery management and conservation needs. Fisheries 26: 23–30.

Salao, C., R. Cola and M. Matillano. 2013. Taytay: taking charge of a critical resource. A case study on the Philippines. WWF-Coral Triangle Initiative on Corals, Fisheries and Food Security pp. 16. wwf.org.ph/wp-content/uploads/2017/11/Taytay-Case-Study-WWF-Philippines-Sep-2013.pdf.

Sattar, S.A. and M.S. Adam. 2005. Review of Grouper Fishery of the Maldives with Additional Notes on the Faafu Atoll Fishery. Marine Research Centre, Maldives.

Scales, H., A. Balmford, M. Liu, Y. Sadovy and A. Manica. 2006. Keeping bandits at bay? Science 313: 612–613.

Shideler, G., D.W. Carter, C. Liese and J.E. Serafy. 2015. Lifting the goliath grouper harvest ban: angler perspectives and willingness to pay. Fisheries Research 161: 156–165.

Shideler, G. and B. Pierce. 2016. Recreational diver willingness to pay for goliath grouper encounters during the months of their spawning aggregation off eastern Florida, USA. Ocean Coastal Management 129: 36–43.

Smith, W.L. and M.T. Craig. 2007. Casting the percomorph net widely: the importance of broad taxonomic sampling in the search for the placement of serranid and percid fishes. Copeia 2007: 35–55.

Southwick, R., D. Maycock and M. Bouaziz. 2016. Recreational fisheries economic impact assessment manual and its application in two study cases in the Caribbean: Martinique and The Bahamas. FAO Fisheries and Aquaculture Circular No. 1128. Bridgetown, Barbados. pp. 118.

Sumaila, U.R., W.W.L. Cheung, L.S.L. Teh, A.H.Y. Bang, T. Cashion, Z. Zeng et al. 2021. Sink or Swim: The future of fisheries in the East and South China Seas. ADM Capital Foundation, Hong Kong. https://www.admcf.org/wp-content/uploads/2021/11/Sink-or-Swim-Full-Report_171121.pdf.

Teh, L.S.L., L.C.L. Teh and U.R. Sumaila. 2013. A global estimate of the number of coral reef fishers. PLOS ONE 8(6): e65397.

Thiao, D., C. Chaboud, A. Samba, F. Laloë and P.M. Cury. 2012. Economic dimension of the collapse of the 'false cod' *Epinephelus aeneus* in a context of ineffective management of the small-scale fisheries in Senegal. African Journal of Marine Science 34: 305–311.

Thrush, S.F. 1999. Complex role of predators in structuring soft-sediment macrobenthic communities: implications of changes inspatial scale for experimental studies. Australian Journal of Ecology 24: 344–354.

To, A.W.L. and Y. Sadovy de Mitcheson. 2009. Shrinking baseline: the growth in juvenile fisheries, with the Hong Kong grouper fishery as a case study. Fish and Fisheries 10: 396–407.

Usseglio, P., A.M. Friedlander, H. Koike, J. Zimmerhackel, A. Schuhbauer, T. Eddy and P. Salinas-de-León. 2016. So long and thanks for all the fish: Overexploitation of the regionally endemic Galapagos grouper *Mycteroperca olfax* (Jenyns, 1840). PLOS ONE 11(10): e0165167.

Wabnitz, C.C.C., A.M. Cisneros-Montemayor, Q. Hanich and Y. Ota. 2018. Ecotourism, climate change and reef fish consumption in Palau: benefits, trade-offs and adaptation strategies. Marine Policy 88: 323–332.

Waldie, P.A., G.R. Almany, T.H. Sinclair-Taylor, R.J. Hamilton, T. Potuku, M.A. Priest, K.L. Rhodes, J. Robinson, J.E. Cinner and M.L. Berumen. 2016. Restricted grouper reproductive migrations support community-based management. Royal Society Open Science 3(3): 150694.

Walsh, S.M., S.L. Hamilton, B.I. Ruttenberg, M.K. Donovan and S.A. Sandin. 2012. Fishing top predators indirectly affects condition and reproduction in a reef-fish community. Journal of Fish Biology 80(3): 519–537.

Warren-Rhodes, K., Y. Sadovy and H. Cesar. 2003. Marine ecosystem appropriation in the Indo-Pacific: a case study of the live reef fish food trade. Ambio 32: 481–488.

Waterhouse, L., S.A. Heppell, C.V. Pattengill-Semmen, C. McCoye, P. Bush, B. Johnson and B.X. Semmens. 2020. Recovery of critically endangered Nassau grouper (*Epinephelus striatus*) in the Cayman Islands following targeted conservation actions. Proceedings of the National Academy of Sciences 117: 1587–1595.

Watson, R. and D. Pauly. 2001. Systematic distortions in world fisheries catch trends. Nature 414: 534–536.

Wenger, A.S., K.E. Fabricius, G.P. Jones and J.E. Brodie. 2015. Effects of sedimentation, eutrophication, and chemical pollution on coral reef fishes. pp. 145–153. *In*: Mora, C. (ed.). Ecology of Fishes on Coral Reefs. Cambridge University Press, UK.

Wong, C.K., I.A.P. Pak and X.J. Liu. 2013. Gill damage to juvenile orange-spotted grouper *Epinephelus coioides* (Hamilton, 1822) following exposure to suspended sediments. Aquatic Research 44: 1685–1695.

Wringe, B.F., N.W. Jeffery, R.R.E. Stanley et al. 2018. Extensive hybridization following a large escape of domesticated Atlantic salmon in the Northwest Atlantic. Communications Biology 1: 108.

Young, J.L., Z.B. Bornik, M.L. Marcotte, K.N.M. Charlie, G.N. Wagner, S.G. Hinch and S.J. Cooke. 2006. Integrating physiology and life history to improve fisheries management and conservation. Fish and Fisheries 7: 262–283.

Yulianto, I., C. Hammer, B. Wiryawan and H. Palm. 2015. Fishing-induced groupers stock dynamics in Karimunjawa National Park, Indonesia. Fisheries Science 81: 417–432.

Zacharia, P.U., A.C. Gupta and H.S. Mahadevaswamy. 1995. Exploitation of juveniles of the spinycheek grouper, *Epinephelus diacanthus* by the multi-day trawlers along Dakshina Kannada coast. Marine Fisheries Information Service, Technical and Extension Series 139: 5–8.

Zhang, W., M. Liu, Y. Sadovy de Mitcheson, L. Cao, D. Leadbitter, R. Newton, D.C. Little, S. Li, Y. Yang, X. Chen and W. Zhou. 2020. Fishing for feed in China: facts, impacts and implications. Fish and Fisheries 21(1): 47–62.

Website Citations

ABC. 2020. www.abc.net.au/news/2020-06-04/chang-lonely-fish-falls-hook-line-sinker-for-childrens-letters/12321964.

https://fish.net/english/fisheries_information/wholesale_prices.php?year=2016.

https://www.4fj.org.fj.

http://www.cep.unep.org/content/about-cep/spaw.

http://www.exportersindia.com/indian-suppliers/grouper-fish.htm.

http://www.fao.org/fi/oldsite/FCP/en/VNM/profile.htm.

http://www.fao.org/fishery/statistics/global-aquaculture-production/en.

http://www.fao.org/fishery/statistics/global-capture-production/en.

http://www.iucnredlist.org.

http://www.iucnredlist.org/species/64409/42691809.

http://www.mingbo-aquatic.com.

http://www.nmfs.noaa.gov.

http://www.SCRFA.org.

http://www.shuichan.cc/news_view-318052.html.

http://www.thefishsite.com/fishnews/3629/researchers-breed-first-hybrid-grouper.

http://www.tianjinhaifa.com.

http://www.youdive.tv/Australia-Dive-with-the-biggest-groupers-of-the-world-at-Cod-Hole_v81.html.

Personal Communications

Gendron, S. 2016. Ocean Park, Hong Kong SAR.

Li, J.L. 2021. Hainan University, China.

Nair, R. 2020. ICAR-Central Marine Fisheries Research Institute, India.

Pet, J. 2016. The Nature Conservancy, Indonesia.

Index

A

Anthropogenic effects 40
Aquaculture 166–169, 171, 172, 174–176, 179–184
Assessment 206, 211–213, 215, 218, 220

B

Bi-directional sex change 55, 59

C

Cephalopods 74, 76–83, 86, 87, 94
Citizen science 128, 139, 145
Co-management 143
Conservation 191, 192, 195, 202, 215, 216, 219, 220
Cooperative hunting 110, 111
Crustaceans 74, 76–83, 86–88, 91, 92, 94
Cryptic speciation 13

D

Data-poor 120, 121, 124, 125, 130, 142
Diet composition 86, 88
Dispersal distance 25, 26, 28, 35–37, 40

E

Ecosystem modeling 121, 128–130
Epinephelidae 4–7, 17, 54, 56, 57, 66, 108, 191
Epinephelinae 4–6, 9, 13, 17, 74

F

Fish 74–83, 86–94
Fish behaviour 75
Fisheries 193, 197, 198, 200–202, 204, 206, 208–213, 217–221
Functional hermaphroditism 54, 55, 61

G

Genetics 180, 182, 183
Global production 167, 168, 198
Gonochorism 54, 55, 58, 59
Groupers 1, 3–14, 166–185, 191–221
Grouper's eggs and larvae 30

H

Habitat of settlement 36, 39
Harvest tags 133, 143
Hermaphroditism 192, 194
High-level predators 103–105, 107, 108, 110

I

Interspecific competition 102, 109

L

Larval behaviour 24, 33, 35, 37, 42
Larval culture 176, 178–180, 183
Larval dispersal 137, 141
Larval replenishment 37
Live food fish trade 193
Local ecological knowledge 125, 139

M

Management 191, 194, 195, 197, 203, 206, 208, 209, 211–220
Management measures 121, 124, 131–134, 139, 140, 143, 144
Mariculture 197, 200, 201, 203, 204, 210, 219
Marine protected areas 131, 134
Mating system 54, 61, 64, 65
Mesopredator release hypothesis 106
Mesopredators 91
Migration 64
Molecular phylogeny 6, 7

Monitoring 191, 206, 210–212, 217, 218, 220
Moratorium 132
Morphology 3–6, 17
Mutualistic interactions 103

O

Ontogenetic movement 25, 26, 28, 39
Ontogenetic variations 86

P

Pelagic larval duration 31, 43
Population dynamics 23, 30, 38, 40
Post-settlement survival 37
Prey types 74, 76, 77, 90, 91
Protection effect 37, 40, 42, 105, 108, 110, 121, 130, 132, 134–136, 140–143
Protogyny 55, 58, 59, 61

R

Recreational fishing 128, 131–133, 143
Recruitment 23, 24, 35, 38–41, 43, 105, 122, 123, 134, 139, 144, 194, 216
Reef fish 4
Reproduction 176–178

S

Scenarios 121, 128, 129, 144, 145
Serranidae 108
Settlement success 24, 35
Sexual pattern 54–57, 59–61, 64, 67, 68, 192–194
Social factors 61
Socio-economics 184
Spawning aggregation 65–68, 120, 121, 125, 131, 132, 134, 137–142, 193, 195, 205, 208, 209, 211, 212, 216, 218, 219
Spawning season 24, 25
Spillover 136–138, 144
Stable isotopes 75, 88, 89, 91, 92, 94
Stock assessment 121–124, 142
Stomach contents 87–89, 91, 94
Sustainability 209
Systematic 3, 4, 6

T

Taxonomic revision 198
Taxonomy 4, 13, 14
Teleosts 54, 55, 61–63, 68
Threat 191, 206–208, 211
Top-down control 104
Trophic cascades 103–105, 108
Trophic level 75, 88–94